ENERGY IMPACTS

Society & Natural Resources Book Series

Jill Belsky, Mallika Bose, Ken Caine, Anna Haines,
Rick Krannich, and Marianne Penker
SERIES EDITORS

This series examines the complexity of interrelationships among human societies, biophysical and built environments, and natural resources and engages emergent issues and informs transformations between society and natural resources toward greater social and environmental justice, health, and sustainability and resilience.

A Community Guide to Social Impact Assessment: 2015 Fourth Edition, Rabel J. Burdge

The Community in Rural America, Kenneth P. Wilkinson

The Concepts, Process and Methods of Social Impact Assessment, Rabel J. Burdge and Colleagues

Daydreams and Nightmares: A Sociological Essay on the American Environment, William R. Burch Jr.

Diffusion Research in Rural Sociology: The Record and Prospects for the Future, Frederick C. Fliegel with Peter F. Korsching

Man, Mind and Land: A Theory of Resource Use, Walter Firey

Rural Sociology and the Environment, Donald R. Field and William R. Burch Jr.

Social Assessment: Theory, Process and Techniques, 3rd ed., C. Nicholas Taylor, C. Hobson Bryan, and Colin G. Goodrich

Three Iron Mining Towns: A Study in Cultural Change, Paul H. Landis

ENERGY IMPACTS

A Multidisciplinary Exploration of North American Energy Development

SYNTHESIS ACROSS THE SOCIAL SCIENCES

EDITED BY

Jeffrey B. Jacquet
Julia H. Haggerty
Gene L. Theodori

UNIVERSITY PRESS OF COLORADO
Louisville, Colorado

SOCIETY AND NATURAL RESOURCES PRESS
Logan, Utah

Published by the Society and Natural Resources Press and University Press of Colorado

University Press of Colorado
245 Century Circle, Suite 202
Louisville, Colorado 80027

Manufactured in the United States of America

The University Press of Colorado is a proud member of the Association of University Presses.

The University Press of Colorado is a cooperative publishing enterprise supported, in part, by Adams State University, Colorado State University, Fort Lewis College, Metropolitan State University of Denver, Regis University, University of Colorado, University of Northern Colorado, University of Wyoming, Utah State University, and Western Colorado University.

∞ This paper meets the requirements of the ANSI/NISO Z39.48–1992 (Permanence of Paper)

ISBN: 978-1-64642-029-2 (hardcover)
ISBN: 978-1-64642-026-1 (paperback)
ISBN: 978-1-64642-027-8 (ebook)
https://doi.org/10.5876/9781646420278

Library of Congress Cataloging-in-Publication Data

Names: Jacquet, Jeffrey B. (Jeffrey Bryan), 1979– editor. | Haggerty, Julia H. (Julia Hobson), editor. | Theodori, Gene Louis, editor.
Title: Energy impacts : a multidisciplinary exploration of North American energy development / edited by Jeffrey B. Jacquet, Julia H. Haggerty, Gene L. Theodori.
Other titles: Society and natural resources book series.
Description: Louisville, Colorado : University Press of Colorado, [2020] | Series: Society and natural resources book series | Includes bibliographical references and index.
Identifiers: LCCN 2020029232 (print) | LCCN 2020029233 (ebook) | ISBN 9781646420292 (cloth) | ISBN 9781646420261 (paperback) | ISBN 9781646420278 (ebook)
Subjects: LCSH: Energy development—Social aspects—North America. | Energy development—Economic aspects—North America.
Classification: LCC HD9502.N582 E54 2020 (print) | LCC HD9502.N582 (ebook) | DDC 333.79097—dc23
LC record available at https://lccn.loc.gov/2020029232
LC ebook record available at https://lccn.loc.gov/2020029233

This material is based upon work supported by the National Science Foundation under Grant No. 1528422. Any opinions, findings, and conclusions or recommendations expressed in this material are those of the author(s) and do not necessarily reflect the views of the National Science Foundation. Publication is also supported, in part, by Montana State University.

Contents

ENERGY IMPACTS

Introduction

JEFFREY B. JACQUET, JULIA H. HAGGERTY, AND GENE L. THEODORI

There is growing understanding that energy is a social issue with technical components, not the other way around (Sovacool 2014a). As novel forms of energy development proliferate across the United States and around the world, the need for credible and informed scientific research on the potential environmental and community impacts of energy development becomes more pronounced (National Research Council 2014). Reflecting these changing understandings and priorities, a growing cadre of researchers has gathered to produce important new research on social, economic, and behavioral impacts from large-scale energy development. This collection demonstrates the momentum and dynamism of the present moment in energy impacts research.

The volume developed through the efforts of a network funded by the National Science Foundation to coordinate research on the social impacts of energy development across the diverse academic landscape in which this research occurs. A centerpiece of that network was The Energy Impacts Symposium, a two-day research symposium held July 26–27, 2017, in Columbus, Ohio. In concert with the event, we issued a call for submissions

DOI: 10.5876/9781646420278.c000

to this volume, which read, in part: "Chapters for this book shall compare, synthesize, bridge, or otherwise work to traverse heretofore constraints on coordination of energy impacts research. We are particularly interested in multidisciplinary author teams and papers that bridge academic disciplines."

The language of the call reflects the circumstances for the research coordination network's activities circa 2015–2017. A proliferation of new energy development had inspired a boom in research on energy and society broadly defined, the products of which stood to be mutually informative and potentially transformational. However, common themes and comparability among diverse strands of the energy impacts literature proved elusive. As such, the *specific goals* of this network were to

1. *build a cross-disciplinary research community* around social, economic, socioeconomic, community, governance, and public health impacts from energy development to coordinate research activity across geographies, energy types, and social science disciplines, increasing effectiveness and reducing redundancy;
2. *develop and promote a set of data collection standards, framing concepts, and research design* to allow for comparison of data acquired across energy landscapes; and
3. *develop a shared library of research tools and educational resources* to engage both academics and nonacademic audiences via integrated tools, concepts, findings, and curricula to inform and engage energy research, policy, and impacts.

Chapters submitted to this volume underwent a blinded peer-review process, managed by the editors. Each chapter was reviewed by at least 2 but often 3 or 4 anonymous reviewers, and the chapters underwent a revise-and-resubmit process that concluded in late spring 2018. Unfortunately, not all chapters made it through the peer-review process.

The twelve chapters in this book and the diverse disciplinary orientations of the authors reflect this original call. We have organized the volume into three main sections—"Section One: Theoretical and Conceptual Approaches," "Section Two: Methodological Approaches," and "Section Three: Case Studies and Applications." At the end of each chapter, we have included one-page "Chapter Summaries," indented to provide a brief overview of the chapter content along with "highlights" or "key takeaways" that showcase how the chapter can inform future policy, regulation, or research in energy-impacted communities.

Section One: Theoretical and Conceptual Approaches

Section one features four chapters that encourage scholars as well as stakeholders in energy systems to rethink conceptual approaches to energy impacts and their evaluation. Together, these chapters indicate the potential significance of interdisciplinary communication and challenging scholarly and cultural norms about energy transitions.

In chapter 1 ("From Climax Thinking toward a Non-equilibrium Approach to Public Good Landscape Change"), Kate Sherren introduces readers to the concept of climax thinking. After defining the concept of climax thinking, Sherren describes the pathology of climax thinking, elaborates on the need for a non-equilibrium model for managing public good landscape change, and proposes an action-oriented research agenda for landscape transitions.

In chapter 2, Seven Mattes and Cameron T. Whitley's engaging essay about the "entangled impacts" that link human and animal lives in energy landscapes and systems ("Entangled Impacts: Human-Animal Relationships and Energy Development") introduces readers to a little-considered dimension of energy impacts, that of the human-animal bond. The authors extend to energy landscapes the notion, established in animal studies scholarship, that society is diminished by the failure to acknowledge the extent and importance of human-animal connections. In addition to an interdisciplinary literature review, the authors consider two very different but exemplary case studies: impacts of fracking on human-animal relationships and the Fukushima Daiichi nuclear accident and its impacts.

In chapter 3 ("The Need for Social Scientists in Developing Social Life Cycle Assessment"), Emily Grubert explores the important and often underappreciated links between engineering and the social sciences, exploring the basis of this importance and introducing new tools for the siting of large industrial projects that incorporates the work of social scientists. Life Cycle Assessment (LCA) is a major decision support tool that is increasingly used in policy making, and one of the major applications of LCA is for evaluating energy systems, both standalone and as contributors to other product systems. As noted, utilizers of LCA aspire for it to be a comprehensive and systematic approach to holistically evaluating a product, process, policy, service, or system; however, in practice, LCA has historically been primarily an environmental assessment tool with socially oriented impacts rarely considered. This chapter calls social scientists to action in contributing to the development of LCA and discusses major needs, such as definition and development of social indicators that could be used in such assessment, approaches to defining scope and descriptive approaches, and critical assessment of the full suite of life cycle methods with respect to justice, ethics, application, and others.

In chapter 4 ("Societal Impacts of Emerging Grassroots Energy Communities: A Capabilities-Based Assessment"), author Ali Adil urges a new deliberative framework for evaluating energy transition scenarios. His case study is the case of Grassroots Energy Communities. The framework Adil offers orients to three axes—equity, justice, and sustainability—and is transdisciplinary in its use of scholarship from policy studies, political theory, and environmental studies. Adil proposes that this framework avoids normative "forcing" of specific visions of progress, and he offers it as a tool for use by decision makers in a variety of contexts and scales.

Section Two: Methodological Approaches

The next section of the collection introduces a set of chapters that reflect on or deploy innovations in methodological approaches to evaluating energy impacts. This includes Fernando et al.'s synthetic evaluation of the variety of existing approaches to data collection and analysis and argument for the merits of mixing methods, with specific respect to the study of shale development's social impacts. It also includes examples of the application of Q method (Parkins and Sherren) and Laninga et al.'s discussion of a new capitals-based assessment tool for energy development projects.

In chapter 5 ("Analysis of Research Methods Examining Shale Oil and Gas Development"), Felix N. Fernando, Julia D. Ulrich-Schad, and Eric C. Larson provide an in-depth review of the research methods used by social scientists to gather data on the social impacts of shale energy development. The authors highlight the various benefits and constraints associated with using either a quantitative or qualitative methodological approach to data collection. They then argue for the increased use of mixed methods in social scientific shale energy research.

In chapter 6 ("Identifying Energy Discourses across Scales in Canada with Q Methodology and Survey Research"), John R. Parkins and Kate Sherren use a mixed methods approach to identify variation in discourses on energy development in Canada at the national and regional levels. Combining data from in-person Q-sorts of individuals who were generally informed about energy issues in three Canadian provinces—Alberta, Ontario, and New Brunswick—with a national survey of Canadian citizens, the authors uncovered a high degree of alignment between regional and national discourses about energy development.

In chapter 7 ("A Capitals Approach to Biorefinery Siting Using an Integrative Model"), an interdisciplinary team of scientists and engineers discusses how decision support tools can enhance industry and public understanding, thus aiding site selection decisions. The authors provide their own innovative decision support tool that incorporates the Community Capitals Framework, and they test this tool with case study locations across the Pacific Northwest.

Section Three: Case Studies and Applications

The final section of the book presents a close look at an array of energy landscapes and their experiences of energy infrastructure transitions. This includes a critical look at the overlay of bioenergy and agriculture in Pennsylvania by Eaton, Hinrichs, and Burnham and a qualitative reflection on experiences of local residents with shale development in the Marcellus region (Podeschi et al.). Ratledge and Zachary offer a comparative exploration of the quantitative and qualitative data describing how public schools fared in US localities that hosted intensive shale development. A complementary chapter led by Myra Moss and colleagues discusses the dimensions of effective community engagement in the context of shale development. Finally, a large, interdisciplinary team from Virginia Tech offers a synthetic model for improving the process of wind facility siting, drawing on extensive local experience.

Chapter 8 ("Cultural Counterpoints for Making Sense of Changing Agricultural and Energy Landscapes: A Pennsylvania Case Study") by Weston M. Eaton, C. Clare Hinrichs, and Morey Burnham examines how cultural resources and social representations for working landscapes influence landowner responses to an emerging bioenergy industry. The authors draw from a case study of bioenergy development in western Pennsylvania's Crawford County, a place undergoing a slow and uneven transition away from traditional agricultural and energy extraction industries, toward new forms of land use that include bioenergy and bioindustry development. In this chapter, the authors show how two cultural counterpoints influence landowner decisions about land use including growing bioenergy crops on their land. Landowners experience, on the one hand, a sense of obligation toward their family's previous generations, and on the other, a yearning for autonomy in the form of flexibility and control over their own land use decisions. Considered are how these cultural counterpoints inform the social-psychological processes through which landowners make sense of changes in land use and community life. The chapter concludes by highlighting the important role cultural meanings for rural landscapes play in social responses to renewable energy technology development that is premised on land use change.

Chapter 9 ("The Wider Array: A Qualitative Examination of the Social and Individual Impacts of Hydraulic Fracturing in the Marcellus Shale") by Christopher W. Podeschi, Lisa Bailey-Davis, Heather Feldhaus, John Hintz, Ethan R. Minier, and Jacob Mowery provides a qualitative analysis of residents' experiences with shale energy development in three communities in the Marcellus Shale region of Pennsylvania. Podeschi et al. use these data to emphasize the voices of local residents and illustrate the wide array of individual and social impacts associated with shale energy development.

In chapter 10 ("Drilling Impacts: A Boom or Bust for Schools? A Mixed Methods Analysis of Public Education in Six Oil and Gas States"), Nathan Ratledge and Laura Zachary examine whether public school districts that experienced the shale energy boom in Pennsylvania, Ohio, West Virginia, North Dakota, Montana, and Colorado fared better or worse in terms of financial and educational outcomes than comparable school districts that did not experience the shale boom in those states. Through the use of secondary data and over seventy predominantly in-person interviews, the authors elaborate on both the short-term benefits and long-term consequences experienced by school districts in the shale boom regions.

In chapter 11 ("Effective Community Engagement in Shale-Impacted Communities in the United States"), an experienced team of University Extension educators provides an overview and case study of effective and (often less-than-effective) attempts on industry's part to engage with host communities. This chapter provides a theoretical and applied contribution based the authors' over three decades of observation and experience with shale-impacted community engagement. As Extension professionals, the authors have been working closely with communities in the Marcellus and Utica shale regions of Pennsylvania and Ohio as well as with Extension peers in other shale regions of the United States and throughout the world. The authors share their reflections and identify best practices and needs for further research, contributing to the larger field of energy research by providing unique experience-based insights into lessons learned and best practices for engaged shale-impacted communities, adding depth and direction to future social science research.

In chapter 12, an interdisciplinary team from Virginia Tech argues that to achieve a sustainable energy supply, public support must reflect adequate consideration of both positive and negative environmental, social, and economic impacts of wind energy facilities (WEFs). Their chapter, "A Framework for Sustainable Siting of Wind Energy Facilities: Economic, Social, and Environmental Factors," draws on research and practice in geography, wildlife biology, economics, and other fields to propose a framework for facility siting that responds to the complex issues involved in siting of WEFs.

Taken together, these chapters reflect a decade or more of novel research into energy impacts. The collection lays the groundwork for continued collaboration across social science disciplines through examples of conceptual, practical, and analytical benefits of going beyond the constraints that have historically separated energy impacts scholarship.

Contributions of this Collected Volume

As the amount of new energy development expands across the United States and elsewhere, the growth of new social science research on energy development has been extraordinary. In the United States, for example, the pace, scale, and intensity of new energy development has included hundreds of thousands of oil and gas wells, over 55,000 wind turbines, over 200 large-scale ethanol plants, and over 130 large-scale biodiesel plants that have been built in or near host communities (and has prompted a major expansion in associated pipeline, rail, and transmission infrastructure) (AWEA 2014; RFA 2014; USEIA 2014).

In response, a mountain of new research has been published. In 2018, Darrick Evensen counted "well over 1,000" (2018:417) English-language publications regarding the impacts of shale development, a number that assuredly continues to grow. However, when the Energy Impacts research coordination network was first proposed to the National Science in 2014, there was not a central forum or clearinghouse for social science research on energy development. Since that same time, Benjamin Sovacool published a comment in the journal *Nature* on the need for social science in energy studies (Sovacool 2014b). Elsevier began publishing the journal *Energy Research and Social Science* on this very subject, producing over 500 articles since 2014, as well as hosting a series of international conferences. Other journals such as *The Extractive Industries and Society*, *Energy Policy*, and *Society and Natural Resources* likewise have produced numerous articles on this topic.

This plethora of scholarship brings various nodes of inquiry and debate to address specific social dynamics of these new energy development technologies, such as the social and economic costs and benefits to landowners, residents, and communities; the efficacy of existing regulatory and governance regimes; and the perception of public health risk (for example, see the August 5, 2014, special issue of *Environmental Science and Technology* entitled "Understanding the Risks of Unconventional Shale Gas Development"). However, research discoveries and insights are often compartmentalized into disciplinary territories in ways that limit their transformative and informative potential.

While there will never be singular theory or approach that can incorporate all the varied research on this broad and complex topic, the entire body of research in this topical area benefits from increased coordination and communication across the various geographies, industries, and disciplines. Sending an uncoordinated army of social scientists out to energy communities to study the impacts of development risks redundancy and inefficiency as well as missed opportunities for synthesis, comparison, and innovation.

The primary contribution of this collected volume is to initiate interdisciplinary dialogue, reflection, and comparison. The collection attempts to span disciplinary and geographical boundaries that have previously isolated energy impacts research. The thirty-seven authors who produced the twelve chapters in this collection offer theories, methods, and approaches to understanding the impacts of energy development from an array of disciplines, ranging from economics, geography, sociology, engineering, community development practice, and beyond. Many of the chapters produce theoretical frameworks or methodological approaches that can be applied across numerous locations or energy types.

Limitations of this Collected Volume

In the face of the broad research constraints outlined in this induction, and despite the original lofty goals of this collected volume, we recognize that this volume is still limited in a number of important ways. This book comprises works submitted in response to the open call for submissions—we did not specifically solicit works to address particular issues. The benefits of this process include that we received a number of remarkable and imaginative chapters that we could not have anticipated or solicited, but a downside of this process is that some critical areas are not specifically addressed.

A Focus on North America

The chapters included here focus largely on North America, though the conclusions or applications of these works need not be limited to this continent. We recognize that other areas of the world grapple with impacts from energy development, with the developing world facing a unique and critically important set of circumstances, and we hope the works here can be applied in international contexts and further adapted for use around the globe.

Academic Disciplines

Despite a diverse range of academic disciplines represented within these works (sociology, engineering, anthropology, science and technology studies, geography, communication, political science, community planning, education, and landscape architecture, to name a few)—we recognize that not all disciplines are represented. Economics and history, in particular, are two areas underrepresented in this volume.

Representativeness in Research Subjects and Authors

We recognize that marginalized populations are particularly vulnerable to the impacts of energy development; indeed, our own prior research has focused on the issues of differentiated costs and benefits from energy development. However, the works in this volume largely do not specifically focus on marginalized populations. Although issues of fairness in the distribution of costs and benefits are embedded in several chapters, we recognize issues of environmental justice (indeed, "energy justice") relative to specific populations—the poor, Sovereign Native peoples, people of color—make up an important and underresearched area of energy social science.

These limitations are not unique to this particular volume but rather speak to the limitations and challenges for the discipline as a whole. The works in this collected volume do attempt to bridge a number of the constraints identified in this introductory chapter, and we hope that these concepts, frameworks, and approaches will serve as the basis for new branches of synthetic and interdisciplinary scholarship, helping to inspire a new generation of social scientists.

References

The Academy of Medicine, Engineering and Science of Texas (TAMEST). 2017. *Environmental and Community Impacts of Shale Development in Texas*. Austin: The Academy of Medicine, Engineering and Science of Texas. https://doi.org/10.25238/TAMESTstf.6.2017.

American Wind Energy Association (AWEA). 2014. "Wind Energy Facts at a Glance." Accessed November 4, 2014. https://www.awea.org/wind-101/basics-of-wind-energy/wind-facts-at-a-glance.

Evensen, Darrick. 2018. "Yet more 'fracking' social science: An overview of unconventional hydrocarbon development globally." *The Extractive Industries and Society*. 5 (4): 417–421.

National Research Council. 2014. *Risks and Risk Governance in Shale Gas Development: Summary of Two Workshops*. Washington, DC: National Academies Press. https://doi.org/10.17226/18953.

Renewable Fuels Association (RFA). 2014. Biorefinery Locations. Accessed November 4, 2014. https://ethanolrfa.org/resources/biorefinery-locations/.

Sovacool, Benjamin K. 2014a. "What Are We Doing Here? Analyzing Fifteen Years of Energy Scholarship and Proposing a Social Science Research Agenda." *Energy Research and Social Science* 1 (March): 1–29.

Sovacool, Benjamin K. 2014b. "Diversity: Energy Studies Need Social Science." *Nature* 511 (July): 529–530.

"Understanding the Risks of Unconventional Shale Gas Development." 2014. Special issue, *Environmental Science and Technology* 48 (15): 8287–8934.

US Energy Information Administration (USEIA). 2014. "Number of Producing Gas Wells." Accessed November 4, 2014. http://www.eia.gov/dnav/ng/ng_prod_wells_s1_a.htm.

Section One

Theoretical and Conceptual Approaches

1

From Climax Thinking toward a Non-equilibrium Approach to Public Good Landscape Change

KATE SHERREN

Introduction

This book is dedicated to exploring the opportunities to coordinate across scales, sources, and social science subfields toward better understanding of energy impacts. One such barrier to coordination has been theory to understand public resistance to landscape change. Current global challenges necessitate widespread transitions that will have significant impacts for landscape appearance, function, and meaning and are thus subject to local opposition. Public good landscape changes discussed here include those required for sustainability transitions: renewable energy but also urban densification and climate adaptation. Explanations for this opposition have thus far been fragmented, but may have common roots.

In recent years it has become common to apply ecological concepts to society (e.g., adaptation, resilience). Many of these instances develop into rich interdisciplinary fields of study and application. The application of ecological concepts to society is often initiated by ecologists recognizing familiar patterns. It is less common for a social scientist to reach into ecology,

DOI: 10.5876/9781646420278.c001

especially given the range of social theories that capture specific phenomena as well or better. For instance, resilience has had a strong uptake among social scientists engaged in team social-ecological research led by ecologists, as well as by policy makers, but has been critiqued for its lack of attention to social dimensions and human subjectivity (Cretney 2014; Davidson 2010; Olsson et al. 2015; Stedman 2016). I thus use ecological analogies here cautiously. Scholars have also applied succession to other aspects of human communities, similarly not without controversy (Rudel 2009). Yet I will build on succession concepts to (1) describe a new concept of climax thinking in relation to a range of parallel literatures and theories, (2) deconstruct the concept's pathology and implications for managing landscape change in lived landscapes, and (3) suggest an action and research agenda to ease the process of transformation.

What Is Climax Thinking?

We are all, from time to time, climax thinkers. That is, we seem to believe that the landscape we currently have is the one that is the *intended* end point for our given context. This recalls Frederick Clements's concept of succession, developed in rangelands (Sayre 2017), where a climax plant community was defined as a stable one that dominates in a given site and set of conditions after a series of predictable and progressive stages. In Clements's thinking this equilibrium state is inevitable, almost fated, and will be reliably returned to after disturbance such as grazing. Indeed, that return was an indication of the plant community's vitality. We often perceive our lived landscapes similarly as progressing from "pioneers" on up to what is seen locally as a mature or "climax" state. In ecology, equilibrium theories such as succession have been surpassed by non-equilibrium concepts such as panarchy and resilience and as multiple potential stable states for given social ecological systems (Elmqvist et al. 2003). This sequence of climax to non-equilibrium theories is an important one for us to follow in the context of landscape change as well. This chapter suggests that in lived landscapes, climax is only an illusion. Although ecologists have stepped away from climax thinking, it seems that social thinking is often stuck with notions of climax (steady state landscapes of place and attachment) that are unhelpful in the face of new challenges. Here I describe the phenomenon of climax thinking and its implications more thoroughly, focusing not on the ways that climax thinking may arrest negative landscape change (Hager and Haddad 2015) but on how such thinking can be a barrier to the landscape transformations we need to meet new societal and planetary needs, such as decarbonization or climate adaptation. I recognize how much nuance such a focus excludes—not all change is good, and not all stasis is bad—but such decisions are sometimes critical for generating useful theory (Healy 2017).

First, however, I would like to note that it is harder than it might seem to identify what is a public good landscape change. In a context of climate change, landscape changes for decarbonization and climate regulation create public benefits that by economic classification are nonexcludable and nonrival. Such public goods are underprovided in part because they also impose at least short-term negative externalities on people living nearby (Stokes 2016), driving opposition to such proposals that increases the cost and reduces the likelihood of transition. In general it should be a good thing to have "interest . . . coincident with duty" (Brennan 1996, 256, citing James Madison ca. 1788), but it complicates such proposals that they do not *exclusively* represent public goods but also economic benefits to developers and town councils, both variously trusted (Hess 2018; Parkins et al. 2017). Those with the power to judge that something is in the public good may not reflect the demographics (class, race) of those affected by the decision (Pasternak 2010; Reed and George 2011). There is very real peril in this situation, though I largely set it aside in what follows. Exemplars of integrated landscape planning and transitions are needed that include close attention to power and justice (Newell and Mulvaney 2013; Stenseke 2016).

The idea of climax thinking has repeatedly emerged from my recent social science case study work, as well as more informal readings of local events, bringing explanatory value to observed public responses to proposed landscape change. Residents around a failing hydroelectric dam recently protested its removal, as they protested its construction fewer than fifty years earlier (Sherren et al. 2016). Many locals disagreed with the dyke realignment and wetland restoration necessary to protect coasts from climate-related risks, though most of the agricultural land that the dykes protect is no longer actively farmed (Sherren, Loik, and Debner 2016). Climax thinking is manifest in resistance to landscape change of all kinds, but particularly explored here in relation to public good landscape change, whether a landscape addition, replacement, or removal (Magilligan, Sneddon, and Fox 2017). It is also manifest in debates over reconstruction after "natural disasters" such as hurricanes, where to rebuild as it was (rather than in preparation for what will be) is seen as most heroic (Birch and Wachter 2006). Sometimes climax thinking seems to emerge as result of "sunk costs," where past effort or investment by ourselves or ancestors to build (farm, log) the current landscape makes the possibility of changing that landscape feel like an invalidation. This emotional "lock-in" becomes a sort of social infrastructure that rejects change to retain identity and honor past generations (Sherren et al. 2017).

Climax thinking is easiest to visualize at an individual scale, with that individual in a bubble: while we stand on a layered landscape, we may be only dimly aware of this history (figure 1.1). This mindset drives our ignorance, inability, or lack of willingness to perceive the current landscape as only one in an ongoing sequence. Instead, we see it as the culmination of a sequence, its persistence privileged. We may assume current solutions will meet future needs, when in fact aggregate resistance to change will inevitably cause degradation of it for all. Climax thinking is a luxury, afforded the socially, politically, or economically powerful who can maintain their own climax landscape at the expense of others. Such resistance to accepting change in lived landscape to meet new public needs pushes the provision of those needs and the implications of that provision onto to those more spatially or socially distant. In disaster contexts it also often pushes the cost of landscape stasis (or restoration/rebuilding) onto governments, who are forced for political ends to prolong current uses.

Climax thinking is a significant problem given the scope of land use change that is required to meet current climate challenges: climate adaptation and decarbonization of the economy, our lifestyles, and the energy sector. New landscapes need to be written into this crowded space, such as renewable energies, new urban forms, restoring ecosystem services, and finding space for sea level rise. This progression is sometimes described as landscape transformation, a step change rather than incremental change (Pelling, O'Brien, and Matyas 2015). Instead, climax thinkers pick a winner, a particular time period and "strategy" (Shepheard 1997), in which to arrest the lived landscape and its meaning. This change subjects new land uses to former needs, much as those who seek to maintain landscapes in specific conditions, such as the sheep-managed Cotswolds, rather than rewilding abandoned agricultural

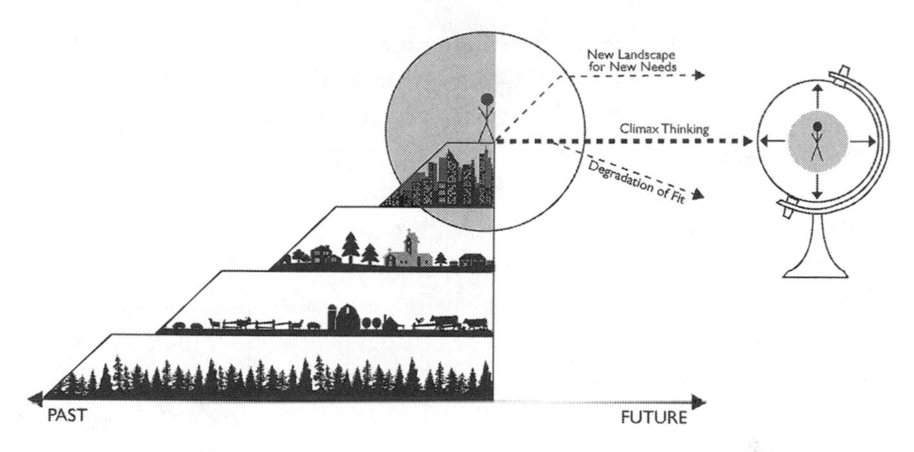

FIGURE 1.1. Climax thinking, illustrated.

land (Monbiot 2014). Silvia Crowe summarized it best sixty years ago in *Landscape of Power*: "The superficial approach to a landscape, seeing only its appearance at the moment, without realizing its past, its essential character or its potential future, can have a stultifying effect at the time we need a broad-minded vision. The humanized landscape is a constantly changing pattern and cannot be arrested at one point in history" (Crowe 1958, 38).

In conservation settings the idea of historical fidelity is being discarded (Higgs et al. 2014), perhaps because it is increasingly clear that future conditions are unlikely to easily support past ecosystems, though the public still prefers, for instance, to support native species conservation rather than immigrant species following shifting climate bands (Lundhede et al. 2014).

Lived Landscapes Are Layered Landscapes

There are few if any places on earth where the hand of humans cannot be seen, but we do not have good descriptors for such places. It is now widely acknowledged that we live in the Anthropocene, the geological time period in which human forces have dominated natural processes, but that term refers to time rather than landscape (Robin and Steffen 2007). The term *landscape* itself is widely acknowledged to encompass the combined outcome of both cultural and natural forces, but these are more commonly referred to as *cultural* landscapes (Council of Europe [CoE] 2000). I introduce here the term lived *landscape* to encompass the range of places where we live, work, and extract natural resources. Most lived landscapes are mundane, yet are no less the combined effort of humans and nature: farms, mill towns, suburbs, hydroelectricity reservoirs, and working forests (through fire suppression and plantation forestry). Lived landscapes may well be the most mature available representations of a given culture as it is currently practiced, but they often do not meet established definitions of cultural landscapes (box 1.1). It frankly doesn't seem that they're "making" any more cultural landscapes as defined by some, but lived landscapes are ubiquitous as we meet our needs for food and fuel, shelter and community, and beauty and inspiration. While the idea of lived landscapes emerges from the "working" landscapes of resource management (Abrams and Bliss 2013), it includes sites of resource consumption as well as production.

Box 1.1. Lived landscapes are not the same as cultural landscapes.

There is a rich vein of research in specifically cultural landscapes. The European Landscape Convention definition of cultural landscapes is oft-cited: an "area, as perceived by people, whose character is the result of the action and interaction of natural and/or human factors" (Council of Europe [CoE] 2000, 2). The cultural landscapes protected since 1992 under the UNESCO World Heritage system are "combined works of nature and man" (Mitchell, Rössler, and Tricaud 2009, 18), typically examples of human landscape modifications for the purposes of aesthetics, spirituality, or livelihood. These landscapes usually but not always represent land uses, meanings, and practices that are now archaic or at least quaint. These landscapes are manifestations of past resources, lifestyles, or spiritual problem-solving. Their protection by UNESCO has often come as a result of their having fallen into disuse or their use being somehow ruptured (Kasemets 2015) and thus "paused" at a specific stage of cultural problem-solving. By contrast, many former cultural landscapes in areas of high pressure for development such as river deltas may be long buried under layers of newer landscape solutions. Any such landscape artifacts serve, as described Dutch writer Willem van Toorn (translated by Martin Drenthen), to "remind us . . . that there is a past, that people who lived in that past had to deal with the world just as we have to; that they had to protect themselves against nature and use its resources" (Drenthen 2015, 66). UNESCO cultural landscapes are thus archetypal and rare, exemplars and celebrations of localized problem-solving.

Lived landscape overcomes limitations of cultural landscape as a concept for the purposes of this chapter.

- First, balance between human and nature is generally a characteristic of cultural landscapes. In lived landscapes such as cities, however, human forces and technologies have sought to overtake nature. This attempt happens despite the fact that many cities today seek to mimic, if not (re)integrate, ecosystems in order to ensure and leverage ecosystem service supply (Depietri, Renaud, and Kallis 2012).
- Second, scale is also an important variable. Cultural landscapes often function as symbols of smaller scale and arguably more sustainable human endeavor, encompassing discrete complexes of human habitation, resource exploitation, and cultural identity. Lived landscapes are more difficult to delineate, representing multiple connections across sometimes large distances for livelihoods, relationships, resources, and meaning.
- Finally, cultural landscapes typically have meaning beyond their boundaries, for instance, representing a cultural group and its traditions, even with "associative" UNESCO category landscapes where there is no material evidence of human use. By contrast, many lived landscapes may seem aesthetically compromised or culturally bereft to an outsider, while for locals they may inspire strong (and likely diverse) senses of personal attachment and cultural identity (Stobbelaar and Pedroli 2011). These lived landscapes are not cultural by most definitions, but they are made by humans of nature and even if they are utilitarian and industrialized, they are variously inhabited, used, perceived, and cherished.

Lived landscapes represent a significant planning challenge (Plieninger et al. 2015). This challenge is partly because of the subjectivity in how they are experienced and interpreted by individuals, which may not be directly connected to the physical meanings or affordances of the place or typical demographic characteristics (Stedman 2016). Individuals in the same physical place may effectively "read" a different landscape "text," depending on any number of personal variables and experiences. Within this diversity, however, there may be clusters, some of which may become dominant and normalized, depending on the power dynamics within the place (Cresswell 1996; Stokowski 2002).

Decades of scholarship have described landscape as a palimpsest (Drenthen 2015): a reused surface upon which a story of current livelihoods is legible, at the same time as evidence of past ones remains visible. Landscapes have many constituencies and thus many ways of being read and thus meaningful (Widgren 2004). D. W. Meinig's famous essay "The Beholding Eye" enumerated ten ways to read landscape: as nature, habitat, artifact, system, problem, wealth, ideology, history, place, and aesthetic (1979). Cosgrove and Daniels (1988) made a smaller list: landscape as scenery/spectacle, institution/rules, and resource. Architect Paul Shepheard would counter these with legible landscape "strategies"—for example, reason, defense, economic exploitation, restoration—that play out at large scales. Shepheard also reminds us that a lack of coherent landscape strategy does not arrest landscape change; it just makes it more emergent, fragmented, and potentially maladaptive: "Incremental changes happen all the time, [and] . . . accrue to big changes in what there is in the world, and whatever you are up to, you will be involved in these already. . . . be aware of the strategy that governs what you do" (Shepheard 1997, 233).

The shared nature of lived landscapes, albeit with their many constituencies over place and time, makes them useful to conceptualize as containers for—as well as outputs of—multiple and overlapping rules and behavioral regimes. Institutions are simply "structures for exchange" (Hotimsky, Cobb, and Bond 2006) and generally refer to the intangible social inventions such as law, education, and markets, more than their physical manifestations (courthouses, schools, banks). Jonathan Turner, however, described institutions in a way that anchors them to problem-solving and resource use: "[as] a complex of positions, roles, norms and values lodged in particular types of social structures and organising relatively stable patterns of human activity with respect to fundamental problems in producing life-sustaining resources, in reproducing individuals, and in sustaining viable societal structures within a given environment" (1997, 6).

Robert Goodin's thinking on institutions echoes Paul Shepheard's, earlier: "[They] can be the product of intentional action without . . . having been literally the intentional product of anyone's action" (Goodin 1998, 28). Anthony Giddens (1979) described structuration cycles that drive and reinforce behaviors. Landscape is thus a physical manifestation of those norms and the way we solve problems in a particular place, and the landscape in turn enables and thus recursively reproduces those patterns. Others have called these sociotechnical landscapes, stable "taken-for-granted backdrop[s]" that do not drive action but "[exert] power and influence . . . provid[ing] deep-structural 'gradients of force' that make some actions easier than others" (Geels and Schot 2007, 403). Changing society means changing landscapes, and vice versa.

Thinking of the interactions between landscapes and institutions is instructive to understanding how change and perhaps transformation should be approached (Pelling, O'Brien, and Matyas 2015). If we can agree that landscapes are institutions in the sense of being commonly held and reproduced by humans, playing out rules and regimes to sustain viable communities in a changing environment (to paraphrase Turner 1997, above), we can more easily see that succession beyond any perceived current "climax" is an obvious outcome. John Handmer and Stephen Dovers identified four approaches to institutional change that can be applied to landscape planning (1996): stability, where the goal is the status quo; incremental or superficial change, often marginal or symbolic; adaptability, where the goal is resilience amidst change; and, flexibility, with the concomitant risk of maladaptation. Given the tendency of future options to be narrowed by past choices (so-called lock-in, or path-dependency [Simmie 2012]), it is appropriate for landscape change processes to face interrogation, as well as some rigidity or resistance. We must balance rigidity and flexibility, consider how much cost or benefit accrues to whom, and find ways to avoid widespread grief and loss from imposed change (Marris 2014; Mels 2016) as well as the solastalgia that results from environmental degradation in cherished places (Albrecht et al. 2007).

A Pathology of Climax Thinking

The attenuation of time and space operate powerfully on us to create climax thinking. Blindnesses or lapses of empathy across temporal and spatial dimensions seem to drive the problem, though whether these emerge as a result of hubris, exceptionalism, ignorance, or uncertainty is unclear (table 1.1). First, climax thinking manifests as an apparent belief that current landscapes are how they are meant to be. This belief may be linked by genuine ignorance of

TABLE 1.1. A multidimensional pathology of climax thinking

Dimension	*Pole*	*Hubris/Exceptionalism*	*Ignorance/Uncertainty*
Time	Past	Previous generations and land uses were paving the way for this one.	There were no previous land uses.
	Future	Future generations matter less than this one.	Current solutions will continue to work in the future.
Space	Self	I should not need to accept landscape change.	I am not able to adapt to landscape change.
	Other	People in other places matter less.	Local landscape decisions do not have implications elsewhere.

former land uses or to a sense that former generations (and their land uses) were more primitive. Second, climax thinking suggests a failure to imagine alternative future landscapes that are equivalently viable and desirable. This belief might be caused by an assumption that current landscapes will continue to meet the needs of current and future generations or simply by a sense that future people matter less: temporal discounting of generations to come. Third, climax thinking may arise from a lack of willingness or ability to adapt to landscape change. Fourth, climax thinking could be linked to the belief that forcing local landscape stasis harms no one, that is, there are no losers. This belief could involve either spatial discounting of distant individuals or true ignorance of our ability to impact faraway places and their people by our landscape decisions. Each dimension is described more fully below.

Specific aspects of this proposed pathology of climax thinking overlap with others that have been advanced to explain mired public processes about landscape change in the years since such processes became commonplace. Most of these have emerged from social science, and they explain defaults to the status quo rather than leverage points for alternatives. Here I try to include links to a wide range of literature that touches upon this concept, without pretenses of being exhaustive. The closest concepts I have found in the literature to date are *immutability*, what Pasqualetti described as "the expectation of landscape permanence" (2011, 914) in his quest to understand drivers of opposition to wind energy; and, *continuity*, whereby adaptive capacity depends in part on whether "places remain continuous and provide same attributes and meet certain needs, giving continuity to identity" (Fresque-Baxter and Armitage 2012, 254). These concepts, however, suggest that landscape is stable state; that is, no change has yet occurred or been experienced.

Past

One driver of climax thinking is a limited awareness of past landscapes. Any individual's time in a given place is limited, and so also are their experiences of it. As Simon Schama (1996, 6–7) described, "Landscape is the work of the mind. It's scenery is built up from strata of memory as much as from layers of rock." That time-in-place may cover periods of slow incremental change such as suburbanization, as well as potentially faster, more significant changes such as hydroelectricity inundation. Changes often involve layering the landscape, from one use to another and consequently one memory to another, as new needs are met. Artifacts of past uses are sometimes still visible to those who know how to see them but can be equally easily ignored by those who do not (Hirsh and Jones 2014). Individuals often demonstrate post hoc adaptability to both kinds of change (e.g., "shifting baselines" with subsequent generations; Keilty, Beckley, and Sherren 2016; Pauly 1995). Before this can happen, however, opposition to public good proposals manifests as efforts to keep the current landscape "solution" intact.

The palimpsest of lived landscapes implies some level of erasure, and this can make the past somewhat illegible to those who did not directly experience it. Paul Shepheard draws a parallel between domesticated dogs, who are happy with their lot without understanding it, and humans: "How many humans are simply domesticated? Living in our civilization—our cultivation—without knowing why it exists?" (1997, 19). He goes on later to explain: "As you retreat in time, the evidence [of past lives] becomes so scanty and so contaminated by the process of being passed down the generations that you can be sure about nothing" (30) and "if it's hard to perceive the ancient, simple landscapes, it's harder still to see the ground beneath the clutter of the modern world" (49).

Our resistance to change in our lived landscapes may thus come in part from simple ignorance of not only *what* has come before but *that* something came before. Our perception that our lived landscape is somehow fated might be weakened by awareness that previous decision makers or inhabitants may have chosen to overwrite previous versions of the same place to fill erstwhile needs.

Better awareness of past landscapes will not necessarily combat climax thinking, however, because of what Carl Sagan called temporal chauvinism

(1997), or what C. S. Lewis called chronological snobbery (1955): past landscape change being dismissed as irrelevant to present-day occupants because of a sense that earlier generations were lesser or primitive. In either instance this past blindness or current hubris serves as a very real barrier to sustainability transitions. This is a failure of local historical knowledge, as well as demonstrating ignorance of the broad strokes of human civilization and our place in it, indigenous and settler. There is a good link to this in succession theory, as Clements defined it. He viewed early successional plant communities as laying the groundwork for later ones, thus facilitating them while being clearly less desirable.

Future

Individuals cannot be faulted for the bias introduced by the "ordinality" of time: simply not knowing what is to come. As Barbara Kingsolver wrote in *The Lacuna*, "The past is all we know of the future" (2009, 240). But it does seem that we are guilty of a kind of paternalistic presentism: assuming that what we have built for current generations will serve future generations. Two fallacies may be associated with this kind of thinking: that current solutions will continue to work in future, for example, in a context of climate change, and that future humans are less important than today's if their needs differ. Futurist Jim Dator has described this phenomenon as temporal crackpot realism: the "fully understandable but quite misleading belief that the world of the present will dominate the future" (qtd. in Candy 2010, 68). Such thinking is a failure of the sociological imaginary (Castoriadis 1987), in that what we collectively hold as possible and desirable is often limited by what already exists. It is also, however, a failure of our capacity to fully consider future generations as equally important to our own (Karlsson 2006). Such temporal discounting is a persistent challenge to implementing the full intentions of sustainable development, that which meets "the needs of the present without compromising the ability of future generations to meet their own needs" (World Commission on Environment and Development [WCED] 1987). Intergenerational equity is thus a cognitive challenge as well as a challenge to hubris. Pop culture commentator Chuck Klosterman explores the limitations of the "informed imagination" in *What If We're Wrong? Thinking about the Present As If It Were the Past* (2016, 259): "Before we can argue that something we currently appreciate deserves inclusion in the world of tomorrow, we must build that future world within our mind. This is not easy (even with drugs). But it's not even the hardest part. The hardest part is accepting that we're building something with parts that don't yet exist" (30).

Uncertainty lies at the heart of some of these concepts. For instance, prospect theory is sometimes used to explain resistance to landscape change: the tendency for loss aversion to outweigh the uncertain possibility of future gain (Holtorf 2015; Rogge, Dessein, and Gulinck 2011). For instance, Kate Reilly and others recently found that locals around a hydroelectric dam headpond were able to map current ecosystem service provision but could not imagine those that would follow scenarios such as dam removal (Reilly, Adamowski, and John 2018). Such uncertainty is generally associated with fear (of change or the unknown) or flawed logic (cognitive dissonance, status quo bias). Some conceptualizations come with implied opprobrium for those said to hold them, so-called NIMBY (not in my backyard) or the more awkward NOOMBY (not out of my backyard) (Fox, Magilligan, and Sneddon 2016), with their echoes of the deficit model (Burningham, Barnett, and Walker 2015). In renewable energy settings these conceptualizations have been superseded by place protection, which builds on a significant literature of sense-of-place and place attachment (Devine-Wright 2009). More informal and perhaps mean-spirited conceptualizations of climax thinking include "last one in, close the door," also called the "gangplank" or "last-settler" syndrome (Graber 1974; Voss 1980).

Ecology gives us numerous analogies for this phenomenon. In resilience terms, climax thinking relates to the desire to unnaturally prolong the "fore loop" of current landscape settings (Allen et al. 2014). A frequent analogy is fire suppression in a forest, which otherwise would naturally recycle nutrients and initiate secondary succession, leading to increased risk of catastrophic fire. In social systems also, extending the fore loop can lead to a more brittle and maladaptive system (Slight, Adams, and Sherren 2016). This is an excellent corollary for climax thinking—the desire to hold in stasis, or force stasis, despite changing conditions and the need to reuse some of the various capitals otherwise locked up. Repeatedly rebuilding after disaster may be another example: a political decision but a poor collective investment. Clements used the ability of a rangeland to return to its identified climax state after disturbance as a critical diagnostic for the health of that system. A non-equilibrium model sees it differently.

Space and Place

Our desire to hold our lived landscape in stasis can force those of others to change. Thanks to a highly interconnected society and economy, local decisions can have far-flung landscape and livelihood implications. The decision to reject hydraulic fracturing for shale gas in one place, for instance, may mean a continued reliance on conventional fuels produced elsewhere, perhaps in places with weaker safety and environmental regulations, and may exacerbate sea level rise in yet another. This type of decision is a failure of intragenerational equity—concern for other members of the same generation elsewhere in the globe—which is another persistent challenge to sustainable development (WCED 1987). To some degree this failure explains why climax thinking is characteristic of developed nations. It is a privileged form of buffering against environmental signals that only those with capital can undertake (Meyfroidt 2013), particularly those with livelihoods outside the primary and secondary economic sectors and thus decoupled from nearby siting and land use decisions.

The scholarship of place has been particularly concerned with the multiplex relationships between people and landscapes and how those can best be sustained. While place theory is useful here for how it explains place attachment, in terms of subjective meanings (Brehm, Eisenhauer, and Stedman 2013; Stedman 2016) and time-in-place (Smaldone, Harris, and Sanyal 2008; Vorkinn and Riese 2001), and though place theory makes reference to iteration, in practice it seems to validate stasis. The difference may simply be semantics; place attachment is positively framed climax thinking. It has been shown, for instance, that place attachment and identity can reduce transformational capacity, such as in changing commodities in the face of climate change in Australia (Marshall et al. 2012). While we may believe we will not be able to adapt to change in our lived landscapes, the opposite has been repeatedly demonstrated as today's landscapes have emerged. Indeed, landscape expectations and preferences can evolve even within the generation that has witnessed quick and dramatic change (e.g., hydroelectricity development) (Keilty, Beckley, and Sherren 2016). Even if current settings or features are perceived as irreplaceable in terms of place attachment or other ecosystem service supply, there may be significant elasticity in the sources of values derived from landscape (Daw et al. 2016).

Leverage Points for Landscape Transition

We need to develop our knowledge, imagination, and empathy to change our landscapes and perhaps our cultural infrastructure in the face of new challenges. Most important, we must do this without cultural obliteration, environmental degradation, and rupture in human-environment relationships (Hourdequin and Havlick 2015). This section proposes three key leverage points to tackle climax thinking, illustrated with examples of research gaps across a range of fields, including social science, environmental assessment, spatial science, "Big Data" and digital technologies (table 1.2). It is significant that none of these leverage points, or specific suggested directions to take action on them, would have the outcome of disempowering valid resistance in the face of development proposals which were not seen to represent a public good; arguably, they would help in identifying such instances.

Heal Our Link with the Past, without Anchoring There

What the above makes clear is that landscape is not ever completed. We need to reveal the fact that landscape is an ongoing trajectory of problem solving, without somehow getting stuck in those past visions. If a blindness to past landscapes drives climax thinking, this is in part because of their erasure. New media may have a role to play. Thanks to archives and online and mobile mapping, improved transparency is well within our ability. We need a better sense of how many times, and in how many ways, our lived landscapes have already been written and rewritten (Hanson 2012). Archival photos and maps can reveal many different versions of a place over time, but their display is limited by cartographic convention. Instead, such resources can be brought to life using online applications such as story maps to remind viewers of past occupants and some of the past viable landscapes they created and/or overwrote as needs changed. Digital archives and social media can also be used to help us track cultural trajectory alongside landscape change, including that resulting from energy transitions (Sherren et al. 2017). Such insight on past land uses at sites facing land use change could be co-created as well as shared with the public via dialogue in more collaborative environmental assessment and stakeholder engagement processes (Eaton et al. 2017). Further research is needed to explore whether exposure to site-specific trajectories of past and

TABLE 1.2. A cross-disciplinary research and action agenda for climax thinking

Field	*Past*	*Future*	*Self and Other*
Spatial science	Exposes past layers and trajectories	Shows challenges and viable scenarios	Reveals how our energy choices propagate globally
Big Data and digital technology	Track cultural trajectories through landscape change	Digitally conserve past solutions to make space for new ones	Reach outside online bubbles to develop empathy
Environmental assessment	Conveys past trajectory	Interrogates "public good" and incorporate end-of-life planning	Documents inter- and intragenerational implications of status quo and proposal
Social science	Do tools to reveal past and possible trajectories ease climax thinking?	How does place reattachment occur, and how can it be facilitated?	How does knowing where your energy comes from, perhaps locally, affect attitudes and consumption?

current landscape solutions eases transitions to new ones. There is some evidence for the former: in research on changing woodland cover in the UK, exposure to maps of past woodland arrangements (as well as literary perceptions of the same) made survey respondents significantly less likely to opt for status quo scenarios (Hanley et al. 2009).

Learn Layering: We Can't Keep Everything

Transition will require learning how to negotiate the editing of lived landscapes. This chapter does not endorse replacing landscape preservation with layering: it is critical that some cultural landscapes are maintained as records of past landscape solutions and associated cultures, but we cannot keep everything. Heritage experts have been grappling with questions about the risk of cultural erasure (Holtorf 2015), as well as the practical need for controlled forgetting (Harrison 2013; Holtorf and Kristensen 2015) to leave space for new culture (e.g., nuclear domes; Holtorf and Högberg 2014). Landscape planners may have things to learn from archaeologists and heritage specialists. Those fields are developing an increasingly wide range of opportunities for "conservation by record" thanks to 3D and immersive technologies (Champion 2017; De Reu et al. 2013; Seif 2009). Techniques such as these have already been used in many landscape change settings to simulate scenarios of future conditions but rarely to conserve or reconstruct those becoming past. Tools such as digital globes and phone apps can also augment our understanding of current landscape values and challenges such as sea level rise to public stakeholders (Bishop 2015; Harwood, Lovett, and Turner 2015). Research is required to understand how such immersive digital archives or renderings might be experienced and perceived by those who see landscapes-at-risk as entangled with their own or ancestors' efforts and identities. Moreover, we need to better understand how to facilitate processes of place reattachment after place disruption (Keilty, Beckley, and Sherren 2016).

Collaborative-planning processes could perhaps also be improved with more transparent end-of-life planning discussions at the proposal stage. It is important to emphasize to local populations the reasonable lifespan of new infrastructure and what might follow. A proposed dam, for instance, has a limited lifespan, so proponents and affected landowners should be able to consider its implications at the outset of the project. We should also be more willing to remove old uses as well as add new ones, which in the context of energy Martin Pasqualetti has called "recycling" landscapes, consistent with nutrient cycling in forests, given the varying permanence and "temporal qualification" of energy landscapes (Pasqualetti and Stremke 2018). More research is needed on whether early and transparent end-of-life planning can ease transition related to specific infrastructure, by casting it as temporary, or whether the specter of future disruption in fact increases resistance. It may also be that certain kinds of baselines (status quo landscapes) operate differently, for instance, in terms of expectations of permanence, or perceptions of naturalness, desirability or "blank slate."

Climax thinking is in part a failure of imagination (Ingold 2012). We may need to reconsider our landscape strategy (*sensu* Shepheard 1997) to recast landscape to meet new challenges. One alternative landscape strategy, for the purposes of illustration only, could be "local energy." Unlike with "local food," in North America there seems a lack of interest in taking similar ownership of and responsibility for energy use, generation choices, and their carry-on effects. Europe is ahead on such thinking, perhaps because of the need to "overwrite" their smaller landmass earlier (e.g., de Waal and Stremke 2014). Rejection of renewable energy proposals, for instance, may prolong reliance on imported fuels and electricity with remote negative externalities. A local energy ethic would expose energy consumers to environmental signals and thus might inspire energy conservation. It might also reduce opposition to local renewable energy infrastructure in contexts where local alternatives (e.g., fracked natural gas) may represent an environmental or a health risk. The negotiation and implementation of such new landscape strategies may be able to ease landscape transition and give meaning to disruption but must be informed by more research. A local energy strategy raises important hypotheses that need testing, for instance, to establish if people generally know where their energy currently comes from and whether places currently supplied with locally produced energy consume less of it. Moreover, we may be able to reduce the impacts of climax thinking if we shift our thinking about renewable energy to considering it as a commodity no different than forestry or agriculture; why should opportunities for export be anathema and drive so much resistance to development?

Build Empathy for Other Lives and Our Impacts on Them

Climax thinking is also in part a failure of empathy. Again, new media has an important role to play in building empathy and system knowledge. So far it has not. As Marshall McLuhan is oft quoted as saying, "We shape our tools and afterwards they shape us." Cosmopolitanism is a philosophy that emphasizes our duty to consider the global community of humans in our actions. Cosmopolitanism is typically associated with calls for wider education in the humanities (Fischer et al. 2007; Sherren 2008), for instance, literature to develop a sense of empathy and the capacity to envision the experience of other lives (Nussbaum 2002). More ubiquitous than literature is that other potential "empathy machine," the internet. Ethan Zuckerman (2013) observes that ready access to global media and culture is making us think we are worldly, despite the fact that we tread relatively worn, familiar paths when we visit there. Such paths delineate the echo chamber, which reinforces prejudices through news feeds fed by algorithms and social networks (Jasny, Waggle, and Fisher 2015). International news coverage in American media has declined with the rise of access to online sources, but our bubbles make sure we do not become exposed to such coverage (Zuckerman 2013). So what might it look like to "do internet differently"? We could learn about what life is like where our energy sources currently come from. Faced with a landscape change, we could seek out a vicarious experience from someone who has faced similar landscape changes. More substantively than recreational internet use, it may be possible to integrate intragenerational considerations into environmental assessment processes (Gibson 2006; Winfield et al. 2010)? Climax thinking may be reduced by understanding how needs (e.g., energy) are currently met, and the impact of that provision on others, as well as how those impacts of new proposals might be distributed.

Conclusion

This chapter draws on the ecological concept of succession to present a new concept, climax thinking, uniting and adding to ideas related to public resistance to landscape change emerging from a range of disciplines. Public good landscape transitions are hampered due to climax thinking, our erroneous perception that our lived landscape is in its peak state. Any perceived summit is only a powerful illusion: land uses must continue to change and layer as new needs and priorities are encountered. That said, this contribution is not meant to provide ammunition to those endorsing unexamined opportunities for large-scale landscape change. This chapter describes lived landscapes as layered landscapes that present us with challenges to adapt for new needs. A multidimensional pathology of climax thinking is proposed that covers time and space, including hubris or ignorance as potential drivers. Three broad leverage points are proposed to help us ease transitions across those dimensions: (1) healing our link with the past, (2) learning layering, and (3) building empathy. A cross-disciplinary mix of directions are proposed to make progress across those leverage points. This research and action agenda should only be a starting point as we learn how to adapt our shared landscapes in the face of significant local and global challenges.

Acknowledgments

Thanks to three anonymous reviewers for their constructive advice, John Parkins and Tom Measham for reading and commenting on an earlier draft, and to Jaya Fahey for graphic design. Thanks also to Dylan Bugden, Nichole Dusyk, Weston Eaton, Josh Fergen, David L. Kay, Leah Stokes, Jeremy Weber, John Whitton, and other attendees for discussions of this work at the NSF-funded Energy Impacts Symposium in Columbus, Ohio, in July 2017, where this work was first presented, and to the symposium organizers for financial support to attend that event. This work was supported in part by grants from the Social Science and Humanities Research Council of Canada to the author, as PI (430-2012-0641) and as co-PI (435-2012-0636, Parkins PI).

References

Abrams, Jesse, and John C. Bliss. 2013. "Amenity Landownership, Land Use Change, and the Re-creation of 'Working Landscapes.'" *Society and Natural Resources* 26 (7): 845–859. https://doi.org/10.1080/08941920.2012.719587.

Albrecht, Glenn, Gina-Maree Sartore, Linda Connor, Nick Higginbotham, Sonia Freeman, Brian Kelly, Helen Stain, Anne Tonna, andGeorgia Pollard. 2007. "Solastalgia: The Distress Caused by Environmental Change." *Australasian Psychiatry* 15 (Supplement 1): S95–S98. https://doi.org/10.1080/10398560701701288.

Allen, Craig R., David G. Angeler, Ahjond S. Garmestani, Lance H. Gunderson, and Crawford S. Holling. 2014. "Panarchy: Theory and Application." *Ecosystems* 17 (4): 578–589. https://doi.org/10.1007/s10021-013-9744-2.

Birch, Eugenie, and Susan Wachter, eds. 2006. *Rebuilding Urban Places after Disaster: Lessons from Hurricane Katrina*. Philadelphia: University of Pennsylvania Press.

Bishop, Ian D. 2015. "Location Based Information to Support Understanding of Landscape Futures." *Landscape and Urban Planning* 142: 120–131. https://doi.org/10.1016/j.landurbplan.2014.06.001.

Brehm, Joan M., Brian W. Eisenhauer, and Richard C. Stedman. 2013. "Environmental Concern: Examining the Role of Place Meaning and Place Attachment." *Society and Natural Resources* 26 (5): 522–538. https://doi.org/10.1080/08941920.2012.715726.

Brennan, Geoffrey. 1996. "Selection and the Currency of Reward." In *The Theory of Institutional Design*, edited by Robert E. Goodin, 256–276. Cambridge: Cambridge University Press.

Burningham, Kate, Julie Barnett, and Gordon Walker. 2015. "An Array of Deficits: Unpacking NIMBY Discourses in Wind Energy Developers' Conceptualizations of Their Local Opponents." *Society and Natural Resources* 28 (3): 246–260. https://doi.org/10.1080/08941920.2014.933923.

Candy, Stuart. 2010. *The Futures of Everyday Life: Politics and the Design of Experiential Scenarios*. PhD diss., University of Hawai'i, Mānoa.

Castoriadis, C. 1987. *The Imaginary Institution of Society*. Translated by Kathleen Blamey. Cambridge: Policy Press.

Champion, Erik. 2017. "The Role of 3D Models in Virtual Heritage Infrastructures." In *Cultural Heritage Infrastructures in Digital Humanities*, edited by Agiatis Benardou, Erik Champion, Costis Dallas, and Lorna Hughes, 15–35. New York: Routledge.

Cosgrove, Denis, and Stephen Daniels, eds. 1988. *The Iconography of Landscape: Essays on the Symbolic Representation, Design and Use of Past Environments*. Vol. 9. Cambridge: Cambridge University Press.

Council of Europe (CoE). 2000. *European Landscape Convention. CETS No. 176*. Florence and Strasbourg: Council of Europe. https://rm.coe.int/1680080621.

Cresswell, Tim. 1996. *In Place/Out of Place: Geography, Ideology, and Transgression*. Minneapolis: University of Minnesota Press.

Cretney, Raven. 2014. "Resilience for Whom? Emerging Critical Geographies of Socio-ecological Resilience." *Geography Compass* 8 (9): 627–640. https://doi.org/10.1111/gec3.12154.

Crowe, Sylvia. 1958. *Landscape of Power*. London: Architectural Press.

Davidson, Debra J. 2010. "The Applicability of the Concept of Resilience to Social Systems: Some Sources of Optimism and Nagging Doubts." *Society and Natural Resources* 23 (12): 1135–1149.

Daw, Tim, Christina Hicks, Katrina Brown, Tomas Chaigneau, Fraser Januchowski-Hartley, William Cheung, and Sérgio Rosendo. et al. 2016. "Elasticity in Ecosystem Services: Exploring the Variable Relationship between Ecosystems and Human Well-Being." *Ecology and Society* 21 (2): 11. https://doi.org/10.5751/ES-08173-210211.

De Reu, Jeroen, Gertjan Plets, Geert Verhoeven, Philippe De Smedt, Machteld Bats, Bart Cherretté, and, Wouter De Maeyer, et al. 2013. "Towards a Three-Dimensional Cost-Effective Registration of the Archaeological Heritage." *Journal of Archaeological Science* 40 (2): 1108–1121. https://doi.org/10.1016/j.jas.2012.08.040.

de Waal, Renée, and Sven Stremke. 2014. "Energy Transition: Missed Opportunities and Emerging Challenges for Landscape Planning and Designing." *Sustainability* 6 (7): 4386.

Depietri, Yaella, Fabrice G. Renaud, and Giorgos Kallis. 2012. "Heat Waves and Floods in Urban Areas: A Policy-Oriented Review of Ecosystem Services." *Sustainability Science* 7 (1): 95–107. https://doi.org/10.1007/s11625-011-0142-4.

Devine-Wright, Patrick. 2009. "Rethinking NIMBYism: The Role of Place Attachment and Place Identity in Explaining Place-Protective Action." *Journal of Community and Applied Social Psychology* 19 (6): 426–441. https://doi.org/10.1002/casp.1004.

Drenthen, Martin. 2015. "Layered Landscapes, Conflicting Narratives, and Environmental Art: Dealing with Painful Memories and Embarrassing Histories of Place." In *Restoring Layered Landscapes: History, Ecology, and Culture*, edited by Marion Hourdequin and David G. Havlick, 239–262. New York: Oxford University Press.

Eaton, Weston M., Morey Burnham, C. Clare Hinrichs, and Theresa Selfa. 2017. "Bioenergy Experts and Their Imagined 'Obligatory Publics' in the United States: Implications for Public Engagement and Participation." *Energy Research and Social Science* 25 (March): 65–75. https://doi.org/10.1016/j.erss.2016.12.003.

Elmqvist, Thomas, Carl Folke, Magnus Nyström, Garry Peterson, Jan Bengtsson, Brian Walker, and Jon Norberg. 2003. "Response Diversity, Ecosystem Change, and Resilience." *Frontiers in Ecology and the Environment* 1 (9): 488–494. https://doi.org/10.1890/1540-9295(2003)001[0488:RDECAR]2.0.CO;2.

Fischer, Joern, Adrian D. Manning, Will Steffen, Deborah B. Rose, Katherine Daniell, Adam Felton, Stephen Garnett, et al. 2007. "Mind the Sustainability Gap." *Trends in Ecology and Evolution* 22 (12): 621–624. https://doi.org/10.1016/j.tree.2007.08.016.

Fox, Coleen A., Francis J. Magilligan, and Christopher S. Sneddon. 2016. "'You Kill the Dam, You Are Killing a Part of Me': Dam Removal and the Environmental Politics of River Restoration." *Geoforum* 70: 93–104. https://doi.org/10.1016/j.geoforum.2016.02.013.

Fresque-Baxter, Jennifer A., and Derek Armitage. 2012. "Place Identity and Climate Change Adaptation: A Synthesis and Framework for Understanding." *Wiley Interdisciplinary Reviews: Climate Change* 3 (3): 251–266. https://doi.org/10.1002/wcc.164.

Geels, Frank W., and Johan Schot. 2007. "Typology of Sociotechnical Transition Pathways." *Research Policy* 36 (3): 399–417. https://doi.org/10.1016/j.respol.2007.01.003.

Gibson, Robert B. 2006. "Sustainability Assessment: Basic Components of a Practical Approach." *Impact Assessment and Project Appraisal* 24 (3): 170–182. https://doi.org/10.3152/147154606781765147.

Giddens, Anthony. 1979. *Central Problems in Social Theory: Action, Structure, and Contradiction in Social Analysis*. Vol. 241. Berkeley: University of California Press.

Goodin, Robert E. 1998. *The Theory of Institutional Design*. Cambridge: Cambridge University Press.

Graber, Edith E. 1974. "Newcomers and Oldtimers: Growth and Change in a Mountain Town." *Rural Sociology* 39 (4): 504.

Hager, Carol, and Mary Alice Haddad. 2015. *NIMBY Is Beautiful: Cases of Local Activism and Environmental Innovation around the World*. New York: Berghahn Books.

Handmer, John W., and Stephen R. Dovers. 1996. "A Typology of Resilience: Rethinking Institutions for Sustainable Development." *Organization and Environment* 9 (4): 482–511.

Hanley, Nick, Richard Ready, Sergio Colombo, Fiona Watson, Mairi Stewart, and E. Ariel Bergmann. 2009. "The Impacts of Knowledge of the Past on Preferences for Future Landscape Change." *Journal of Environmental Management* 90 (3): 1404–1412. https://doi.org/10.1016/j.jenvman.2008.08.008.

Hanson, Lorelei L. 2012. "Changes in the Social Imaginings of the Landscape: The Management of Alberta's Rural Public Lands." In *Social Transformation in Rural Canada: Community, Cultures, and Collective Action*, edited by John R. Parkins and Maureen G. Reed, 148–168. Vancouver: University of British Columbia Press.

Harrison, Rodney. 2013. "Forgetting to Remember, Remembering to Forget: Late Modern Heritage Practices, Sustainability and the 'Crisis' of Accumulation of the Past." *International Journal of Heritage Studies* 19 (6): 579–595. https://10.1080/13527258.2012.678371.

Harwood, Amii R., Andrew A. Lovett, and Jenni A. Turner. 2015. "Customising Virtual Globe Tours to Enhance Community Awareness of Local Landscape Benefits." *Landscape and Urban Planning* 142 (October): 106–119. https://doi.org/10.1016/j.landurbplan.2015.08.008.

Healy, Kieran. 2017. "Fuck Nuance." *Sociological Theory* 35 (2): 118–127. https://doi.org/10.1177/0735275117709046.

Hess, David J. 2018. "Energy Democracy and Social Movements: A Multi-coalition Perspective on the Politics of Sustainability Transitions." *Energy Research and Social Science* 40 (June): 177–189. https://doi.org/10.1016/j.erss.2018.01.003.

Higgs, Eric, Donald A. Falk, Anita Guerrini, Marcus Hall, Jim Harris, Richard J. Hobbs, Stephen T. Jackson, Jeanine M. Rhemtulla, and William Throop. 2014. "The Changing Role of History in Restoration Ecology." *Frontiers in Ecology and the Environment* 12 (9): 499–506. https://doi.org/10.1890/110267.

Hirsh, Richard F., and Christopher F. Jones. 2014. "History's Contributions to Energy Research and Policy." *Energy Research and Social Science* 1 (March): 106–111. https://doi.org/10.1016/j.erss.2014.02.010.

Holtorf, Cornelius. 2015. "Averting Loss Aversion in Cultural Heritage." *International Journal of Heritage Studies* 21 (4): 405–421. https://doi.org/10.1080/13527258.2014.938766.

Holtorf, Cornelius, and Anders Högberg. 2014. "Communicating with Future Generations: What Are the Benefits of Preserving for Future Generations? Nuclear Power and Beyond." *European Journal of Post-classical Archaeologies* 4: 315–330.

Holtorf, Cornelius, and Troels Myrup Kristensen. 2015. "Heritage Erasure: Rethinking 'Protection' and 'Preservation.'" *International Journal of Heritage Studies* 21 (4): 313–317. https://doi.org/10.1080/13527258.2014.982687.

Hotimsky, Samy, Richard Cobb, and Alan Bond. 2006. "Contracts or Scripts? A Critical Review of the Application of Institutional Theories to the Study of Environmental Change." *Ecology and Society* 11(1):41.

Hourdequin, Marion, and David G. Havlick. 2015. *Restoring Layered Landscapes: History, Ecology, and Culture*. New York: Oxford University Press.

Ingold, Tim. 2012. "Introduction." In *Imagining Landscapes: Past, Present and Future*, edited by Monika Janowski and Tim Ingold, 1–18. New York: Routledge.

Jasny, Lorien, Joseph Waggle, and Dana R. Fisher. 2015. "An Empirical Examination of Echo Chambers in US Climate Policy Networks." *Nature Climate Change* 5 (8): 782–786. https://doi.org/10.1038/nclimate2666.

Karlsson, Rasmus. 2006. "Reducing Asymmetries in Intergenerational Justice." *Organization and Environment* 19 (2): 233–250. https://doi.org/10.1177/1086026606288227.

Kasemets, Kadri. 2015. "Affect, Rupture and Heritage on Hashima Island, Japan." In *Ruptured Landscapes: Landscape, Identity and Social Change*, edited by Helen Sooväli-Sepping, Hugo Reinert, and Jonathan Miles-Watson, 97–109. Dordrecht, Netherlands: Springer.

Keilty, Kristina, Thomas M. Beckley, and Kate Sherren. 2016. "Baselines of Acceptability and Generational Change on the Mactaquac Hydroelectric Dam Headpond (New Brunswick, Canada)." *Geoforum* 75: 234–248.

Kingsolver, Barbara. 2009. *The Lacuna*. New York: HarperCollins.

Klosterman, Chuck. 2016. *What If We're Wrong? Thinking about the Present as If It Were the Past*. New York: Blue Rider Press.

Lewis, C. S. 1955. *Surprised by Joy: The Shape of My Early Life*. Vol. 320. New York: Houghton Mifflin Harcourt.

Lundhede, Thomas Hedemark, Jette Bredahl Jacobsen, Nick Hanley, Jon Fjeldså, Carsten Rahbek, Niels Strange, and Bo Jellesmark Thorsen. 2014. "Public Support for Conserving Bird Species Runs Counter to Climate Change Impacts on Their Distributions." *PLOS ONE* 9 (7): e101281. https://doi.org/10.1371/journal.pone.0101281.

Magilligan, F. J., C. S. Sneddon, and C. A. Fox. 2017. "The Social, Historical, and Institutional Contingencies of Dam Removal." *Environmental Management* 59 (6): 982–994. https://doi.org/10.1007/s00267-017-0835-2.

Marris, Peter. 2014. *Loss and Change (Psychology Revivals)*. Rev. ed. New York: Routledge.

Marshall, N. A., S. E. Park, W. N. Adger, Katrina Brown, and S. M. Howden. 2012. "Transformational Capacity and the Influence of Place and Identity." *Environmental Research Letters* 7 (3): 034022.

Meinig, Donald W. 1979. "The Beholding Eye: Ten Versions of the Same Scene." In *The Interpretation of Ordinary Landscapes: Geographical Essays*, edited by Donald W. Meinig, 33–50. New York: Oxford University Press.

Mels, Tom. 2016. "The Trouble with Representation: Landscape and Environmental Justice." *Landscape Research* 41 (4): 417–424. https://doi.org/10.1080/01426397.2016.1156071.

Meyfroidt, Patrick. 2013. "Environmental Cognitions, Land Change, and Social-Ecological Feedbacks: An Overview." *Journal of Land Use Science* 8 (3): 341–367. https://doi.org/10.1080/1747423X.2012.667452.

Mitchell, Nora, Mechtild Rössler, and Pierre Marie Tricaud. 2009. *World Heritage Cultural Landscapes: A Handbook for Conservation and Management*. Paris: UNESCO World Heritage Centre.

Monbiot, George. 2014. *Feral: Rewilding the Land, the Sea, and Human Life*. Chicago: University of Chicago Press.

Newell, Peter, and Dustin Mulvaney. 2013. "The Political Economy of the 'Just Transition.'" *Geographical Journal* 179 (2): 132–140.

Nussbaum, M. 2002. "Education for Citizenship in an Era of Global Connection." *Studies in Philosophy and Education* 21 (4): 289–303. https://doi.org/10.1023/a:1019837105053.

Olsson, Lennart, Anne Jerneck, Henrik Thoren, Johannes Persson, and David O'Byrne. 2015. "Why Resilience Is Unappealing to Social Science: Theoretical and Empirical Investigations of the Scientific Use of Resilience." *Science Advances* 1 (4): e1400217. https://doi.org/10.1126/sciadv.1400217.

Parkins, John R., Thomas Beckley, Louise Comeau, Richard C. Stedman, Curtis L. Rollins, and Anna Kessler. 2017. "Can Distrust Enhance Public Engagement? Insights from a National Survey on Energy Issues in Canada." *Society and Natural Resources* 30 (8): 934–948. https://doi.org/10.1080/08941920.2017.1283076.

Pasqualetti, Martin J. 2011. "Opposing Wind Energy Landscapes: A Search for Common Cause." *Annals of the Association of American Geographers* 101 (4): 907–917. https://doi.org/10.1080/00045608.2011.568879.

Pasqualetti, Martin J., and Sven Stremke. 2018. "Energy Landscapes in a Crowded World: A First Typology of Origins and Expressions." *Energy Research and Social Science* 36 (February): 94–105. https://doi.org/10.1016/j.erss.2017.09.030.

Pasternak, Judy. 2010. *Yellow Dirt: An American Story of a Poisoned Land and a People Betrayed*. New York: Simon and Schuster.

Pauly, Daniel. 1995. "Anecdotes and the Shifting Baseline Syndrome of Fisheries." *Trends in Ecology and Evolution* 10 (10): 430.

Pelling, Mark, Karen O'Brien, and David Matyas. 2015. "Adaptation and Transformation." *Climatic Change* 133 (1): 113–127. https://doi.org/10.1007/s10584-014-1303-0.

Plieninger, Tobias, Thanasis Kizos, Claudia Bieling, Le Dû-Blayo, Marie-Alice Budniok, Matthias Bürgi, and Carole Crumley, et al. 2015. "Exploring Ecosystem-Change and Society through a Landscape Lens: Recent Progress in European Landscape Research." *Ecology and Society* 20 (2): 5.

Reed, Maureen G., and Colleen George. 2011. "Where in the World Is Environmental Justice?" *Progress in Human Geography* 35 (6): 835–842. https://doi.org/10.1177/0309132510388384.

Reilly, Kate, Jan Adamowski, and Kimberly John. 2018. "'Participatory Mapping of Ecosystem Services to Understand Stakeholders' Perceptions of the Future of the Mactaquac Dam, Canada." *Ecosystem Services* 30: 107–123. https://doi.org/10.1016/j.ecoser.2018.01.002.

Robin, Libby, and Will Steffen. 2007. "History for the Anthropocene." *History Compass* 5 (5): 1694–1719. https://doi.org/10.1111/j.1478–0542.2007.00459.x.

Rogge, Elke, Joost Dessein, and Hubert Gulinck. 2011. "Stakeholders Perception of Attitudes towards Major Landscape Changes Held by the Public: The Case of Greenhouse Clusters in Flanders." *Land Use Policy* 28 (1): 334–342. https://doi.org/10.1016/j.landusepol.2010.06.014.

Rudel, Thomas K. 2009. "Succession Theory: Reassessing a Neglected Meta-narrative about Environment and Development." *Human Ecology Review* 16 (1): 84.

Sagan, Carl. 1997. *Pale Blue Dot: A Vision of the Human Future in Space*. New York: Random House.

Sayre, Nathan F. 2017. *The Politics of Scale: A History of Rangeland Science*. Chicago: University of Chicago Press.

Schama, Simon. 1996. *Landscape and Memory*. New York: Knopf.

Seif, Assaad. 2009. "Conceiving the Past: Fluctuations in a Multi-value System." *Conservation and Management of Archaeological Sites* 11 (3–4): 282–295. https://doi.org/10.1179/175355210X12747818485484.

Shepheard, Paul. 1997. *The Cultivated Wilderness, Or, What Is Landscape?* Cambridge, MA: MIT Press.

Sherren, Kate. 2008. "A History of the Future of Higher Education for Sustainable Development." *Environmental Education Research* 14 (3): 238–256. https://doi.org/10.1080/13504620802148873.

Sherren, Kate, Thomas Beckley, Simon Greenland-Smith, and Louise Comeau. 2017. "How Provincial and Local Discourses Aligned against the Prospect of Dam Removal in New Brunswick, Canada." *Water Alternatives* 10 (3): 697–723.

Sherren, Kate, Thomas M. Beckley, John R. Parkins, Richard C. Stedman, Kristina Keilty, and Isabelle Morin. 2016. "Learning (or Living) to Love the Landscapes of Hydroelectricity in Canada: Eliciting Local Perspectives on the Mactaquac Dam via Headpond Boat Tours." *Energy Research and Social Science* 14: 102–110.

Sherren, Kate, Logan Loik, and James A. Debner. 2016. "Climate Adaptation in 'New World' Cultural Landscapes: The Case of Bay of Fundy Agricultural Dykelands (Nova Scotia, Canada)." *Land Use Policy* 51: 267–280.

Sherren, Kate, John R. Parkins, Michael Smit, Mona Holmlund, and Yan Chen. 2017. "Digital Archives, Big Data and Image-based Culturomics for Social Impact Assessment: Opportunities and Challenges." *Environmental Impact Assessment Review* 67: 23–30. https://doi.org/10.1016/j.eiar.2017.08.002.

Simmie, James. 2012. "Path Dependence and New Technological Path Creation in the Danish Wind Power Industry." *European Planning Studies* 20 (5): 753–772. https://doi.org/10.1080/09654313.2012.667924.

Slight, Penny, Michelle Adams, and Kate Sherren. 2016. "Policy Support for Rural Economic Development Based on Holling's Ecological Concept of Panarchy." *International Journal of Sustainable Development and World Ecology* 23 (1): 1–14. https://doi.org/10.1080/13504509.2015.1103801.

Smaldone, David, Charles Harris, and Nick Sanyal. 2008. "The Role of Time in Developing Place Meanings." *Journal of Leisure Research* 40 (4): 479.

Stedman, Richard C. 2016. "Subjectivity and Social-Ecological Systems: A Rigidity Trap (and Sense of Place as a Way Out)." *Sustainability Science* 11 (6): 891–901.

Stenseke, Marie. 2016. "Integrated Landscape Management and the Complicating Issue of Temporality." *Landscape Research* 41 (2): 199–211. https://doi.org/10.1080/01426397.2015.1135316.

Stobbelaar, Derek Jan, and Bas Pedroli. 2011. "Perspectives on Landscape Identity: A Conceptual Challenge." *Landscape Research* 36 (3): 321–339. https://doi.org/10.1080/01426397.2011.564860.
Stokes, Leah C. 2016. "Electoral Backlash against Climate Policy: A Natural Experiment on Retrospective Voting and Local Resistance to Public Policy." *American Journal of Political Science* 60 (4): 958–974. https://doi.org/10.1111/ajps.12220.
Stokowski, Patricia A. 2002. "Languages of Place and Discourses of Power: Constructing New Senses of Place." *Journal of Leisure Research* 34 (4): 368–382.
Turner, Jonathan H. 1997. *The Institutional Order: Economy, Kinship, Religion, Policy, Law, and Education in Evolutionary and Comparative Perspective*. New York: Longman Publishing Group.
Vorkinn, Marit, and Hanne Riese. 2001. "Environmental Concern in a Local Context." *Environment and Behavior* 33 (2): 249–263. https://doi.org.10.1177/00139160121972972.
Voss, Paul R. 1980. "A Test of the 'Gangplank Syndrome' among Recent Migrants to the Upper Great Lakes Region." *Community Development* 11 (1): 95–111.
Widgren, Mats. 2004. "Can Landscapes Be Read?" *European Rural Landscapes: Persistence and Change in a Globalising Environment*, edited by Hannes Palang, Helen Sooväli, Marc Antrop, and Gunhild Setten, 455–465. Dordrecht, The Netherlands: Kluwer Academic Publishers.
Winfield, Mark, Robert B. Gibson, Tanya Markvart, Kyrke Gaudreau, and Jennifer Taylor. 2010. "Implications of Sustainability Assessment for Electricity System Design: The Case of the Ontario Power Authority's Integrated Power System Plan." *Energy Policy* 38 (8): 4115–4126. https:// doi.org/10.1016/j.enpol.2010.03.038.
World Commission on Environment and Development (WCED). 1987. *Our Common Future ('The Brundtland Report')*. New York: Oxford University Press.
Zuckerman, Ethan. 2013. *Digital Cosmopolitans: Why We Think the Internet Connects Us, Why It Doesn't, and How to Rewire It*. New York: W.W. Norton and Company.

Chapter 1 Summary

In the absence of strong political will, public resistance to landscape change is a significant challenge to the kinds of transitions needed to make to achieve long-term sustainability. Such resistance happens across the urban-rural gradient: from protesting the condo development next door that will house more people in a smaller area and reduce our need for cars, to residents opposing a large-scale wind farm to reduce our dependency on fossil fuels. Such arguably public good landscape changes are always challenged by those living nearby, a phenomenon well explored in energy research. This chapter draws upon succession theory to describe the phenomenon of "climax thinking," the sense we have that the landscape we currently live in is in its ideal, perhaps even intended, state. The pathology of climax thinking is dissected into temporal and spatial dimensions of ignorance or egotism. The past dimension is being unaware of any previous land uses or, if aware, seeing those past landscapes as lesser, along with past residents who had to suffer change for today's landscape to emerge. The future dimension is assuming current land uses will continue to work in future or feeling that we have no duty to anticipate the needs of future residents. The spatial dimensions are anchored in the self—our feeling that we should not need to, or cannot, accept change—toward incomprehensible others, elsewhere, whose landscapes change precisely because we seek to hold ours static. The pathology, once described, is tackled by outlining a potential set of leverage points for easing each dimension: healing our link with the past, learning layering, and building empathy for other lives. Returning to succession theory, the chapter advocates for a multidisciplinary research and action agenda across the social and computational sciences to facilitate a non-equilibrium way of thinking about landscapes in the face of sustainability transitions.

KEY TAKEAWAYS

- Landscapes must change in line with new societal needs, but such change is politically difficult.
- Climax thinking is fallacious thinking, but near ubiquitous in Western settings.
- Climax thinking is the privileged mobilization of ignorance and hubris across time and space.
- Forcing landscape stasis despite changing conditions and needs pushes impacts to those less able to resist.
- Leverage points to reduce climax thinking may include improving awareness of past landscape changes and landscape changes elsewhere that our decisions may cause.
- As in succession theory, climax thinking should be challenged by a non-equilibrium approach to thinking about landscapes that acknowledges a range of viable futures exist beyond the status quo.

2

Entangled Impacts

Human-Animal Relationships and Energy Development

SEVEN MATTES AND CAMERON T. WHITLEY

A growing area of inquiry addresses the place of animals in energy development with a particular focus on history, the use of sentinels (animals used to detect risks to humans such as the canary in a coal mine), and impacts to animals and the human-animal bond (for review see, e.g., Whitley 2017). Much of this work recognizes that because animals are so intertwined with human lives, the place of animals in energy development and associated impacts to animals may elicit significant psychological, social, and community repercussions for humans. Risk and resilience studies suggest that we should take the human-animal bond seriously. For instance, how an individual responds after a disaster (e.g., energy development disasters) can be significantly related to their connection to a companion animal and whether this connection is honored in disaster management practices (Irvine 2009).

Joining sociological and anthropological approaches, we examine two case studies of the entangled impacts of humans and animals in energy development. The first case study looks at the place of animals in the context of the high-volume hydraulic fracturing (HVHF) "fracking" boom in the United

DOI: 10.5876/9781646420278.c002

States. We suggest that little attention has been paid to the human-animal bond in terms of HVHF but that this bond is critically important not only to understanding general impacts, but also to understanding public perception of HVHF. The second case study looks at the impact of an energy development disaster, the Fukushima Daiichi nuclear disaster in Japan in 2011, in which only humans were permitted in evacuation centers and how disaster response policy ignored the plethora of agricultural and companion animals left in the no-go zone (Mattes 2017). This lack of recognition of the human-animal bond continues to have far-reaching consequences for society. Both cases explore the manner in which human-animal entanglement is significant in the context of energy development, the damage ignoring these relationships can wield, and approaches for building resilience for the humans and animals within each society. These cases further highlight the importance of considering animals in energy development because they have inherent value and deserve access to an environment that is not contaminated. Within human society, additionally, animals have economic value (e.g., food, tourism). Finally, regardless of an animal's role, humans form strong bonds with animals that may alter their attitudes and perceptions of development practices (such as in disasters), and animals are already being used as unintentional sentinels, indicators of current and future risks to humans.

Introduction

Humans have perpetually tried to harness energy from natural resources to power society, a process known as energy development. Because energy development is crucial to sustaining social systems, much research and political attention have been paid to understanding its social, political, and economic dynamics. Although rarely discussed, animals have been, and remain, an important part of energy development. Before the burning of fossil fuels became a popular mechanism to harness energy, people relied on the movement of animal workers to produce energy. Animal-powered engines (e.g., Major 1978) and animal-powered machines (Major 2008) drove animals around a center post, transferring kinetic energy from the animal through a task-oriented machine to complete a chore. The use of animals as energy producers systematically transformed society. For example, the harnessing of animal power drove paper production, especially during the Chinese golden age (Lucas 2005). As industrialization took hold, animal bodies used for the transfer of kinetic energy were repurposed into intentional sentinels—indicators of environmental stress and risk to humans.

Historically, birds and small mammals were selected for this task because they respond to environmental stress by showing physical and cognitive symptoms well before humans are impacted. The classic example of a sentinel animal is the canary in the coal mine. In the late 1890s, John Scott Haldane suggested that canaries and other small animals should be used as risk assessors in mining practices (Acott 1999). He was the first to do so, making a recommendation that was backed by his examination of mining disasters and laboratory experimentation with animals who could detect noxious gas (Acott 1999). Haldane suggested that animals could be used in place of technology at lower cost and with greater accuracy (Duin and Sutcliffe 1992; Goodman 2008). Canaries were specifically chosen because of their rapid breathing rate, their small size, their high metabolism, and the ease with which they could be handled. The importance of sentinels to mining operations and the resulting organizational changes are articulated throughout historical documents.

For instance, George A. Burrell wrote, "In the author's opinion the use of birds and mice is superior to chemical tests for carbon monoxide in that the test is quickly made, requires no technical experience, and is sufficiently exact" (1914, 1251). What is noticeably absent from this book, and most historical accounts of animals in energy development, is an analysis of the human-animal relationship or, more specifically in this case, how miners related to their sentinel companions. At first glance, one might assume that there was limited emotional connection between humans and their sentinel air-monitoring systems. However, Kat Eschner asserts that canaries "were so ingrained in the culture, miners report whistling to the birds and coaxing them as they worked, treating them as pets" (2016, 1). This interaction is perhaps unsurprising, given that coal mining has always been a lonely and dangerous job, not just for workers but for families as well (Giesen 2014). It was not until 1986 that canaries were replaced in British mining pits with electronic gas detectors (Goodman 2008).

Although the proverbial canary is now gone, the expansion of new energy technologies has given rise to the repurposing of animals once again, this time with less intention. Contemporary energy technologies such as hydraulic fracturing and horizontal drilling and energy disasters such as the Fukushima Daiichi nuclear power plant accident once again place animals in the role of sentinel, but with far less intention. Unlike animals placed in coal mines for a stated purpose, the unintentional animal sentinels of contemporary development are stakeholders who, like humans, are at risk of being harmed by energy production. This chapter's exploration of the place of animals and the human-animal bond in two contemporary energy development events—the energy boom in horizontal drilling and hydraulic fracturing happening in the United States and the Fukushima Daiichi nuclear power plant disaster in Japan in 2011—reveals that though these events are radically different, their examination leads to the same conclusion: animals continue to be a significant part of energy development but are rarely a focal point in energy development discussions. This lack of recognition of the human-animal bond, once again, continues to have far-reaching consequences for society. Both scenarios explore the way human-animal entanglement is significant in the context of energy development, the damage ignoring these relationships can wield, and approaches for building resilience for humans and animals within each society. Ultimately, we argue that people should care about animals in energy development because humans and animals share emotional bonds, animals are economic assets (specifically in production and tourism), animals signal danger to humans and ecosystem collapse, and animals have a right to equal consideration in terms of resources such as clean air and water. In the context of energy development, animals are not only companions, workers, or otherwise accompaniments to humans, but stakeholders—their lives and homes at risk in the shadow of energy development.

Literature Review

Energy Development and the Impact to Ecosystems

Most energy development comes from harnessing natural resources, a process that does not come without consequences, something that is true across energy development forms—conventional, unconventional and renewable. Constructing wind turbines, fortifying hydroelectric dams, building nuclear power plants, and using technology to extract previously unattainable oil and gas stores have all been shown to have varied effects on surrounding ecosystems. For instance, renewable source developments such as wind turbines can lead to loss of habitat, landscape disturbances, population dispersion, fragmentation, and collision risks (see, e.g., Drewitt and Langston 2006; Fox et al. 2006; Stewart, Pullin, and Coles 2004, 2007; Thomsen et al. 2006). Since wind turbine development also occurs offshore, the impacts to marine ecosystems have become increasingly significant. Researchers find that noise from turbines may restrict or "mask communication and orientation signals" among fish (Wahlberg and Westerberg 2005, 295) and that marine animals (e.g., harbor seals, porpoises) may experience similar effects (Carstensen, Henriksen, and Teilmann 2006; Koschinski et al. 2003; Madsen et al. 2006; K. Thompson 2013). The impacts of hydroelectric dams also have unfortunate impacts on the ecosystem in which they are built. The reduction or alteration of freshwater fish populations following the construction of dams has been studied around the world (Anderson, Freeman, and Pringle 2006; Fukushima et al. 2007; Ziv et al. 2012), leading in some cases to their removal (e.g., Bednarek 2001). These studies consider potential trade-offs inherent in dam building, such as a reduction of certain species who lose the ability to migrate and/or return to their original spawning habitat (e.g., salmon, steelhead trout).

Conventional oil and gas extraction (techniques that do not involve hydraulic fracturing) can have similar impacts as wind turbines, depending on where infrastructure is sited. The added dimension of hydraulic fracturing creates new potential problems. Although many studies have addressed how individuals in extractive communities perceive the costs and benefits of hydraulic fracturing, researchers have largely ignored how expansion affects animals and human-animal relationships (for exceptions, see, e.g., Bamberger and Oswald 2012, 2014, 2015). Elizabeth Royte notes this oversight and suggests that cattle and other farm animals are the new proverbial canaries in the coal mine, unintentional victims of environmental destruction and contamination caused by human hands (2012). Cameron Thomas Whitley (2017, 2018) labels these unintentional victims as "unintentional sentinels" and suggests that unlike coal mining, where canaries and other small animals were deliberately placed in energy extraction environments to expose potential environmental harms to humans, animals in high-production zones are being watched as their environments are inundated with extraction technologies from the current energy boom. In many cases, these animals are monitored in their natural habitats to determine potential environmental risk and subsequent risk to human health. A similar process happens when energy disasters occur.

Nuclear power energy is statistically safer than other forms of energy development, having the lowest accident and immediate fatality rates (Markandya and Wilkinson 2007). Peter Burgherr and Stefan Hirschberg found the "coal chain accounted for 65.3% of all accidents, with oil a distant 2nd at 21.2% . . . natural gas (7.2%) and LPG (5.6%) chains were much smaller, while both hydro and nuclear account for less than 1% each" (2008, 958). Between 1969 and 2000, only thirty-one confirmed deaths occurred due to nuclear energy—all of which occurred during Chernobyl, whereas coal causes 25,107, and oil causes 20, 218 (Burgherr and Hirschberg 2008). The nuclear energy number does not include the latent deaths that result from radiation exposure, which are estimated between the hundreds to the thousands for both Chernobyl and Fukushima. Large accidents such as Three Mile Island (1979), Chernobyl (1986), and Fukushima Daiichi (2011) have painted a dangerous image of nuclear energy. Although all energy development forms impact animals, nuclear energy is unique in that its effects on animal life are largely constrained to accidents. When accidents do occur, however, they can be catastrophic for domestic, wildlife and agriculture animals. While the immediate loss of human life in nuclear accidents is relatively minor in comparison to other energy technologies, the economic, community, and environmental impacts are staggering (see, e.g., Betzer, Doumet, and Rinne 2013; Houts, Cleary, and Hu 2010).

Whereas much of this research studies species and ecosystems independently of humans, it is important to note these entities' inseparability from human populations. Their vulnerability is inherently tied to ours. When a marine population struggles or collapses, there are consequences throughout the surrounding ecosystem. Whether we rely on this species for tourism or food, or merely as a piece of the biodiversity chain, in the Anthropocene, population loss can have far-reaching consequences. Thus, we are progressively understanding that building resiliency includes considering our nonhuman entanglements whether in the form of companion, agriculture, or wildlife.

The Human-Animal Bond, Risk and Resilience

Scholars are increasingly interested in how the relationships humans form with animals impact attitudes and behaviors. Humans have lived among, relied on, and cared for animals since their first appearance as hunter-gatherers 100,000 years ago (Clutton-Brock 1999). Learning from animals as hunters, integrating their bodies and movements as symbols, including them in our cosmology, as our kin, and in our homes—we have formed relationships with them that have been complicated, dynamic, and culturally significant (Hurn 2012; Kalof and Montgomery 2011; Mullin 2002; Noske 1989). Incorporating the modern human-animal bond, recent studies have considered animal subjects in a number of new forms, including identities as kin (Cormier 2003; Hansen 2013), as citizens of interspecies communities (Donaldson and Kymlicka 2011), or as sentient property or legal persons (Favre 2010; Francione 2004). These emergent considerations of animals are significant in the realm of energy development, risk assessment, and disaster response policy.

Although not specific to energy development, studies by disaster scholars have shown that companion animal ownership is an important factor in determining how individuals react to disaster warnings and whether they choose to evacuate (Trigg et al. 2016). Recognizing this importance, numerous government organizations such that the Federal Emergency Management Agency (FEMA) and the Centers for Disease Control and Prevention (CDC) have created disaster plans and resources that include animal family members (Centers for Disease Control and Prevention 2020; Federal Emergency Management Agency 2015). This recognition is consistent with past research that suggests humans form strong and enduring relationships with companion animals that influence how they process information and behave (see, e.g., Blouin 2013; Irvine 2009; Sanders 2003; Triebenbacher 2000; Walsh 2009). Kirrilly Thompson, for example, states that "the willingness of people to risk their own lives during disasters to save those of animals has been well documented" (2013, 123). In an early account assessing how horse owners identify their priorities during a disaster, researchers found they rated family safety first, followed closely by the safety of their horses. In addition, 75 percent said the ability to be assured the welfare of their horses would dictate their decision to evacuate (Linnabary et al. 1993).

Although research established this relationship in the early 1990s, when Hurricane Katrina hit in 2005 there were few disaster preparation and evacuation systems that acknowledged the human-animal bond (see, e.g., Hunt, Al-Awadi, and Johnson 2008; Irvine 2009). Because of the lack of understanding and limited policy support to address human-animal relationships, a substantial number of people chose not to evacuate (see, e.g., Hunt, Al-Awadi, and Johnson 2008; Irvine 2009). It became clear in the aftermath of Hurricane Katrina that animals needed to be an important part of risk assessment and decision management processes. Since then, government agencies have made a number of systematic changes, including the development of organizations to temporarily house animals, and an increased recognition that risk and disaster management plans must honor human and animal relationships (see, e.g., Irvine 2009). Although less is known regarding how relationships with animals inform perceptions of energy development, there is some evidence to suggest that having a close relationship with a companion animal does increase perception of risk, particularly in terms of hydraulic fracturing (Whitley 2017). In the next section, we use a "multispecies interdisciplinary perspective" to present two case studies highlighting the ways in which the comingled lives of humans and companion animals are impacted by energy development. In this process, we assert that that recognizing the place of animals and human-animal relationships in energy development is important not only because animals have inherent value but also because animals are important economically, their suffering can signal risks to humans and ecosystem services, and humans share bonds with animals with the effect that the suffering or loss of animals can have psychological impacts on individuals and communities.

Case Study 1: Assessing the Impacts of Hydraulic Fracturing on Animals and the Human-Animal Relationship

Hydraulic fracturing, or "fracking," is a well-stimulation process that uses large volumes of pressurized fluid pumped into wells to induce rock fractures that release previously unattainable oil and gas stores, a process that has spurred an energy boom in the United States. The Department of Energy estimates that over 2 million wells in the United States have been hydraulically fractured (US Department of Energy 2013). Of the new wells that are being drilled, upwards of 95 percent are being fracked (US Department of Energy 2013). In late 2016, it was estimated that fracking procedures had occurred in at least twenty-one states (Hirji and Song 2016). Although not all states have experienced a fracking boom, many states support the boom by providing sand and chemicals or are impacted by the boom truck transportation and oil and gas pipeline siting. Several localities have instituted limited moratoriums. In March 2017, Maryland banned fracking (Wood 2017). Drilling was banned in New York and Vermont in 2012, though Vermont has no known gas reserves. Most regulation of fracking is done at the state or local level, meaning that impacts are likely to vary across jurisdictions. In fact, hydraulic fracturing is exempt from the federal Clean Air Act; Clean Water Act; Safe Drinking Water Act (Halliburton Loophole); National Environmental Policy Act; Emergency Planning and Community Right-to-Know Act; Resource Conservation and Recovery Act; and Comprehensive Environmental Response, Compensation and Liability Act (CERCLA or Superfund) among others.

Although many have heard of the Halliburton Loophole, few understand the history or importance. Former chief executive of Halliburton Vice President Dick Cheney encouraged the Halliburton Loophole to be inserted in the Energy Policy Act of 2005. It is known as the Halliburton Loophole because "fracking" was patented by Halliburton, and this exemption was pushed by a former chief executive. Ultimately, the loophole allows for companies to use any chemicals they deem necessary without regard to the impacts those chemicals and practices have on drinking water safety. It also allows for companies to not disclose the types and quantities of chemicals used citing that they are trademark secrets. It effectively stripped authority from the EPA to regulate hydraulic fracturing processes. At the time of this book's publication, there remains little understanding of the extent to which animals and people interact with fracking fluids and what chemical compositions are most harmful (Ralston and Kalmbach 2018).

Research on the impacts of hydraulic fracturing on animals and human-animal relationships is limited due to varied geographies and regulatory procedures across the United States. However, scientists have identified several risks animals living near wells may experience. These risks include land fragmentation, exposure to toxic fracking fluids, increased air pollution, reduced access to water sources, truck traffic, noise pollution, and light pollution. Although not all animals face all risks, scholars argue that all of those residing near fracked wells could be exposed to increased air pollution and fracking fluid contamination (see, e.g., C. Davis 2012; J. Davis and Robinson 2012; Entrekin et al. 2011; Gillen and Kiviat 2012; Kiviat 2013; Rozell and Reaven

2012). Michelle Bamberger and Robert Oswald (2012) imply that one of the leading sources of concern for animals is exposure to toxins through contaminated water. Animals exposed to fracking fluid can experience reproductive, neurological, urological, gastrointestinal, dermatological, upper respiratory, and musculoskeletal impacts, as well as death. In addition, hydraulic fracturing is shown to increase air pollution, which poses a concern for human and animal health (see, e.g., Colborn et al. 2014; Garti 2012). Below we detail the known impacts to companion animals, agriculture animals, and wildlife. What is limited in this discussion is social science research exploring how these impacts affect human-animal relationships. Scholars are only beginning to explore important questions such as how humans think about the risks hydraulic fracturing poses to animals, whether closeness to animal influences risk perceptions, or how individuals and communities respond to and manage risks to animal populations (but cf. Whitley 2017 and 2018). Since public perception of risk is a critical part of mitigating harm, understanding how connections between humans and animals influence opinion becomes particularly critical, especially when there are shared short- and long-term impacts from hydraulic fracturing across species.

Companion Animals

Identifying the risk hydraulic fracturing poses to companion animals is critically important for understanding broader human health and well-being concerns. Companion animals and humans likely experience shared risks by residing in the same home, breathing the same air, and drinking from the same water source. In addition, companion animal owners may face tough decisions if their animals develop health symptoms. The cost of treating companion animals for respiratory problems or toxic chemical exposure could induce financial strain and create emotional and mental distress, especially for those on fixed incomes. To date, the only published article addressing companion animal health used a community health survey with questions about companion and livestock health to assess how the distance to the nearest fracked well relates to reported human and animal health (Slizovskiy et al. 2015). The authors find that the reported health of dogs is significantly lower for people living within 1 km of a gas well compared to those living farther away. By following participants from their 2012 study,

Bamberger and Oswald (2015) demonstrated that negative health impacts decreased for families and animals who moved away from hydraulic fracturing areas, while health impacts remained the same or increased for those continuing to reside in high-volume drilling areas. Although not speaking to health-related risks, Cameron Whitley shows that in general, people are very concerned about the risks to physical and ecological well-being that hydraulic fracturing poses to animals and having a companion animal makes individuals more likely to be concerned about the risk hydraulic poses to all animals (2017).

Agriculture Animals

Like companion animals, limited research has been done to assess how hydraulic fracturing impacts agricultural animals and what this means for human-animal relationships and social systems. In the only systematic study of livestock, Madelon Finkel and colleagues (Finkel et al. 2013) assess how hydraulic fracturing affects cow and milk production over a five-year period, finding that production decreased as development increased. Bamberger and Oswald (2012) provide a glimpse of the impacts hydraulic fracturing have on animals by looking at a selection of case studies. In one case, the authors describe how wastewater leaked into a cattle pasture, causing direct exposure. Within one hour of the leak, 17 cows died. In another case, a leak created a natural experiment whereby 60 cattle had access to a contaminated creek, while a second group of 36 cattle was at pasture and did not have access to the creek. Of the 60 cattle who had access to the contaminated water, 21 died and 16 failed to produce calves the following spring; the 36 who were not exposed to that water showed no symptoms or abnormal health problems. The authors note that because animals have restricted movement and continuous exposure, they "especially livestock, are sensitive to the contaminants released into the environment by drilling and by its cumulative impacts" (Bamberger and Oswald 2012, 72). Not only is accidental exposure a problem, but several studies show that livestock will ingest crude oil and other petroleum products when faced with dehydration, have a lack of clean water, are fed contaminated or poor feeds, need to increase their salt intake, or are bored (Coppock et al. 1995, 1996; W. C. Edwards 1989; Edwards, Coppock, and Zinn 1979). Although studies have not been conducted, companion animals and wildlife may drink or bathe in contaminated sources or in wastewater pools for similar reasons. There are few studies that look at how contaminant exposure might enter into the human food chain through the consumption of animals. Of the few studies that have been conducted, Luisa Torres, Om Prakash Yadav, and Eakalak Khan provide the most promising early insight (2018). They model the possibility of exposure to radium-226 (Ra-226) from wastewater (water that returns to the surface after a well is fracked). They suggest that contamination is not only possible, but that it could be very likely if plants and animals that have been exposed to spills are consumed. There have been no published studies that interview or survey farmers to assess how well development has impacted animal health or their relationships with their animals.

Wildlife

Of all animal groups, most studies have examined the impacts of hydraulic fracturing on wildlife, though research is still limited. As with other forms of development, land use changes can create fragmentation leading to species dispersal and biodiversity reduction (Entrekin et al. 2018; Gillen and Kiviat 2012; Kiviat 2013). As John Davis and George Robinson note, forest fragmentation contributes to range restriction and species extinction (2012). Animals avoid roadways and other structural changes that fragment the landscape, a trend documented in grassland bird species (S. Thompson et al. 2015), salamanders (Brand, Wiewel, and Grant 2014), mule deer (Lendrum et al. 2012), and river otters (Godwin et al. 2015). In addition, two studies show that fracking fluids have endocrine-disrupting properties that may impact wildlife (Kassotis et al. 2016; Kassotis et al. 2013). Beyond water contamination, the depletion of water reserves may have long-term consequences for animal populations by reducing wetlands (Kiviat 2013). In addition, compared to humans, wildlife are particularly sensitive to noise and light pollution (Gillen and Kiviat 2012; Kiviat 2013). Erik Kiviat suggests that "continuous loud noise from, for example, transportation networks, motorized recreation, and urban development can interfere with acoustic communication for frogs, birds, and mammals and cause hearing loss, elevated stress hormone levels, and hypertension in various animals" (2013, 5). Lucas Habib, Erin Bayne, and Stan Boutin (2007) found that noise pollution from drilling in Alberta, Canada, reduced bird-pairing success. However, Clinton Francis, Catherine Ortega, and Alexander Cruz found that noise effects increased breeding success in birds near wells, likely because of reduced predation (2009). So, while fracking showed a positive outcome for bird species in this study, it remains to be seen how this activity could impact predation and ecosystem equilibrium. Not only is onshore noise a problem, but offshore drilling noises have been found to disrupt underwater animal communication channels (Schlossberg 2016). Although the studies are limited, Erik Kiviat asserts that light pollution may impact reproduction and foraging (2013). None of these studies assess whether there are differences in how people use and engage with the natural environment when animals are being impacted or how changed engagement relates to physical and mental health. Ultimately, some species such as the woodland caribou in Alberta, Canada, show signs of decline and are now classified as endangered due to petroleum exploration (Bradshaw, Boutin, and Herbert 1997).

Case Study 2: Assessing the Impacts of the Fukushima Daiichi Nuclear Power Plant Disaster on Animals and the Human-Animal Relationship

The triple disasters that occurred in Northeastern Japan on March 11, 2011, shifted the perception of nuclear power. The tsunami that followed a 9.0 earthquake caused the cooling systems of the Fukushima Daiichi Nuclear Power Plant to shut down, resulting in the meltdown of three reactors. Radioactive material was released, and evacuation of the nearby area had to be ordered. The fear of nuclear fallout and of further explosions and damage was an international concern for months following the disaster, with Germany and France committing to reduce their reliance on nuclear power. Within Japan, a nation already victim to a tumultuous nuclear history, social movements to end nuclear power gained significant momentum. Using this disaster as a lens, this case study examines the impact of nuclear energy disasters on (1) companion animals, (2) agricultural animals, and (3) wildlife.

Companion Animals

When evacuation was announced for residents surrounding the malfunctioning nuclear plant, people were told to leave their animals behind. Temporary shelters could not immediately accommodate pets. Furthermore, evacuation was to be swift, and human safety was a priority (Ito 2013). Tragically, the residents were told they would only be absent for a few days, when in fact they would not be permitted to return to their homes, and their pets, for weeks. After this long absence, they were then only permitted to return for two-to-five-hour visits to collect select family belongings. Those who had pets within the zone returned to find them lost, dead, or otherwise suffering from the neglect. As companion animals are largely dependent on their human counterparts, mortality was high. Starvation, dehydration, and other ailments took the lives of the majority. This region of Japan was previously home to over 6,000 dogs, and potentially an even greater number of cats (Ito 2013).

Those animals who survived the initial evacuation were nonetheless barred from most temporary shelters. Therefore, pet owners might drop off large bags of food and water for an animal chained within the evacuation zone during short visits. Those who removed their pets from the no-go zone had to decontaminate them, which simply involved shampooing; find a location that would house both of them and their pet; or identify an animal rescue to care for the animal. Some people kept their pets in their cars, or even chose to live in their cars with the animals (Mattes 2017). A study conducted by Ross Mouer and Hazuki Kajiwara interviewed evacuees who struggled with having pets postevacuation (2016). They reported many referred to their animals as "children" or "kid" and found their pets came to signify all they had left behind: "their animals served as a bridge to the past that allowed their owners to maintain a life narrative that incorporated their lost worlds" (207). The human-first disaster policy made life following the disaster aftermath especially trying for evacuees.

In Japan, the human-pet bond is significant. A beloved cat may be the only friend to an elderly woman in Japan's aging society, a couple of chihuahuas the "children" to a childless couple (Hamano 2013; Hansen 2013). As many as 80 percent of owners keep dogs indoors, with 30 percent allowing them to sleep in their bed (Kakinuma 2008). For others, the dog chained in the evacuation zone stands loyally as their guard dog—alone and now competing with local wildlife and strays for the food resources left behind. In each case, the nuclear disaster left the human-animal bonds torn apart—adding significant emotional distress to an already trying circumstance.

Those companion animals who survived, in both Fukushima and the wider Tōhoku region, and were not rescued by intervening nonprofit organizations, currently roam the deserted area searching for food, becoming increasingly feral. As many pets were not neutered/spayed prior to the evacuation, they continue to reproduce. Individuals and local nonprofit organizations have set up feeding stations within the no-go zone. They leave and collect traps to take as many companion animals as possible out and get them into shelters or homes. Those animals who were rescued by nonprofit organizations have an uphill battle in either waiting for their people to be resettled or be adopted out to a new family. As these nonprofit animal rescue shelters are scant, not well funded or staffed, and often operating at capacity, the lives of the animals and the human rescuers will be negatively impacted by the disaster for years to come (Mattes 2017).

Agriculture Animals

Whereas farmers share a wide variety of bonds with agricultural animals, their bodies are valued as products within a broader economic system. Prior to this disaster, there were around 3,400 cattle, 31,500 pigs, and 630,000 chickens (Ito 2013) in what became the no-entry zone surrounding the Fukushima nuclear site. The Kitasato University report "The Study on the Impact of the Fukushima Nuclear Accident on Animals" details the widespread mortality and causes within the evacuation zone (Ito 2013). Whereas radiation contamination was not a direct cause of death for agricultural animals, it was an indirect cause. Neglect occurred due to the inability of farmers to reach and provide for their animals. Furthermore, because radiation—real or perceived—made animals' bodies unsellable, those who survived lost their value as economic assets. According to the report "National authorities provided animal keepers with two choices for their livestock: death by starvation or euthanasia. The euthanasia option did not progress smoothly and more animals were left to die of starvation" (Ito 2013, 28). Furthermore, the crops grown within the zone were also deemed unsafe due to radioactivity. Before ordinances to prevent their shipment and use were implemented, the radioactive feed had already been distributed throughout Japan, leading to a halt in beef sales and a need to compensate farmers. Many of these farmers lost their entire livelihoods, leading to economic troubles, unemployment, and suicide. More so, for some farmers, the loss of cattle that had been bred in their family for generations was a deep emotional burden.

Those agricultural animals who still survive in the evacuation zone roam the area. As of October 5, 2013, the Kitasato Report found as many as 2,560 cattle living in the 20 km area (Ito 2013). Pigs and chickens also traverse the land, alongside the usual assortment of Japanese wildlife. As the bulls are capable of reproducing, veterinarians have been sent in to castrate them as a preventative measure. Whereas smaller agricultural animals, such as chickens, were rescued and removed from the 20 km zone, the larger surviving animals remained. Some continue to be fed by their former owners, or with help from local individuals or nonprofit organizations.

Wildlife

This disaster displaced approximately 160,000 people, leaving a 20 km region with few human residents. Within this zone, cleanup workers, animal rescuers, and other employees behind the checkpoints report an abundance of wildlife. Deer, wild boar, and the increasingly feral stray companion animals are changing the ecosystem—all unchecked by hunting. Macaques and other wildlife are said to have come into farms, eating the crops in the fields and living among abandoned homes. Without farmers, however, weeds are dominating and pollination of crops will not occur—causing the food supply to dwindle for these new residents (Ito 2013). A small number of pigs who survived and escaped enclosures following the evacuation have bred with wild boar, raising concern over the transfer of infectious disease to domestic pigs outside of the exclusion zone (von Wehrden et al. 2012). Through consumption and interaction (digging, etc.) with the physical environment, wildlife within the fallout landscape becomes contaminated.

The long-term consequences of contamination in this unanticipated natural preserve are only beginning to undergo research. The effects of contamination within the landscape are known to be long term, but the details on impacts to the ecosystem over time are yet unknown. The best comparison study to Fukushima's evacuation zone is the 1,600 sq. mi. exclusion zone surrounding the Chernobyl nuclear meltdown site. However, the environmental conditions in Chernobyl are quite different from that of Japan—which is a wet, monsoon-prone area scattered with rice paddies (von Wehrden et al. 2012). The evacuated area surrounding Chernobyl is now acting as one of Europe's largest wildlife sanctuaries. Despite the open, uninhibited space, radioactivity has had devastating genetic consequences surrounding Chernobyl and limits healthy development of individuals and entire species populations (Møller and Mousseau 2015; Møller et al. 2010; Wendle 2016). Similarities in Fukushima have already been recorded regarding the ecological impact of radiation in species populations, including white spots on bird feathers, signaling radiation-induced oxidative stress and population decline (e.g., Møller et al. 2013; Møller et al. 2015). As consequences are multigenerational, it may be years before we have a full understanding of how radiation will impact the species within this ecosystem.

Discussion

Historically, animals have been an important part of energy development, with little acknowledgment of their role. Increasingly, scientists and policy makers are recognizing the importance of human-animal bonds, but as these two cases demonstrate, there is much work needed. Although the two case studies we present are extremely different, both highlight the importance of taking animals and human-animal relationships into account when considering energy development and the potential for energy-development-related disasters. To date, most research on energy development is conducted from a human-centered perspective or human exceptionalism perspective. This viewpoint situates the human as the most important part of the equation. Although on the surface this method makes sense for articulating impacts to humans, it has the potential to miss interconnected impacts, such as those between humans and animals. This "missing animal connection" is what was observed during Hurricane Katrina, leading to catastrophe.

The reality that people lost their lives because of their unwillingness to leave their companion animals sparked widespread policy change. Following Hurricane Katrina, in accordance with the PETS Act, pet-friendly shelters are increasingly becoming the norm. These shelters take a variety of forms, including separate centers to hold animals, centers that allow the companion animal's family to care for them, and shelters in which humans can live with their pets directly. The presence of pets not only physically relaxes those near them, including lowering blood pressure, but also provides "ice breakers," encouraging communication and entertainment (Hurn 2012). In a space such as a temporary shelter, such distractions from the reality of the present dismal situation is welcome. Despite this domestic progress, on the world stage there was a repeat of the "missing the animal" scenario when a dual natural and energy disaster hit Japan. In Japan, the locals frequently suggested having shelters for both those with pets and for those without so that no one would be inconvenienced by a dog or cat in temporary housing (Mattes 2017) These plans must be made in accordance with local animal rescue and veterinary services where possible and should be established prior to the disaster occurring. Having this protocol in place could bring peace of mind to those in an area potentially impacted by an energy-related disaster.

These events and more recent nonenergy disasters consistently demonstrate how important animals are to our social systems. As discussed previously, disaster policies that disregard the human-animal bond jeopardize the safety of humans and animals (Irvine 2009). We see this trend not only in disasters but in research on domestic violence as well. The abuse of pets in a home not only serves as a marker of potential human-focused violence (Ascione and Arkow 1999), but research has found that battered women may choose to not seek safety unless their companion animal is permitted to escape to a shelter alongside them (Fitzgerald 2005). This example further showcases the importance of the human-animal bond, highlighting how individuals may remain in horrific or dangerous situations when this bond is not acknowledged.

Furthermore, when agricultural animals are left out of disaster planning, entire family livelihoods are lost, increasing the length and difficulty of the aftermath. Resiliency plans must take animals into consideration. For example, Steve Glassey, an expert on animals in emergency (Potts and Gadenne 2014), presents a fiasco following a disaster in Myanmar, where the goal was to swiftly reestablish the rice fields, preventing famine and economic stress. Oxen were used to plow the fields, but they had perished in the floods. An outside agency brought in oxen from a different region, but they all died within a year—unvaccinated from local diseases. This situation was preventable with proper resiliency plans in place by local actors and qualified veterinarians (Potts and Gadenne 2014). Although much work has been done to address risk and resiliency in disasters, this work needs to extend into thinking about energy development disasters. Plans need to be in place to assess how all animals (companion, agriculture, and wild) are cared for as a component of broader human health and well-being.

The first case study begins the discussion about the impacts of energy development on animals and the human-animal relationship by delving into the rapid expansion of hydraulic fracturing across the United States. Like the case study, this analysis shows that animals are important to the discussion of risk and resiliency but that they have largely been excluded, functioning today as unintentional sentinels. These unintentional sentinels give us insight into the potential health impacts that humans may experience; however, little research has been conducted to assess how impacts to animals may affect the human-animal relationship and lead to changes in society. The limited existing research suggests that humans and animals experience similar health symptoms when living close to wells (Slizovskiy et al. 2015) and that humans with companion animals consider their animals when thinking about hydraulic fracturing risk (Whitley 2017); additional social science studies are needed to investigate these relationships. For instance, researchers might want to examine how social systems have changed in response to animals' exposure to hydraulic fracturing. Scholars may do this by asking innovative questions, for example, whether veterinarians located in high-volume hydraulic fracturing areas alter their practices to deal with nondisclosure policies when treating animals that have encountered fracking fluids. Or, how do individuals on fixed incomes who experience shared health symptoms with companion animals make decisions about who gets medical care first? Although studies investigating how humans process hydraulic fracturing's harm to animals are lacking, broad research shows that watching a companion animal suffer or die can cause extended grief (see, e.g., Eckerd, Barnett, and Jett-Dias 2016; Pierce 2013; Testoni et al. 2017) and elicit PTSD symptoms (see, e.g., Adrian and Stitt 2017). Another avenue for future research lies in energy boomtowns. Rapid expansion of energy development often produces boomtowns. These instantaneous communities lack social cohesion and can breed violence. Several studies have documented increases in crime rates among shale gas boomtowns, but these studies do not investigate crime against animals (see, e.g., Archbold 2015; James and Smith 2017; Riggs 2013). There is a wealth of literature suggesting that violence against animals is often connected to violence against humans (see, e.g., Flynn 2011; McPhedran 2009; Piper 2003). As a result, future studies should explore the relationship between unconventional development boomtowns, crime, and human-animal well-being. It should be noted that even in studies of violence related to energy development boomtowns, the animal remains an unintentional sentinel—giving us a glimpse into what may be underneath the surface in a community.

The second case study illustrates how the impacts of energy development disasters are further compounded because they involve toxic exposure. In the case of the Fukushima Daiichi meltdown, the effects of radiation on companion, agriculture, and wild animals are only beginning to be understood. So far, studies have revealed declining populations, genetic mutations, and weakening of individuals within the radioactive zone. These effects have long-lasting and far-reaching consequences for the overall ecosystem and can lead to ecosystem collapse. Although there is likely much resistance to creating energy development disaster plans, such procedures should be considered, especially given that climate change is likely to exacerbate the severity of natural disasters, which could pose a problem for energy development infrastructure.

Although not extensive, this chapter situates the animal back into discussions about the impacts of energy development, arguing that more attention needs to be paid, by social scientists specifically, to assessing how energy development impacts animals and human-animal relationships with recognition that how animals are impacted may inform how people respond to energy development processes. As energy development expands, as does the potential of more energy development disasters, it is essential to avoid using animals as tools for assessing potential impacts on human health and well-being without taking the human-animal relationship into account. Animals are an integral part of human society. Societies across the world rely on them for food, entertainment, economic resources, and companionship. When animals are impacted, humans are also at risk. As Royte (2012) notes, animals are the new proverbial canaries in the coal mine. As Whitley (2017, 2018) refines this metaphor, in contemporary energy development animals have become unintentional sentinels, no longer deliberate canaries in the coal mine but situational stakeholders being used as canaries without proper recognition of the social implications. They are the unintentional victims of environmental destruction and contamination caused by human hands in our quest for energy.

References

Acott, Chris. 1999. "Haldane, JBS Haldane, L Hill and A Siebe: A Brief Resume of Their Lives." *SPUMS Journal* 29 (3): 161–165.

Adrian, Julie Ann Luiz, and Alexander Stitt. 2017. "Pet Loss, Complicated Grief, and Post-traumatic Stress Disorder in Hawaii." *Anthrozoös* 30 (1): 123–133.

Anderson, Elizabeth P., Mary C. Freeman, and Catherine M. Pringle. 2006. "Ecological Consequences of Hydropower Development in Central America: Impacts of Small Dams and Water Diversion on Neotropical Stream Fish Assemblages." *River Research and Applications* 22 (4): 397–411.

Archbold, Carol A. 2015. "Established-Outsider Relations, Crime Problems, and Policing in Oil Boomtowns in Western North Dakota." *Criminology, Criminal Justice Law and Society* 16 (3): 19–40.

Ascione, Frank R., and Phil Arkow, eds. 1999. *Child Abuse, Domestic Violence, and Animal Abuse: Linking the Circles of Compassion for Prevention and Intervention*. West Lafayette, IN: Purdue University Press.

Bamberger, Michelle, and Robert E. Oswald. 2012. "Impacts of Gas Drilling on Human and Animal Health." *New Solutions: A Journal of Environmental and Occupational Health Policy* 22 (1): 51–77.

Bamberger, Michelle, and Robert E. Oswald. 2014. "Unconventional Oil and Gas Extraction and Animal Health." *Environmental Science: Processes and Impacts* 16 (8): 1860–1865.

Bamberger, Michelle, and Robert E. Oswald. 2015. "Long-Term Impacts of Unconventional Drilling Operations on Human and Animal Health." *Journal of Environmental Science and Health, Part A* 50 (5): 447–459.

Bednarek, Angela T. 2001. "Undamming Rivers: A Review of the Ecological Impacts of Dam Removal." *Environmental Management* 27 (6): 803–814.

Betzer, André, Markus Doumet, and Ulf Rinne. 2013. "How Policy Changes Affect Shareholder Wealth: The Case of the Fukushima Daiichi Nuclear Disaster." *Applied Economics Letters* 20 (8): 799–803.

Blouin, David D. 2013. "Are Dogs Children, Companions, or Just Animals? Understanding Variations in People's Orientations toward Animals." *Anthrozoös* 26 (2): 279–294.

Bradshaw, Corey J. A., Stan Boutin, and Daryll M. Hebert. 1997. "Effects of Petroleum Exploration on Woodland Caribou in Northeastern Alberta." *Journal of Wildlife Management* (1, October): 1127–1133.

Brand, Adrianne B., Amber N.M. Wiewel, and Evan H. Campbell Grant. 2014. "Potential Reduction in Terrestrial Salamander Ranges Associated with Marcellus Shale Development." *Biological Conservation* 180 (December): 233–240.

Burgherr, Peter, and Stefan Hirschberg. 2008. "A Comparative Analysis of Accident Risks in Fossil, Hydro, and Nuclear Energy Chains." *Human and Ecological Risk Assessment: An International Journal* 14 (5): 947–793.

Burrell, George Arthur. 1914. *The Use of Mice and Birds for Detecting Carbon Monoxide after Mine Fires and Explosions*. Washington, DC: US Department of the Interior, Bureau of Mines.

Carstensen, J., O. D. Henriksen, and J. Teilmann. 2006. "Impacts of Offshore Wind Farm Construction on Harbour Porpoises: Acoustic Monitoring of Echo-Location Activity Using Porpoise Detectors (T-PODs)." *Marine Ecology-Progress Series* 321 (September): 295–308.

Centers for Disease Control and Prevention. 2020. *Pet Safety in Emergencies*. https://www.cdc.gov/healthypets/emergencies/index.html.

Clutton-Brock, Juliet. 1999. *A Natural History of Domesticated Mammals*. Cambridge: Cambridge University Press.

Colborn, Theo, Kim Schultz, Lucille Herrick, and Carol Kwiatkowski. 2014. "An Exploratory Study of Air Quality near Natural Gas Operations." *Human and Ecological Risk Assessment: An International Journal* 20 (1): 86–105.

Coppock, R., M. Mostrom, A. Khan, and S. Semalulu. 1995. "Toxicology of Oil Field Pollutants in Cattle: A Review." *Veterinary and Human Toxicology* 37 (6): 569–576.

Coppock, R., M. Mostrom, E. Stair, and S. Semalulu. 1996. "Toxicopathology of Oilfield Poisoning in Cattle: A Review." *Veterinary and Human Toxicology* 38 (1): 36–42.

Cormier, Loretta A. 2003. *Kinship with Monkeys: The Guajá Foragers of Eastern Amazonia*. New York: Columbia University Press.

Davis, Charles. 2012. "The Politics of 'Fracking': Regulating Natural Gas Drilling Practices in Colorado and Texas." *Review of Policy Research* 29 (2): 177–191. https://doi.org/10.1111/j.1541-1338.2011.00547.x.

Davis, John, and George Robinson. 2012. "A Geographic Model to Assess and Limit Cumulative Ecological Degradation from Marcellus Shale Exploitation in New York, USA." *Ecology and Society* 17 (2): 25.

Donaldson, Sue, and Will Kymlicka. 2011. *Zoopolis: A Political Theory of Animal Rights*. New York: Oxford University Press.

Drewitt, Allan L., and Rowena H. Langston. 2006. "Assessing the Impacts of Wind Farms on Birds." *Ibis* 148 (March): 29–42.

Duin, Nancy, and Jenny Sutcliffe. 1992. *A History of Medicine: From Pre-history to the Year 2000*. New York: Simon and Schuster.

Eckerd, Lizabeth M., James E. Barnett, and Latishia Jett-Dias. 2016. "Grief following Pet and Human Loss: Closeness Is Key." *Death Studies* 40 (5): 275–282.

Edwards, V. C., R. W. Coppock, and L. L. Zinn. 1979. "Toxicoses Related to the Petroleum Industry." *Veterinary and Human Toxicology* 21 (5): 328–337.

Edwards, William C. 1989. "Toxicology of Oil Field Wastes: Hazards to Livestock Associated with the Petroleum Industry." *Veterinary Clinics of North America: Food Animal Practice* 5 (2): 363–374.

Entrekin, Sally, Anne Trainor, James Saiers, et al. 2018. "Water Stress from High-Volume Hydraulic Fracturing Potentially Threatens Aquatic Biodiversity and

Ecosystem Services in Arkansas, United States." *Environmental Science and Technology* 52 (4): 2349–2358.

Entrekin, Sally, Michelle Evans-White, Brent Johnson, and Elisabeth Hagenbuch. 2011. "Rapid Expansion of Natural Gas Development Poses a Threat to Surface Waters." *Frontiers in Ecology and the Environment* 9 (9): 503–511.

Eschner, Kat. 2016. "The Story of the Real Canary in a Coal Mine." *Smithsonian*, December 30, 2016. http://www.smithsonianmag.com/smart-news/story-real-canary-coal-mine-180961570/.

Favre, David. 2010. "Living Property: A New Status for Animals within the Legal System." *Marquette Law Review* 93: 1021–1072.

Federal Emergency Management Agency. 2015. Include Your Pets in Disaster Preparedness Planning. Release no. 037. Washington, DC: Federal Emergency Management Agency.

Finkel, Madelon L., Jane Selegean, Jake Hays, and Nitin Kondamudi. 2013. "Marcellus Shale Drilling's Impact on the Dairy Industry in Pennsylvania: A Descriptive Report." *NEW SOLUTIONS: A Journal of Environmental and Occupational Health Policy* 23 (1): 189–201.

Fitzgerald, Amy J. 2005. *Animal Abuse and Family Violence*. Lewiston, NY: Edwin Mellen Press.

Flynn, Clifton P. 2011. "Examining the Links between Animal Abuse and Human Violence." *Crime, Law and Social Change* 55 (5): 453–468.

Fox, A. D., Mark Desholm, Johnny Kahlert, Thomas Kjaer Christensen, and I. B. Krag Petersen. 2006. "Information Needs to Support Environmental Impact Assessment of the Effects of European Marine Offshore Wind Farms on Birds." *Ibis* 148 (March): 129–144.

Francione, Gary L. 2004. "Animals: Property or Persons?" *Rutgers Law School Faculty Papers*. Working Paper 21. Newark, NJ.

Francis, Clinton D., Catherine P. Ortega, and Alexander Cruz. 2009. "Noise Pollution Changes Avian Communities and Species Interactions." *Current Biology* 19 (16): 1415–1419.

Fukushima, Michio, Satoshi Kameyama, Masami Kaneko, Katsuya Nakao, and E. Ashley Steel. 2007. "Modelling the Effects of Dams on Freshwater Fish Distributions in Hokkaido, Japan." *Freshwater Biology* 52 (March): 1511–1524. doi:10.1111/j.1365-2427.2007.01783.x.

Garti, Anne Marie. 2012. "The Illusion of the Blue Flame: Water Law and Unconventional Gas Drilling in New York." *Environmental Law New York* 23 (11): 159–165.

Giesen, Carole A. B. 2014. *Coal Miners' Wives: Portraits of Endurance*. Lexington: University Press of Kentucky.

Gillen, Jennifer L., and Erik Kiviat. 2012. "Environmental Reviews and Case Studies: Hydraulic Fracturing Threats to Species with Restricted Geographic Ranges in the Eastern United States." *Environmental Practice* 14 (4): 320–331.

Godwin, Braden L., Shannon E. Albeke, Harold L. Bergman, Annika Walters, and Merav Ben-David. 2015. "Density of River Otters (*Lontra canadensis*) in Relation to Energy Development in the Green River Basin, Wyoming." *Science of The Total Environment* 532 (November): 780–790. https://doi.org/10.1016/j.scitotenv.2015.06.058.

Goodman, Martin. 2008. *Suffer and Survive: Gas Attacks, Miners' Canaries, Spacesuits and the Bends: The Extreme Life of Dr. JS Haldane*. New York: Pocket Books.

Habib, Lucas, Erin Bayne, and Stan Boutin. 2007. "Chronic Industrial Noise Affects Pairing Success and Age Structure of Ovenbirds *Seiurus aurocapilla*." *Journal of Applied Ecology* 44 (1): 176–184.

Hamano, Sayoko. 2013. "Katei Dobutsu to no Tsukiai" [Relations with the family animals]. In *Nihon no Dobutsu Kan: Hito to Dobutsu no Kankei Shi* [Japanese attitudes toward animals: A history of human-animal relations], edited by Osamu Ishida, Sayoko Hamano, Makoto Hanazono, and Akihisa Setoguchi, 19–35. Tokyo: University of Tokyo Press.

Hansen, Paul. 2013. "Urban Japan's 'Fuzzy' New Families: Affect and Embodiment in Dog–Human Relationships." *Asian Anthropology* 12 (2): 83–103.

Hirji, Zahra, and Lisa Song. 2016. "Map: The Fracking Boom, State by State." *InsideClimate News*, April 28, 2016. https://insideclimatenews.org/news/20150120/map-fracking-boom-state-state.

Houts, Peter S., Paul D. Cleary, and Tei-Wei Hu. 2010. *Three Mile Island Crisis: Psychological, Social, and Economic Impacts on the Surrounding Population*. University Park: Pennsylvania State University Press.

Hunt, Melissa, Hind Al-Awadi, and Megan Johnson. 2008. "Psychological Sequelae of Pet Loss following Hurricane Katrina." *Anthrozoös* 21 (2): 109–121.

Hurn, Samantha. 2012. *Humans and Other Animals: Cross-cultural Perspectives on Human-Animal Interactions*. Chicago: Pluto Press.

Irvine, Leslie. 2009. *Filling the Ark: Animal Welfare in Disasters*. Philadelphia: Temple University Press.

Ito, Nobuhiko, ed. 2013. *Study on the Impact of the Fukushima Nuclear Accident on Animals*. Tokyo: Kitasato University.

James, Alexander, and Brock Smith. 2017. "There Will Be Blood: Crime Rates in Shale-Rich US Counties." *Journal of Environmental Economics and Management* 84 (July): 125–152.

Kakinuma, Miki. 2008. "Hattatsu Shinrigaku Kara Mita Kainushi to Inu no Kankei–Hito no Migatte na Yokyu ni Honrosareru Inu" [The relationship between dogs and their owners: Spoilt by human interaction]. In *Petto to Shakai: Hito to Dobutsu no Kankeigaku* 3 [Pets and society: The study of human animal relations 3], edited by in Y. Hayashi, Y. Mori, F. Akishinonomiya, K. Ikeya, and T. Okuno, 76–99. Tokyo: Iwanami Shoten.

Kalof, Linda, and Georgina M. Montgomery, eds. 2011. *Making Animal Meaning*. East Lansing: Michigan State University Press.

Kassotis, Christopher D., Luke R. Iwanowicz, Denise M. Akob, Isabelle M. Cozzarelli, Adam C. Mumford, William H. Orem, and Susan C. Nagel. 2016. "Endocrine Disrupting Activities of Surface Water Associated with a West Virginia Oil and Gas Industry Wastewater Disposal Site." *Science of the Total Environment* 557 (July): 901–910.

Kassotis, Christopher D., Donald E. Tillitt, J. Wade Davis, Annette M. Hormann, and Susan C. Nagel. 2013. "Estrogen and Androgen Receptor Activities of Hydraulic Fracturing Chemicals and Surface and Ground Water in a Drilling-Dense Region." *Endocrinology* 155 (3): 897–907.

Kiviat, Erik. 2013. "Risks to Biodiversity from Hydraulic Fracturing for Natural Gas in the Marcellus and Utica Shales." *Annals of the New York Academy of Sciences* 1286 (1): 1–14.

Koschinski, Sven, Boris M. Culik, Oluf Damsgaard Henriksen, Nick Tregenza, Graeme Ellis, Christoph Jansen, and Günter Kathe. 2003. "Behavioural Reactions of Free-Ranging Porpoises and Seals to the Noise of a Simulated 2 MW Wind-power Generator." *Marine Ecology Progress Series* 265 (December): 263–273.

Lendrum, Patrick E., Charles R. Anderson, Ryan A. Long, John G. Kie, and R. Terry Bowyer. 2012. "Habitat Selection by Mule Deer during Migration: Effects of Landscape Structure and Natural-Gas Development." *Ecosphere* 3 (9): 1–19.

Linnabary, Robert D., John C. New, Barbara M. Vogt, Carol Griffith-Davies, and Lee Williams. 1993. "Emergency Evacuation of Horses: A Madison County, Kentucky Survey." *Journal of Equine Veterinary Science* 13 (3): 153–158.

Lucas, Adam Robert. 2005. "Industrial Milling in the Ancient and Medieval Worlds: A Survey of the Evidence for an Industrial Revolution in Medieval Europe." *Technology and Culture* 46 (1): 1–30.

Madsen, Peter T., Magnus Wahlberg, Jakob Tougaard, Klaus Lucke, and P. Tyack. 2006. "Wind Turbine Underwater Noise and Marine Mammals: Implications of Current Knowledge and Data Needs." *Marine Ecology Progress Series* 309 (March): 279–295.

Major, John Kenneth. 1978. *Animal-Powered Engines*. London: Batsford.

Major, John Kenneth. 2008. *Animal-Powered Machines*. London: Bloomsbury.

Markandya, Anil, and Paul Wilkinson. 2007. "Electricity Generation and Health." *Lancet* 370 (9591): 979–990.

Mattes, Seven. 2017. "The Shared Vulnerability and Resiliency of the Fukushima Animals and Their Rescuers." In *Responses to Disasters and Climate Change: Understanding Vulnerability and Fostering Resilience*, edited by Michele Companion and Miriam S. Chaiken, 103–116. Philadelphia: Taylor and Francis.

McPhedran, Samara. 2009. "Animal Abuse, Family Violence, and Child Wellbeing: A Review." *Journal of Family Violence* 24 (1): 41–52.

Møller, A. P., J. Erritzøe, F. Karadas, and T. A. Mousseau. 2010. "Historical Mutation Rates Predict Susceptibility to Radiation in Chernobyl Birds." *Journal of Evolutionary Biology* 23 (10): 2132–2142.

Møller, Anders Pape, and Timothy A. Mousseau. 2015. "Strong Effects of Ionizing Radiation from Chernobyl on Mutation Rates." *Scientific Reports* 5 (8363): 1–6.

Møller, Anders Pape, Timothy A. Mousseau, Isao Nishiumi, and Keisuke Ueda. 2015. "Ecological Differences in Response of Bird Species to Radioactivity from Chernobyl and Fukushima." *Journal of Ornithology* 156 (1): 287–296.

Møller, Anders Pape, Isao Nishiumi, Hiroyoshi Suzuki, Keisuke Ueda, and Timothy A. Mousseau. 2013. "Differences in Effects of Radiation on Abundance of Animals in Fukushima and Chernobyl." *Ecological Indicators* 74 (January): 75–81.

Mouer, Ross, and Hazuki Kajiwara. 2016. "Strong Bonds: Companion Animals in Post-tsunami Japan." In *Companion Animals in Everyday Life: Situating Human-Animal Engagement within Cultures*, edited by Michał Piotr Pręgowski, 201–115. Warsaw: Palgrave Macmillan.

Mullin, Molly. 2002. "Animals and Anthropology." *Society and Animals* 10 (4): 387–393.

Noske, Barbara. 1989. *Humans and Other Animals: Beyond the Boundaries of Anthropology*. London: Pluto Press.

Pierce, Jessica. 2013. "The Dying Animal." *Journal of Bioethical Inquiry* 10 (4): 469–478.

Piper, Heather. 2003. "The Linkage of Animal Abuse with Interpersonal Violence a Sheep in Wolf's Clothing?" *Journal of Social Work* 3 (2): 161–177.

Potts, Annie, and Donelle Gadenne. 2014. *Animals in Emergencies: Learning from the Christchurch Earthquakes*. Christchurch, NZ: Canterbury University Press.

Ralston, Jonah J., and Jason A. Kalmbach. 2018. "Regulating under Conditions of Uncertainty and Risk: Lessons Learned from State Regulation of Hydraulic Fracturing." *Environmental Practice* 20 (2–3): 1–12.

Riggs, Mike. 2013. "Why Energy Boomtowns Are a Nightmare for Law Enforcement." *Atlantic*, October 18, 2013. http://www.citylab.com/politics/2013/10/why-energy-boomtowns-are-nightmare-law-enforcement/7274/.

Royte, Elizabeth. 2012. "Fracking Our Food Supply: Are Dying Cattle the Canaries in the Coal Mine? Farmers and Ranchers Are Sounding Alarms about the Risks to Human Health of Hydraulic Fracturing." *Nation*, November 28, 2012. http://www.thenation.com/article/fracking-our-food-supply/.

Rozell, Daniel J., and Sheldon J. Reaven. 2012. "Water Pollution Risk Associated with Natural Gas Extraction from the Marcellus Shale." *Risk Analysis* 32 (8): 1382–1393.

Sanders, Clinton R. 2003. "Actions Speak Louder than Words: Close Relationships between Humans and Nonhuman Animals." *Symbolic Interaction* 26 (3): 405–426.

Schlossberg, Tatiana. 2016. "A Plan to Give Whales and Other Ocean Life Some Peace and Quiet." *New York Times*, June 3, 2016. https://www.nytimes.com/2016/06/03/science/noise-pollution-oceans-noaa-roadmap.html.

Slizovskiy, Ilya B., Lisa A. Conti, Sally J. Trufan, John S. Reif, Vanessa T. Lamers, Meredith H. Stowe, J. Dziura, and Peter M. Rabinowitz. 2015. "Reported Health Conditions in Animals Residing near Natural Gas Wells in Southwestern Pennsylvania." *Journal of Environmental Science and Health, Part A* 50 (5): 473–481.

Stewart, Gavin B., Andrew S. Pullin, and Christopher F. Coles. 2004. "Effects of Wind Turbines on Bird Abundance." *Systematic Review No.* 4. Birmingham, UK: University of Birmingham. https://www.researchgate.net/profile/Andrew_Pullin/publication/33375945_Effects_of_wind_turbines_on_bird_abundance/links/0fcfd507eab6c425b9000000/Effects-of-wind-turbines-on-bird-abundance.pdf.

Stewart, Gavin B., Andrew S. Pullin, and Christopher F. Coles. 2007. "Poor Evidence-Base for Assessment of Windfarm Impacts on Birds." *Environmental Conservation* 34 (1): 1–11.

Testoni, Ines, Loriana De Cataldo, Lucia Ronconi, and Adriano Zamperini. 2017. "Pet Loss and Representations of Death, Attachment, Depression, and Euthanasia." *Anthrozoös* 30 (1): 135–148.

Thompson, Kirrilly. 2013. "Save Me, Save My Dog: Increasing Natural Disaster Preparedness and Survival by Addressing Human-Animal Relationships." *Australian Journal of Communication* 40 (1): 123–136.

Thompson, Sarah J., Douglas H. Johnson, Neal D. Niemuth, and Christine A. Ribic. 2015. "Avoidance of Unconventional Oil Wells and Roads Exacerbates Habitat Loss for Grassland Birds in the North American Great Plains." *Biological Conservation* 192 (December): 82–90.

Thomsen, Frank, Karin Lüdemann, Rudolf Kafemann, and Werner Piper. 2006. *Effects of Offshore Wind Farm Noise on Marine Mammals and Fish*. Hamburg, DE: Biola.

Torres, Luisa, Om Prakash Yadav, and Eakalak Khan. 2018. "Risk Assessment of Human Exposure to Ra-226 in Oil Produced Water from the Bakken Shale." *Science of the Total Environment* 626 (June): 867–874.

Triebenbacher, Sandra Lookabaugh. 2000. "The Companion Animal within the Family System: The Manner in Which Animals Enhance Life within the Home." In *Handbook on Animal-Assisted Therapy: Theoretical Foundations and Guidelines for Practice*, edited by Aubrey Fine, 357–374. San Diego: Academic Press.

Trigg, Joshua L., Kirrilly Thompson, Bradley Smith, and Pauleen Bennett. 2016. "A Moveable Beast: Subjective Influence of Human-Animal Relationships on Risk Perception, and Risk Behaviour during Bushfire Threat." *Qualitative Report* 21 (10): 1881–1903.

US Department of Energy. 2013. *How Is Shale Gas Produced?* Washington, DC. https://energy.gov/sites/prod/files/2013/04/f0/how_is_shale_gas_produced.pdf.

von Wehrden, Henrik, Joern Fischer, Patric Brandt, Viktoria Wagner, Klaus Kümmerer, Tobias Kuemmerle, Anne Nagel, Oliver Olsson, and Patrick Hostert. 2012. "Consequences of Nuclear Accidents for Biodiversity and Ecosystem Services." *Conservation Letters Banner* 5 (2): 81–89.

Wahlberg, Magnus, and Håkan Westerberg. 2005. "Hearing in Fish and Their Reactions to Sounds from Offshore Wind Farms." *Marine Ecology Progress Series* 288 (March): 295–309.

Walsh, Froma. 2009. "Human-Animal Bonds I: The Relational Significance of Companion Animals." *Family Process* 48 (4): 462–480.
Wendle, John. 2016. "Animals Rule Chernobyl Three Decades after Nuclear Disaster." *National Geographic*. April 18, 2016. https://www.nationalgeographic.com/news/2016/04/060418-chernobyl-wildlife-thirty-year-anniversary-science/.
Whitley, Cameron Thomas. 2017. *Altruism, Risk, Energy Development and the Human-Animal Relationship*. PhD diss., Michigan State University, Lansing.
Whitley, Cameron Thomas. 2018. Animals, Humans and Energy Development. In *Fractured Communities: Risk, Impacts, and Protest against Hydraulic Fracking in U.S. Shale Regions*, edited by Anthony E. Ladd, 128–148. New Brunswick: Rutgers University Press.
Wood, Pamela. 2017. "Maryland General Assembly Approves Fracking Ban." *Baltimore Sun*, March 27, 2017.
Ziv, Guy, Eric Baran, So Nam, Ignacio Rodríguez-Iturbe, and Simon A. Levin. 2012. "Trading-Off Fish Biodiversity, Food Security, and Hydropower in the Mekong River Basin." *PNAS* 109 (15): 5609–5614.

Chapter 2 Summary

This chapter addresses the place of animals in energy development, focusing on the use of sentinels, and highlighting the significance of the human-animal bond. Recognizing the social, psychological, and economical entanglement of humans and animals, the study of risk and resilience is increasingly exploring the significance of animals—especially domesticated animals—in energy development and disaster. We suggest this multispecies bond is critically important in not only understanding general impacts, but also in understanding public perception. To engage these points, this chapter introduces two case studies: (1) the high-volume hydraulic fracturing (HVHF) boom in the United States, and (2) a nuclear power disaster in Japan. Within each case study, we discuss how companion animals, agricultural animals, and wildlife interact with the energy form at hand. Broadly, these case studies demonstrate the consequences of excluding animals in both contexts, while highlighting the key insights that are viewable when we consider both the animal and the human-animal bond as significant. Regarding HVHF, animals are functioning as unintentional sentinels, or serving as canaries in the coal mine, providing insight into the potential health impacts that humans may experience. The second case study introduces the short- and long-term effects of a nuclear disaster, illustrating the importance of planning for both human and nonhuman resilience to radiation exposure.

KEY TAKEAWAYS

- The human-animal bond is critically important not only to understanding general impacts, but also to understanding public perception of HVHF.
- This lack of recognition of the human-animal bond continues to have far-reaching consequences for society—especially during disaster situations.
- Building resiliency for animals simultaneously builds resilience for the humans who rely upon them economically, emotionally, or otherwise.
- Within human society, animals are already being used as unintentional sentinels, indicators of current and future risks to humans.

3

The Need for Social Scientists in Developing Social Life Cycle Assessment

EMILY GRUBERT

Introduction

Life Cycle Assessment and Energy Systems

Life Cycle Assessment (LCA) is a decision support tool focused on multi-criteria evaluation of the impacts of a system, historically heavily focused on environmental impacts. The primary innovation of LCA relative to other analytical methods is to recognize that the overall impact of a given object of analysis cannot be described accurately without considering all the stages implicated in its existence, from inception through retirement. Although LCA is used to evaluate the environmental impacts of extremely diverse products, processes, systems, and services, it has been particularly relevant to thc study of energy systems. Not only does the LCA literature reflect explicit interest in certain types of energy systems, such as biofuels (Chen et al. 2014), but also one of the most commonly investigated environmental impacts, climate change pollution (Grubert 2017a), is heavily correlated with energy use (IPCC Working Group III 2014). Even for LCA

DOI: 10.5876/9781646420278.c003

studies investigating nonenergy topics, therefore, energy is frequently an important part of LCA.

As LCA work and other analytical efforts continue to investigate questions about the most appropriate energy systems for a given time, place, and goal (Arvesen et al. 2018; Arvidsson and Svanström 2015; Gagnon, Bélanger, and Uchiyama 2002; Gibon, Arvesen, and Hertwich 2017; Jones et al. 2017; Li, Wang, and Yan 2018; Raadal et al. 2011; Stamford and Azapagic 2012; Venkatesh et al. 2011; Xie et al. 2018), the relevance of understanding impacts of energy development and use beyond greenhouse gas emissions has grown. Particularly given the often unequal spatial distribution of energy infrastructure and the associated localized impacts of such infrastructure (see, e.g., Bell 2013; Clough and Bell 2016; Liu, Liu, and Zhang 2014; Malin and DeMaster 2016; Ottinger 2013; Walker 2010), social impacts of energy systems are often significant and long term. These social impacts are often relevant not only because of the implications they bear for receiving communities, but also because societal acceptance or opposition can have major influence on a project's success.

It has become increasingly clear that the sustainability of an energy system cannot be effectively assessed using environmental metrics alone. In the LCA world, this realization is often addressed with proposals to expand life cycle methods to include more criteria, often through the addition of concepts of Life Cycle Costing (LCC) and Social Life Cycle Assessment (S-LCA) to the more common environmental LCA (E-LCA or just LCA). While LCC is relatively straightforward because of its use of monetary units as a metric, S-LCA shares the multicriteria nature of E-LCA. However, S-LCA is much less common than E-LCA, and it remains relatively immature and unformalized. This chapter argues that S-LCA cannot become a viable method for assessing social impact without the participation of social scientists, particularly those who investigate nonmonetary issues. Further, given LCA's connection with energy and the deep literature on social impacts associated with energy systems, this chapter argues that the social scientific energy research community likely has the greatest opportunity to engage either critically or constructively with S-LCA in a way that could dramatically improve the understanding and treatment of social issues within socioenvironmental assessment more broadly.

S-LCA likely has extensive applications across topics of life cycle–oriented studies. However, robust S-LCA is particularly important for energy policy and project evaluation. Energy systems are deeply embedded in society, and that embedding suggests two major motivations for understanding that relationship in a robust, policy-integrated manner. First, energy policy and projects affect people at a community to societal scale. Although scholarship on the social impacts of energy development has long debated the exact nature and magnitude of these effects (e.g., Wilkinson et al. 1982), they clearly exist. A second, instrumental reason for understanding the energy-society relationship is the inverse relationship: community and societal attitudes affect energy policy and projects. Communities can invite or prevent development (Pathak 2020; Wood 2016), and societal attitudes can reflect support of or opposition to projects and policy (Jacquet 2012; Larson and Krannich 2016; Liebe, Bartczak, and Meyerhoff 2017). A further influence is that community and societal reaction to energy can engender social conflict—a metasocial impact that might be independent of any material changes to the energy system (Colvin, Witt, and Lacey 2016; Grubert and Skinner 2017). The energy-oriented social science community is well positioned to comment on how social impacts might be addressed in a life cycle framework in part also because of the extensive work on social impacts throughout the life cycle of a project, from entrance through exit, for example, through the social license to operate and just transitions literatures (Haggerty et al. 2018; Jasanoff 2018; Smith and Richards 2015).

Social Life Cycle Assessment and the Case for Social Scientific Engagement

S-LCA is immature but not new (O'Brien, Doig, and Clift 1996). Over its twenty-year history, only about 120 papers indexed in Web of Knowledge with the topic "social life cycle assessment" have been published (using the same approach described in Grubert [2016]), including 66 published between 2015 and September 2017.[1] Although S-LCA has been infrequently applied, it is currently gaining traction and attention from the scientific community. Appetite for standardization and comparability across case studies is also evident in the increasing use of the S-LCA methodological sheets proposed by (Benoît-Norris et al. 2011). That is, there is increasing attention to S-LCA proposals that indicate both that there is opportunity to meaningfully alter the method's design and that this window of opportunity may be closing soon.

One noticeable characteristic of the recent rapid growth of S-LCA literature is the very low participation by social scientists, even relative to other socially oriented assessment literatures such as social impact assessment (SIA) and corporate social responsibility (CSR). Based on the listed affiliations of corresponding authors, five of the roughly 120 S-LCA articles in the literature as of late 2017 were led by social scientists, all economists or from economics-adjacent fields. Although some noncorresponding authors might be social scientists, and though some corresponding authors in interdisciplinary or non-social-science departments might be social scientists, this absence is striking. A similar search for "social impact assessment" returns over 400 documents, about a quarter of which are categorized as being from sociology, anthropology, psychology, and related noneconomics social science fields (compared to only three documents—2%—from the S-LCA literature). "Corporate social responsibility" returns over 10,000 documents, about 10 percent of which are similarly categorized (the vast majority—over 75%—are categorized as business, management, and similar). These observations support the assertion that social science participation in S-LCA is not only limited but substantially more limited than for SIA and CSR, even assuming that some literature outside Web of Knowledge exists.

This chapter focuses on S-LCA for two major reasons. First, S-LCA is of interest largely because of a broader goal of simultaneous financial, environmental, and social assessment within the life cycle family of methods (see, e.g., Grubert 2017b; Jørgensen, Herrmann, and Bjørn 2013; Kloepffer 2008; Zamagni 2012; Zamagni et al. 2012), which means it is likely to gain substantial traction within the broader socioenvironmental assessment community if it is developed well. Second, S-LCA is still in active development, which means that there is limited established practice and therefore room for potentially radical change. The major goal of this chapter is to make the case that S-LCA should not be developed and deployed without significant engagement by the social science community. LCA more generally is primarily an engineering method, and despite calls for engagement by social scientists in developing S-LCA (see, e.g., Hunkeler and Rebitzer 2005) and critiques of LCA as biased against social science engagement (see, e.g., Hertwich and Pease 1998 and the response by Marsmann et al. 1999 framing social science as primarily value-driven rather than analytical[2]), most of the academic discourse on LCA—including S-LCA—has taken place in journals that generally do not solicit participation from social scientists.[3] This chapter represents an attempt to introduce S-LCA in an explicitly social scientific forum, focused on the particularly relevant topical area of energy. Given the depth of research on the social impacts of energy development in particular, focusing on energy systems as a test bed for meaningful S-LCA that is consistent with social scientific theory and practice is likely to be productive.

Given the goal of introducing S-LCA as a target for engagement, the remainder of this chapter aims to fulfill three goals. First, it describes socioenvironmental assessment broadly, placing S-LCA within the history of adjacent methods to contextualize how S-LCA might be different from similar and often critiqued social assessment methods. Second, it describes constraints on S-LCA associated with being a life cycle assessment method. Finally, it proposes specific areas for social scientific engagement in methodological development.

Socioenvironmental Assessment

Socioenvironmental assessment is a broad term for a group of multicriteria approaches to evaluating or predicting the direct socioenvironmental consequences of creating or implementing a product, process, policy, service, or system. Often, these consequences (which may be positive or negative) are classified in three realms: financial, environmental, and nonfinancial social areas. Generally, assessments are quantitative. One of the hallmarks of socioenvironmental assessment is its multicriteria approach, where even a single-realm (e.g., environmental) assessment generally investigates multiple types of impacts, such as air pollution and water pollution, that cannot be directly compared. Given this incommensurability, one of the key challenges of socioenvironmental assessment is to effectively collect, analyze, and present information about these multiple issues of interest.

Many socioenvironmental assessment methods are explicitly designed as decision support tools, indicating that findings from these assessments are both intended for a nonacademic audience and intended to influence outcomes that affect people and the environment. That is, the goal of socioenvironmental assessment is not primarily theoretical. Therefore, the types of information gathered, how the information is analyzed, and how the information is presented are meaningful and should be held to a relatively higher standard of robustness, accuracy, and justice than might otherwise be the case, particularly when socioenvironmental assessment is used for policy analysis.

This type of decision- and policy-oriented analysis is extremely common in energy and other natural resource management contexts, largely due to the broad scope of potential impacts associated with these often expensive and long-term projects. Almost all energy-related decisions are ultimately multicriteria decisions, where no single option is typically strictly better than all alternatives across all management criteria of interest. These decisions are made more challenging by the fact that decisions must be made even with incomplete data, a "central issue in natural resources management" (Owens 2014). Socioenvironmental assessment—which can be mandated, voluntary, or externally introduced to a decision process—represents an effort to more systematically identify relevant information and make it available before action is taken.

Given the purported equal emphasis on economy, environment, and society suggested by common business and management frameworks such as the "triple bottom line" or "people, planet, profit"—and given the goal of socioenvironmental assessment to identify, organize, and provide decision-relevant data that are as complete as practical—an immediate and serious gap in the socioenvironmental assessment sphere becomes clear. That is, in general, socially oriented assessment methods are far less developed than financial and environmental assessment methods,[4] even though the concepts are approximately the same age. One potential driver of this relative immaturity is that socioenvironmental assessment has historically been situated in natural science and engineering realms, in part because of a persistent demand for quantitative methods. This chapter addresses the history and status of socioenvironmental assessment and argues that more social scientific engagement with the development of socially oriented assessment methods, particularly S-LCA, is warranted and necessary.

History of Socioenvironmental Assessment

The three major quantitative socioenvironmental assessment frameworks (impact assessment, corporate social responsibility, and life cycle methods) all began to develop around the same time, though in different domains, largely in tandem with the emergence of the environmental movement in mid-twentieth-century United States. Government policy, primarily in the form of the National Environmental Policy Act of 1969, effectively mandated socioenvironmental assessment for certain projects under federal jurisdiction (US Congress 1970) and fostered the development of impact assessment (environmental impact assessment, EIA, and social impact assessment, SIA). Corporate social responsibility (CSR) began to emerge around 1974 as a way for businesses to enhance their reputations and long-term profitability by tracking and publicizing voluntary corporate actions with positive socioenvironmental benefits (Carroll 2008).

The origins of LCA are different in three major ways. First, LCA was formalized later than EIA/SIA and CSR, though it originated around the same time (see Guinée et al. 2011 for a more detailed history). Second, and critically, LCA has always been situated in a more scientific context than EIA/SIA and CSR. Although some early life cycle analyses were performed on products for companies or the government, they were often performed by research institutes, consultants, or other organizations at least nominally independent of the producer (Guinée et al. 2011). Life cycle assessment differs from EIA/SIA and CSR in that it is neither a fully governmental nor fully corporate approach to socioenvironmental assessment. In general, LCA attracts more academic engagement than does EIA/SIA or CSR,[5] with a greater focus on methodological development and refinement. The third major difference between LCA and impact assessment or CSR, and perhaps the one most relevant to this chapter, is that social issues were not included in LCA from the beginning.

Socially Oriented Socioenvironmental Assessment Approaches

Nonfinancial, nonhealth social impacts were not part of the original formulation of LCA, though interest in a socially focused form or element of LCA has long been expressed (see, e.g., Dreyer, Hauschild, and Schierbeck 2006; Fava et al. 1993; O'Brien Martin, Doig, and Clift 1996). This lack of social focus at inception differs from the National Environmental Policy Act (NEPA)-based EIA/SIA (see, e.g., Title I, Sec. 101 and 102 [US Congress 1970]) or CSR and might explain the relative lack of social scientific involvement with LCA and S-LCA specifically. This chapter argues that because S-LCA is under active development, there is still opportunity for engagement to change the method and specifically to potentially avoid some of the major problems with SIA or CSR. For example, SIA's project-level focus and CSR's business-level focus limit their value for assessing the broader social context and full range and timing of impacts. Similarly, the rigidity of SIA as a NEPA process and the comparative informality of CSR as a voluntary process (though see standardization frameworks, e.g., GRI 2017) pose challenges to genuine inquiry. These weaknesses could also develop in S-LCA. Given that core issues such as the definition of a functional unit, the nature of impact categories and indicators, and the approach to impact assessment are not yet formalized, however, engagement building on the experience of SIA and CSR could be extremely valuable.

What Does It Mean for S-LCA to Be Life Cycle Assessment?

Given this chapter's goal of encouraging social scientific engagement in S-LCA development, this section describes the basic characteristics of life cycle assessment approaches. For a social assessment method to be S-LCA, it likely must inherit these broad characteristics, and so these methodological elements also represent areas of potential engagement. Understanding what it means for a method to be LCA—perhaps most relevantly, quantitative and focused on measurable impacts associated with specific processes—is also relevant to answering the outstanding question of whether S-LCA can or should be effectively developed as a companion method to E-LCA.

Formally, based on the governing international standard (ISO 14044), LCA has four major steps: goal and scope definition, inventory analysis, impact assessment, and interpretation (ISO 14044 2006). Goal and scope definition involves some of the most important choices of the assessment, including many of the most result-critical value judgments in the process. Specifically, goal and scope include the LCA's functional unit, or object of analysis; the boundaries of the system, which include the processes to be analyzed,[6] and sometimes the temporal and geographic scope of analysis; assumptions that will be used during analysis, including issues such as efficiencies, life times, and maintenance needs for relevant processes; allocation approaches associated with impacts from processes that are shared; and choice of impact categories.

Inventory analysis, sometimes referred to as life cycle inventory, is the data collection step. Here, flows between the object of analysis (the functional unit) and the environment are cataloged and quantitatively measured or estimated. For example, this stage is where the total amount of CO_2

emitted during fuel combustion in a boiler would be assessed. For social impacts—where flows are generally not physical and impacts generally do not occur in discrete, additive units—defining what parameters to inventory has been a major challenge.

Once the inventory is available, impact assessment is used to assess the significance of inventoried flows. The major goal is to analyze how inventory data map to impact categories, then to assess impacts within these categories on a common basis. Characterization factors within impact categories normalize impacts from different environmental flows for within-category comparison. For example, one common characterization factor is global warming potential (GWP), which allows analysts to compare the climate change impacts of different greenhouse gases. Impact assessment thus takes inventory data about, for example, how much methane, CO_2, and N_2O are emitted from various processes, then uses GWP as a characterization factor to provide an overall estimate of greenhouse gas impact in CO_2-equivalents. The last stage, interpretation, can take many different forms, and it is generally used to summarize LCA information in a way that is useful to its users.

Both qualitative and quantitative input is needed on topics ranging from core issues, for instance, whether S-LCA is possible or worthwhile to implement, to more technical issues, for instance, how to define goal and scope in a social context, how to measure social impact, and how to aggregate impact indicators into meaningful impact categories. Social impacts are different from environmental impacts in some major ways that would almost certainly require adaptation of current LCA practice. For example, LCA currently defines a life cycle primarily based on material flow, but social effects can begin before material extraction and persist well beyond material disposal. If S-LCA can exist as a quantitative, impact-category-based method, guidance on how to quantify social impact requires careful consideration by both qualitative and quantitative means, including critical assessment of measurement techniques and trials of those techniques. Qualitative research is likely to be particularly valuable in informing choices of impact categories and measured indicators by identifying mechanisms by which a given measurable outcome contributes to the impact of interest. For example, CO_2 emissions are an indicator of climate change impact because there is a well-understood mechanism by which CO_2 contributes to climate change. Similarly well-defined indicators would be needed for S-LCA.

The Status of Social Life Cycle Assessment

The S-LCA literature has been reviewed several times (see, e.g., Chhipi-Shrestha, Hewage, and Sadiq 2014; Grubert 2016; Jørgensen et al. 2008; Petti, Serreli, and Cesare 2016), with repeated observation that S-LCA is still immature and requires substantial work before it will be seen as a consistent, rigorous method with well-defined approaches to common questions. Currently, S-LCA does not have consistent impact assessment methodologies, accepted impact categories, or accepted functional unit boundaries, though some attempts to standardize S-LCA have been proposed and used (see, e.g., Benoît et al. 2010; Benoît-Norris et al. 2011; Norris, Norris, and Aulisio 2014). S-LCA is "an ideological hybrid" (Sakellariou 2016), which both poses challenges and creates opportunities for collaborative consideration of what S-LCA should be and why.

One of the most significant established frameworks guiding S-LCA is ISO 26000, *Guidance on Social Responsibility*, which establishes both the "Seven Key Principles" of accountability, transparency, ethical behavior, respect for stakeholder interests, respect for the rule of law, respect for international norms of behavior, and respect for human rights, and also the "Seven Core Subjects" of organizational governance, human rights, labor practices, environment, fair operating practices, consumer issues, and community involvement and development. As with CSR and E-LCA, ISO 26000 does not draw a completely mutually exclusive line between environmental and social concerns, in part because of the importance of environmentally mediated health and justice outcomes. Another highly influential framework, the UNEP/SETAC Guidelines for Social Life Cycle Assessment of Products and the associated methodological sheets, suggests categories of human rights, working conditions, health and safety, cultural heritage, governance, and socioeconomic repercussions. Although this approach has been critiqued given challenges associated with integrating stakeholder participation in impact category selection and other issues (Mathe 2014), the use of consistent general impact categories rather than locally defined impact categories is typical across life cycle approaches. Notably, these frameworks do not easily accommodate some of the impacts identified in the social science energy literature, such as the effects of changes to social capital, demographic changes, infrastructural strain, identity impacts, and others (Benham 2016; Groves et al. 2017). Even in contexts where impacts are not seen as usefully quantifiable, identifying the scope of what social impacts can be is a valuable point of engagement.

Some of the major challenges for S-LCA are similar in character to challenges experienced in E-LCA (see Grubert 2016). For example, data collection in S-LCA (as with E-LCA) can be exceptionally challenging, particularly in culturally specific and socially dynamic contexts. Other challenges are specific to S-LCA, such as the idea that whether a social outcome is a "harm" or a "benefit" can be much more contextually specific than for E-LCA, or the similar idea that an indicator might be subject to nonmonotonic benefit curves where having more of an impact (e.g., working hours) is desirable in some settings and not in others (Arvidsson, Baumann, and Hildenbrand 2014; Baumann et al. 2013). Other challenges that differentiate S-LCA from E-LCA include the meaning of the life cycle itself. Social impacts can precede a project (Grubert and Skinner 2017), whereas environmental impacts generally do not, and S-LCA frequently does not address the use phase of a functional unit's life cycle (Chhipi-Shrestha, Hewage, and Sadiq 2014).

Perhaps the most serious challenge for S-LCA is that by nature of its status as an LCA method, S-LCA leaves little space for assertions that quantification of social impacts is inappropriate or not meaningful. This potentially poses major problems for its rigor and long-term survival. Given the momentum behind the approach, however, it is arguably more important for the social science community to address S-LCA if the consensus is that S-LCA is not valid. The likely alternative to social scientific engagement is that S-LCA will continue to develop without input from disciplinary and epistemological traditions that house the most knowledge about social impact. As Nicholas Sakellariou writes: "Yet, regardless of the fact that the translation of quantitative data into engineering decisions is no less problematic in LCA than in SLCA, most engineers view LCAs as more robust and technically sound. SLCA is then an attempt by sustainability practitioners to make *social impacts* more 'sound' in engineering cultures" (2016).

The desire to quantify social impacts (which in part reflects the common privileging of quantitative data in many decision contexts, appropriate or not) is derivative of this interest in fitting social impacts into an epistemological understanding of truth as fundamentally measurable and quantitative, despite the fact that fields devoted to studying social impacts do not universally share that understanding. Although this mismatch and the nature of interdisciplinary research more generally pose genuine barriers to engagement (Schuitema and Sintov 2017), even establishing a literature of critique, as exists for SIA and CSR, would be valuable for the socioenvironmental assessment community. An aspiration would be to define more valid approaches, building on what has been learned from social assessment experience to date.

Social Science in Life Cycle Assessment

As earlier sections of this chapter have addressed, S-LCA requires substantial development effort before it might be viewed as valid and useful. Further, LCA more generally faces challenges that engage areas of study more common in the social sciences, such as justice, ethics, and meaningful engagement of value systems. This section therefore describes areas of potential engagement for social scientists both in S-LCA specifically and in LCA broadly. For the sake of discussion, this section assumes that S-LCA is worth developing within an ISO 14000 framework.

Direct Development of S-LCA: Scope and Functional Unit

Understanding how the scope of a study might change when S-LCA is conducted alongside E-LCA would benefit greatly from existing research on psychological and community effects of development. For example, E-LCA does not generally consider impacts that occur before material flows between the functional unit system and the environment begin or after they end. However, social impacts are not typically material in the way that environmental impacts are, in the sense that there is not typically a mass of substance moving from one environmental sphere to another. For S-LCA to both adequately address the full life cycle of social impacts associated with a functional unit and remain compatible with E-LCA, guidelines on scope choice should likely consider specific life cycle stages for social impacts that might not match the traditional E-LCA life cycle stages.

Impact Categories

One critique of S-LCA is that to date, it has not always highlighted decision-relevant information (see, e.g., Ekener-Petersen and Moberg 2013;

Freudenburg 1986). A major task, then, is to ensure that the set of issues being investigated in S-LCA, called impact categories, addresses topics that are decision relevant as well as important. As William Freudenburg notes, a goal of focusing on sociological variables rather than available data is particularly important (1986), especially in the context of a quantitative method being used in a setting where quantitative data are privileged. Defining impact categories should rely on both empirical and theoretical social science work, drawing on findings from social impact assessment, case studies, and other sources. One possible approach to identifying potential impact categories is to use large-scale metareview approaches to identify relevant impact categories and the relationships among them. An application of this approach to the adaptive capacity literature shows that computer-aided review in particular can identify useful patterns from large literatures (Siders 2017).

Indicators, Measurement Strategies, and Impact Assessment

Once impact categories have been chosen, LCA requires some way to quantify the extent of impact within those categories. First, relevant indicators within each category must be chosen. For example, for the impact category of ecotoxicity, indicators include emissions of a variety of toxic materials to different media (e.g., lead emissions to soil or mercury emissions to water). These same-category indicators are then compared to one another using characterization factors to provide a single value for impact in each category.

For S-LCA, indicators and measurement strategies are not well defined. Health concerns are sometimes measured and characterized using Disability-Adjusted Life Years (DALYs) and Quality-Adjusted Life Years (QALYs), but these are controversial. Substantial effort is required toward identifying appropriate indicators in social impact categories and approaches to quantitative measurement of these indicators in a way that is translatable across contexts. Issues of differentiated ways of knowing, cultural understanding of the relative goodness or badness of an outcome, value judgments related to within-category equivalency, and others are particularly relevant in social assessment given that meaningful social impacts are often related to emotional response or other highly personal or cultural outcomes in ways that are less common for environmental impact. For example, the sociocultural context of infrastructural development mediates both the magnitude (big or small) and direction (good or bad) of the impact (see, e.g., Fergen and Jacquet 2016; Liebe, Bartczak, and Meyerhoff 2017). Characterization factors might be site, time, and culture specific, which would require a consistent approach to determining these characterization factors rather than using consistent characterization factors themselves.

Interpretation

Interpretation is challenging in E-LCA and is likely to also be challenging in S-LCA for similar reasons. Two of the most significant challenges are cross-category normalization and prioritization. Cross-category normalization refers to a process analogous to within-category characterization using characterization factors, in that it is an approach to determining how severe problems in one impact category are relative to problems in another, incommensurable impact category. For example, how might one compare 18 units of human rights violation with 42 units of socioeconomic loss? Can, or should, that comparison even be done? Prioritization represents the next step: once one understands the harm inflicted in one impact category in normalized relationship to the harm inflicted in another, how does an analyst incorporate the idea that one issue might be more important than another? That is, if an equal amount of human rights harm and socioeconomic harm exist, but there is a societal understanding that human rights are inherently more important than socioeconomic harm or vice versa, how is that included in analysis? Although these are major questions for S-LCA, they are also important in E-LCA and have not yet been solved. See Grubert (2017b) for a deeper discussion.

Challenges in Developing Social Life Cycle Assessment

Standardizing approaches to S-LCA so that various approaches can be used, tested, and improved upon requires the identification and deployment of best practices. One of the most challenging problems in LCA generally, and S-LCA specifically, is ensuring analytical robustness. LCA is a data-heavy framework that relies heavily on case studies, so it is important to understand where existing data can be applied in other work. Although it can be argued that data sharing is not fundamental to the success of LCA given that LCA is an analytical method rather than a tool for generating theory, a lack of shared social data would pose a significant barrier to uptake of S-LCA in exploratory settings. Related challenges include assuring consistency in S-LCA results in comparative studies, the need to be quantitative in nearly all contexts, and the costs and effort associated with data collection and maintenance.

Another challenge is the likely difference in preferred scope for S-LCA versus E-LCA or LCC. Although differences in which life cycle stages are relevant are fairly easily mitigated by expanding the definition of life cycle to include both pre- and postimplementation phases, differences in the scope of the functional unit or analytical boundary are more difficult to overcome. For example, an S-LCA might more reasonably focus on an entire factory or company than on a single product, but evaluations with unreconcilable scopes will not easily contribute to a goal of integrating social and environmental assessment. Similarly, allocating impacts to a company rather than to a product can be challenging (Grubert 2016).

A third challenge is that S-LCA has so far more frequently allowed for the idea that impacts can be positive than has E-LCA (Di Cesare et al. 2016). For example, the impact category "working conditions" can include indicators of improved working conditions. In E-LCA, by contrast, a product system rarely contributes positively to the impact category "climate change" (the challenge of performing E-LCA on functional units designed as a beneficial environmental measure is substantial and still under active development, e.g., Petit-Boix et al. 2017).

Beyond Social Life Cycle Assessment

Beyond S-LCA, social science attention to and engagement with LCA methods generally could be fruitful in a variety of areas. One of the most significant areas is the question of how to apply value judgments to socio-environmental assessment. For example, any decision based on multicriteria data involves prioritizing across multiple incommensurable categories. As mentioned above, this question of prioritization addresses the point that in a situation where harms in two areas are equivalent, people might still prefer to address one area over the other. Questions of what these priorities might be, whose priorities matter, and how to measure and apply such priorities are topics with wide relevance in decision analysis and multicriteria assessment generally, not just LCA. Other, more meta-analytic questions for social scientists engaged with life cycle methods are similar to other areas of science and technology studies inquiry. For example, how just are life cycle methods as designed, and how just are they as implemented? What types of ethical underpinnings do life cycle methods currently have, and how might they need to change? LCA could also be productively critiqued on the grounds of how topics of analysis are chosen, how data are collected and measured, and who controls the generation and interpretation of knowledge. In general, it is understood that LCA includes value judgments, but the implications of these value judgments, their role, and their future role are not especially rigorously engaged or studied. Social scientific attention and experience would likely strengthen the methods as a whole.

Notes

1. Although the Web of Knowledge collections do not perfectly reflect the social science literature, contrast this absence with the environmental LCA literature, with about 15,000 academic papers with the topic "life cycle assessment" indexed in Web of Knowledge as of September 2017.
2. Clearly, social science research can be analytical or value-driven, and there are standards for robustness and rigor in either case. In LCA, the distinction is largely relevant for distinguishing between social science research on impacts and social science research on how to use data to make decisions.
3. Based on a September 2017 search in Web of Knowledge, the topic "life cycle assessment" appears most frequently in the *Journal of Cleaner Production* (12% of about 15,000 articles), the *International Journal of Life Cycle Assessment* (8%), and a variety of other engineering and natural science journals. Similarly, "social life cycle assessment" appears most frequently in the *International Journal of Life Cycle Assessment* (37% of about 120 articles), *Sustainability* (10%), and the *Journal of Cleaner Production* (10%).
4. Note, however, that both financial and environmental assessment methods often include elements with social significance, such as costs, ecotoxicity influences on human health, and environmental impacts with aesthetic and other social implications: smog, acid rain, and so on.
5. The academic CSR literature is relatively large, with about 10,000 documents indexed in Web of Science versus LCA's 14,000 as of September 2017. However, whereas the LCA literature contains mostly LCA case studies and methodological comments, the CSR literature contains mostly analyses of corporate CSR case

studies and critiques of CSR. That is, companies control the generation of CSR information, and academics are far more engaged in generating LCA than CSR reports. The academic EIA and SIA literatures are smaller, with primary documents typically taking the form of government compliance documents.

6. One of the major challenges of LCA is that, ultimately, essentially every process in the economy is implicated (perhaps in minute quantities) in a complex system. Deciding where to draw the line on which indirectly involved processes are important is a meaningful choice. LCA practitioners occasionally joke about this; for example, is it environmentally significant whether the farmers who grow the hay that feeds the cattle that provides beef to the workers at the steel factory were vegetarians or not?

References

Arvesen, Anders, Gunnar Luderer, Michaja Pehl, Benjamin Leon Bodirsky, and Edgar G. Hertwich. 2018. "Deriving Life Cycle Assessment Coefficients for Application in Integrated Assessment Modelling." *Environmental Modelling and Software* 99: 111–125. https://doi.org/10.1016/j.envsoft.2017.09.010.

Arvidsson, Rickard, Henrikke Baumann, and Jutta Hildenbrand. 2014. "On the Scientific Justification of the Use of Working Hours, Child Labour and Property Rights in Social Life Cycle Assessment: Three Topical Reviews." *International Journal of Life Cycle Assessment*: 1–13. https://doi.org/10.1007/s11367-014-0821-3.

Arvidsson, Rickard, and Magdalena Svanström. 2015. "A Framework for Energy Use Indicators and Their Reporting in Life Cycle Assessment." *Integrated Environmental Assessment and Management* 12 (3): 429–436. https://doi.org/10.1002/ieam.1735.

Baumann, Henrikke, Rickard Arvidsson, Hui Tong, and Ying Wang. 2013. "Does the Production of an Airbag Injure More People than the Airbag Saves in Traffic?: Opting for an Empirically Based Approach to Social Life Cycle Assessment." *Journal of Industrial Ecology* 17 (4): 517–527. https://doi.org/10.1111/jiec.12016.

Bell, Shannon Elizabeth. 2013. *Our Roots Run Deep as Ironweed: Appalachian Women and the Fight for Environmental Justice*. Urbana: University of Illinois Press.

Benham, Claudia. 2016. "Change, Opportunity and Grief: Understanding the Complex Social-Ecological Impacts of Liquefied Natural Gas Development in the Australian Coastal Zone." *Energy Research and Social Science* 14 (April): 61–70. https://doi.org/10.1016/j.erss.2016.01.006.

Benoît, Catherine, Gregory A. Norris, Sonia Valdivia, Andreas Ciroth, Asa Moberg, Ulrike Bos, Siddharth Prakash, Cassia Ugaya, and Tabea Beck. 2010. "The Guidelines for Social Life Cycle Assessment of Products: Just in Time!" *International Journal of Life Cycle Assessment* 15 (2): 156–163. https://doi.org/10.1007/s11367-009-0147-8.

Benoît-Norris, Catherine, Gina Vickery-Niederman, Sonia Valdivia, Juliane Franze, Marzia Traverso, Andreas Ciroth, and Bernard Mazijn. 2011. "Introducing the UNEP/SETAC Methodological Sheets for Subcategories of Social LCA." *International Journal of Life Cycle Assessment* 16 (7): 682–690. https://doi.org/10.1007/s11367-011-0301-y.

Carroll, Archie B. 2008. "A History of Corporate Social Responsibility." In *The Oxford Handbook of Corporate Responsibility*, edited by Andrew Crane, Dirk Matten, Abagail McWilliams, Jeremy Moon, and Donald S. Siegel, 19–46. New York: Oxford University Press. https://doi.org/10.1093/oxfordhb/9780199211593.003.0002.

Chen, Haibin, Yu Yang, Yan Yang, Wei Jiang, and Jingcheng Zhou. 2014. "A Bibliometric Investigation of Life Cycle Assessment Research in the Web of Science Databases." *International Journal of Life Cycle Assessment* 19 (10): 1674–1685. https://doi.org/10.1007/s11367-014-0777-3.

Chhipi-Shrestha, Gyan Kumar, Kasun Hewage, and Rehan Sadiq. 2014. "'Socializing' Sustainability: A Critical Review on Current Development Status of Social Life Cycle Impact Assessment Method." *Clean Technologies and Environmental Policy* 17 (3): 579–596. https://doi.org/10.1007/s10098-014-0841-5.

Clough, Emily, and Derek Bell. 2016. "Just Fracking: A Distributive Environmental Justice Analysis of Unconventional Gas Development in Pennsylvania, USA." *Environmental Research Letters* 11 (2): 025001. https://doi.org/10.1088/1748-9326/11/2/025001.

Colvin, R. M., G. Bradd Witt, and Justine Lacey. 2016. "How Wind Became a Four-Letter Word: Lessons for Community Engagement from a Wind Energy Conflict in King Island, Australia." *Energy Policy* 98 (supplement C): 483–494. https://doi.org/10.1016/j.enpol.2016.09.022.

Di Cesare, Silvia, Federica Silveri, Serenella Sala, and Luigia Petti. 2016. "Positive Impacts in Social Life Cycle Assessment: State of the Art and the Way Forward." *International Journal of Life Cycle Assessment* 23 (3): 1–16. https://doi.org/10.1007/s11367-016-1169-7.

Dreyer, Louise, Michael Hauschild, and Jens Schierbeck. 2006. "A Framework for Social Life Cycle Impact Assessment." *International Journal of Life Cycle Assessment* 11 (2): 88–97. https://doi.org/10.1065/lca2005.08.223.

Ekener-Petersen, Elisabeth, and Åsa Moberg. 2013. "Potential Hotspots Identified by Social LCA—Part 2: Reflections on a Study of a Complex Product." *International Journal of Life Cycle Assessment* 18 (1): 144–154. https://doi.org/10.1007/s11367-012-0443-6.

Fava, James, Frank Consoli, Richard Denison, Kenneth Dickson, Tim Mohin, and Bruce Vigon, eds. 1993. *A Conceptual Framework for Life-Cycle Impact Assessment*. Pensacola, FL: Society of Environmental Toxicology and Chemistry and SETAC Foundation for Environmental Education. http://trove.nla.gov.au/version/10066843.

Fergen, Joshua, and Jeffrey B. Jacquet. 2016. "Beauty in Motion: Expectations, Attitudes, and Values of Wind Energy Development in the Rural U.S." *Energy Research and Social Science* 11: 133–141. https://doi.org/10.1016/j.erss.2015.09.003.

Freudenburg, William R. 1986. "Social Impact Assessment." *Annual Review of Sociology* 12 (1): 451–478. https://doi.org/10.1146/annurev.so.12.080186.002315.

Gagnon, Luc, Camille Bélanger, and Yohji Uchiyama. 2002. "Life-Cycle Assessment of Electricity Generation Options: The Status of Research in Year 2001." *Energy Policy* 30 (14): 1267–1278. https://doi.org/10.1016/S0301-4215(02)00088-5.

Gibon, Thomas, Anders Arvesen, and Edgar G. Hertwich. 2017. "Life Cycle Assessment Demonstrates Environmental Co-benefits and Trade-offs of Low-Carbon Electricity Supply Options." *Renewable and Sustainable Energy Reviews* 76 (September): 1283–1290. https://doi.org/10.1016/j.rser.2017.03.078.

GRI. 2017. GRI Standards Download Center. Accessed October 2, 2017. https://www.globalreporting.org/standards/gri-standards-download-center/.

Groves, Christopher, Karen Henwood, Fiona Shirani, Gareth Thomas, and Nick Pidgeon. 2017. "Why Mundane Energy Use Matters: Energy Biographies, Attachment and Identity." *Energy Research and Social Science* 30: 71–81. https://doi.org/10.1016/j.erss.2017.06.016.

Grubert, Emily. 2016. "Rigor in Social Life Cycle Assessment: Improving the Scientific Grounding of SLCA." *International Journal of Life Cycle Assessment* 23 (3): 481–491.

Grubert, Emily. 2017a. "Implicit Prioritization in Life Cycle Assessment: Text Mining and Detecting Metapatterns in the Literature." *International Journal of Life Cycle Assessment* 22 (2): 148–158. https://doi.org/10.1007/s11367-016-1153-2.

Grubert, Emily. 2017b. "The Need for a Preference-Based Multicriteria Prioritization Framework in Life Cycle Sustainability Assessment." *Journal of Industrial Ecology* 21 (6): 1522–1535. https://doi.org/10.1111/jiec.12631.

Grubert, Emily, and Whitney Skinner. 2017. "A Town Divided: Community Values and Attitudes towards Coal Seam Gas Development in Gloucester, Australia." *Energy Research and Social Science* 30 (August): 43–52. https://doi.org/10.1016/j.erss.2017.05.041.

Guinée, Jeroen B., Reinout Heijungs, Gjalt Huppes, Alessandra Zamagni, Paolo Masoni, Roberto Buonamici, Tomas Ekvall, and Tomas Rydberg. 2011. "Life Cycle Assessment: Past, Present, and Future." *Environmental Science and Technology* 45 (1): 90–96. https://doi.org/10.1021/es101316v.

Haggerty, Julia H., Mark N. Haggerty, Kelli Roemer, and Jackson Rose. 2018. "Planning for the Local Impacts of Coal Facility Closure: Emerging Strategies in the U.S. West." *Resources Policy*. https://doi.org/10.1016/j.resourpol.2018.01.010.

Hertwich, Edgar G., and William S. Pease. 1998. "ISO 14042 Restricts Use and Development of Impact Assessment." *International Journal of Life Cycle Assessment* 3 (4): 180–181.

Hunkeler, David, and Gerald Rebitzer. 2005. "The Future of Life Cycle Assessment." *International Journal of Life Cycle Assessment* 10 (5): 305–308. https://doi.org/10.1065/lca2005.09.001.

IPCC Working Group III. 2014. *Chapter 7: Energy Systems*. http://www.ipcc.ch/report/ar5/wg3/.

ISO 14044. 2006. *Environmental Management—Life Cycle Assessment—Requirements and Guidelines*. https://www.iso.org/standard/38498.html.

Jacquet, Jeffrey B. 2012. "Landowner Attitudes toward Natural Gas and Wind Farm Development in Northern Pennsylvania." *Energy Policy* 50 (November): 677–688. https://doi.org/10.1016/j.enpol.2012.08.011.

Jasanoff, Sheila. 2018. "Just Transitions: A Humble Approach to Global Energy Futures." *Energy Research and Social Science* 35 (January): 11–14. https://doi.org/10.1016/j.erss.2017.11.025.

Jones, Christopher, Paul Gilbert, Marco Raugei, Sarah Mander, and Enrica Leccisi. 2017. "An Approach to Prospective Consequential Life Cycle Assessment and Net Energy Analysis of Distributed Electricity Generation." *Energy Policy* 100 (January): 350–358. https://doi.org/10.1016/j.enpol.2016.08.030.

Jørgensen, Andreas, Ivan T. Herrmann, and Anders Bjørn. 2013. "Analysis of the Link between a Definition of Sustainability and the Life Cycle Methodologies." *International Journal of Life Cycle Assessment* 18 (8): 1440–1449. https://doi.org/10.1007/s11367-013-0617-x.

Jørgensen, Andreas, Agathe Le Bocq, Liudmila Nazarkina, and Michael Hauschild. 2008. "Methodologies for Social Life Cycle Assessment." *International Journal of Life Cycle Assessment* 13 (2): 96–103. https://doi.org/10.1065/lca2007.11.367.

Kloepffer, Walter. 2008. "Life Cycle Sustainability Assessment of Products." *International Journal of Life Cycle Assessment* 13 (2): 89–95. https://doi.org/10.1065/lca2008.02.376.

Larson, Eric C., and Richard S. Krannich. 2016. "A Great Idea, Just Not Near Me! Understanding Public Attitudes about Renewable Energy Facilities." *Society and Natural Resources* 29 (12): 1436–1451. https://doi.org/10.1080/08941920.2016.1150536.

Li, Jiao, Yuan Wang, and Beibei Yan. 2018. "The Hotspots of Life Cycle Assessment for Bioenergy: A Review by Social Network Analysis." *Science of the Total Environment* 625 (June): 1301–1308. https://doi.org/10.1016/j.scitotenv.2018.01.030.

Liebe, Ulf, Anna Bartczak, and Jürgen Meyerhoff. 2017. "A Turbine Is Not Only a Turbine: The Role of Social Context and Fairness Characteristics for the Local Acceptance of Wind Power." *Energy Policy* 107 (August): 300–308. https://doi.org/10.1016/j.enpol.2017.04.043.

Liu, Lee, Jie Liu, and Zhenguo Zhang. 2014. "Environmental Justice and Sustainability Impact Assessment: In Search of Solutions to Ethnic Conflicts Caused by Coal Mining in Inner Mongolia, China." *Sustainability* 6 (12): 8756–8774. https://doi.org/10.3390/su6128756.

Malin, Stephanie A., and Kathryn Teigen DeMaster. 2016. "A Devil's Bargain: Rural Environmental Injustices and Hydraulic Fracturing on Pennsylvania's Farms." *Journal of Rural Studies* 47 (Part A): 278–290. https://doi.org/10.1016/j.jrurstud.2015.12.015.

Marsmann, Manfred, Sven Olaf Ryding, Helias Udo de Haes, James Fava, Willie Owens, Kevin Brady, Konrad Saur, and Rita Schenck. 1999. "In Reply to Hertwich & Pease, Int. J. LCA 3 (4) 180–181, 'ISO 14042 Restricts Use and Development of Impact Assessment.'" *International Journal of Life Cycle Assessment* 4 (2): 65. https://doi.org/10.1007/BF02979402.

Mathe, Syndhia. 2014. "Integrating Participatory Approaches into Social Life Cycle Assessment: The SLCA Participatory Approach." *International Journal of Life Cycle Assessment* 19 (8): 1–9. https://doi.org/10.1007/s11367-014-0758-6.

Norris, Catherine, Gregory Norris, and Deana Aulisio. 2014. "Efficient Assessment of Social Hotspots in the Supply Chains of 100 Product Categories Using the Social Hotspots Database." *Sustainability* 6 (10): 6973–6984. https://doi.org/10.3390/su6106973.

O'Brien, Martin, Alison Doig, and Roland Clift. 1996. "Social and Environmental Life Cycle Assessment (SELCA)." *International Journal of Life Cycle Assessment* 1 (4): 231–237. https://doi.org/10.1007/BF02978703.

Ottinger, Gwen. 2013. *Refining Expertise: How Responsible Engineers Subvert Environmental Justice Challenges*. New York: New York University Press.

Owens, Hannah L. 2014. "Closing the Gap Between Researchers and Policymakers: Lessons from the History of Fisheries Management in the United States." *Society and Natural Resources* 27 (12): 1339–1345. https://doi.org/10.1080/08941920.2014.933930.

Pathak, Ruchie. 2020. "Sharing the Sun: Community Solar in Ohio." Master's thesis, Ohio State University Columbus.

Petit-Boix, Anna, Eva Sevigné-Itoiz, Lorena A. Rojas-Gutierrez, et al. 2017. "Floods and Consequential Life Cycle Assessment: Integrating Flood Damage into the Environmental Assessment of Stormwater Best Management Practices." *Journal of Cleaner Production* 162 (September): 601–608. https://doi.org/10.1016/j.jclepro.2017.06.047.

Petti, Luigia, Monica Serreli, and Silvia Di Cesare. 2016. "Systematic Literature Review in Social Life Cycle Assessment." *International Journal of Life Cycle Assessment* 23 (3): 1–10. https://doi.org/10.1007/s11367-016-1135-4.

Raadal, Hanne Lerche, Luc Gagnon, Ingunn Saur Modahl, and Ole Jørgen Hanssen. 2011. "Life Cycle Greenhouse Gas (GHG) Emissions from the Generation of Wind and Hydro Power." *Renewable and Sustainable Energy Reviews* 15 (7): 3417–3422. https://doi.org/10.1016/j.rser.2011.05.001.

Sakellariou, Nicholas. 2016. "A Historical Perspective on the Engineering Ideologies of Sustainability: The Case of SLCA." *International Journal of Life Cycle Assessment* 1–11. https://doi.org/10.1007/s11367-016-1167-9.

Schuitema, Geertje, and Nicole D. Sintov. 2017. "Should We Quit Our Jobs? Challenges, Barriers and Recommendations for Interdisciplinary Energy Research." *Energy Policy* 101: 246–250. https://doi.org/10.1016/j.enpol.2016.11.043.
Siders, Anne R. 2017. "Collocations and Network Structure as Insights to Functional Elements of Building Adaptive Capacity." Presented at the Digital Humanities 2017, Montreal, Canada: Alliance of Digital Humanities Organizations. https://dh2017.adho.org/abstracts/451/451.pdf.
Smith, Don C., and Jessica M. Richards. 2015. "Social License to Operate: Hydraulic Fracturing–Related Challenges Facing the Oil and Gas Industry." *Oil and Gas, Natural Resources, and Energy Journal* 81 (posted April; updated October): 1–56. http://papers.ssrn.com/abstract=2591988.
Stamford, Laurence, and Adisa Azapagic. 2012. "Life Cycle Sustainability Assessment of Electricity Options for the UK: Life Cycle Sustainability Assessment of Electricity Options for the UK." *International Journal of Energy Research* 36 (14): 1263–1290. https://doi.org/10.1002/er.2962.
US Congress. 1970. *The National Environmental Policy Act of 1969*. https://www.fws.gov/r9esnepa/RelatedLegislativeAuthorities/nepa1969.PDF.
Venkatesh, Aranya, Paulina Jaramillo, W. Michael Griffin, and H. Scott Matthews. 2011. "Uncertainty Analysis of Life Cycle Greenhouse Gas Emissions from Petroleum-Based Fuels and Impacts on Low Carbon Fuel Policies." *Environmental Science and Technology* 45 (1): 125–131. https://doi.org/10.1021/es102498a.
Walker, Gordon. 2010. "Environmental Justice, Impact Assessment and the Politics of Knowledge: The Implications of Assessing the Social Distribution of Environmental Outcomes." *Environmental Impact Assessment Review* 30 (5): 312–318. https://doi.org/10.1016/j.eiar.2010.04.005.
Wilkinson, Kenneth P., James G. Thompson, Robert R. Reynolds Jr., and Lawrence M. Ostresh. 1982. "Local Social Disruption and Western Energy Development: A Critical Review." *Pacific Sociological Review* 25 (3): 275–296. https://doi.org/10.2307/1388767.
Wood, Roberta. 2016. "Protesters' Pipeline Victory." *Guardian (Sydney)*, October 12, 2016, issue no. 1752.
Xie, Minghui, Jiuli Ruan, Weinan Bai, et al. 2018. "Pollutant Payback Time and Environmental Impact of Chinese Multi-crystalline Photovoltaic Production Based on Life Cycle Assessment." *Journal of Cleaner Production* 184 (May): 648–659. https://doi.org/10.1016/j.jclepro.2018.02.290.
Zamagni, Alessandra. 2012. "Life Cycle Sustainability Assessment." *International Journal of Life Cycle Assessment* 17 (4): 373–376. https://doi.org/10.1007/s11367-012-0389-8.
Zamagni, Alessandra, Paolo Masoni, Patrizia Buttol, Andrea Raggi, and Roberto Buonamici. 2012. "Finding Life Cycle Assessment Research Direction with the Aid of Meta-Analysis." *Journal of Industrial Ecology* 16 (s1): S39–S52. https://doi.org/10.1111/j.1530–9290.2012.00467.x.

Chapter 3 Summary

Life Cycle Assessment (LCA) is a common quantitative tool for evaluating environmental outcomes. LCA is widely used to address questions about the multicriteria effects of creating and deploying products, policies, and services. Although life cycle methods primarily address environmental outcomes and costs (as life cycle costing, or LCC), theorists and practitioners have long articulated a desire for life cycle methods to address social impacts as well. A socially oriented version of LCA, known as social life cycle assessment (S-LCA), is emerging, but it remains undertheorized and immature. Major questions about S-LCA remain, including both fundamental questions such as "Should a quantitative social life cycle assessment tool exist?" and implementation issues relevant across life cycle methods such as "How should life cycle methods include value judgments?" Despite growing attention to S-LCA, however, it remains notably isolated from the social science community. This chapter is intended to both introduce LCA to social scientists who might not be familiar with it and to argue that social science perspectives are badly needed in LCA. Energy systems represent a particularly fruitful opportunity for engagement. Energy systems are a very common modern target for environmental LCA, given the close relationship between energy systems and issues such as climate change, air quality, water quality, and waste streams. Simultaneously, extensive and high-quality recent social science scholarship on energy systems demonstrates that energy systems also have a close relationship with numerous social outcomes. Drawing on this knowledge to evaluate whether S-LCA can be productively developed, and how, is a major opportunity for both the social science and LCA communities.

KEY TAKEAWAYS

- Life Cycle Assessment (LCA) is a common decision support tool that aims to systematically and holistically evaluate impacts of human activity.
- Energy systems are commonly evaluated using LCA because of their large and multicriteria impacts on the environment and society.
- Social impacts are rarely considered in LCA, despite frequent and long-term calls to integrate them.
- Social science scholarship on energy systems is extensive, and social impacts from energy systems are meaningful.

- Social Life Cycle Assessment (S-LCA) is immature and rapidly developing but to date has limited contact with the social science community.
- Social science perspectives on questions such as whether and how S-LCA should be developed are needed.
- The immaturity of S-LCA means that there is a major opportunity to shape and contribute to its development.
- The depth of social scientific knowledge on energy systems, and the LCA community's existing attention to energy, suggests that energy social scientists could be especially effective participants in S-LCA dialogues.

4

Societal Impacts of Emerging Grassroots Energy Communities

A Capabilities-Based Assessment

ALI ADIL

Introduction

Incumbent utility efforts to undermine and marginalize distributed solar installations are a well-documented phenomena in transitions scholarship (Hess 2015; see also Bulkeley, Castán Broto, and Maassen 2014; Geels 2014). At the same time, exponential growth of large-scale solar installations represent incumbent attempts to preserve the conventional centralized approach to electricity generation and, to this end, co-opt solar technology (Hess 2015; REN21 2017; SEIA 2017). Amidst ongoing struggles between incumbent utilities and technologically empowered solar consumers to control local energy transitions, the rise of Grassroots Energy Communities (GECs) is increasingly touted as a potent countervailing and emancipatory force against the prevailing status quo (Hufen and Koppenjan 2015; Pahl 2012; Stirling 2014). I use the term *GECs* to describe communities of citizens undertaking local energy initiatives, typically based on residential and community-scale initiatives, and ranging in their organization from simple group-buy initiatives and full-fledged consumer-owned and -run cooperatives to formal nonprofit

DOI: 10.5876/9781646420278.c004

organizations. However, despite the growing popularity and potential of GECs in the United States and abroad (ILSR 2016; Klein and Coffey 2016), there has not been much critical assessment of how GECs may impact the energy sector and a just energy transition.

To this end, a three-axis analytical framework is proposed to assess the nature of GECs in terms of their capacity to actualize a *fundamental* energy transition—a technological transition that accompanies transformation in the underlying centralized and accumulative logics governing the energy sector (Fairchild and Weinrub 2017; Stirling 2014). This chapter contributes to scholarship on energy transition governance as well as to the interdisciplinary field of energy research social science by drawing on a non-Western conception of (energy) justice by Amartya Sen (2009)—identified as a gap in this scholarship (Sovacool and Dworkin 2015)—to conceptualize the proposed framework. Furthermore, in engaging with scholarship from policy studies, political theory, and energy engineering toward developing the axes of the proposed framework, I envision it as a tool for GEC organizers, local policy makers, and planners to encourage self-reflexive assessments of grassroots civic action towards more radical enactments.

This chapter begins with a brief review of transitions scholarship to locate GECs and highlight a critical gap in the literature in regards to assessing their capacity to overcome the incumbent status quo. In the following section, I identify key dimensions of a *fundamental* energy transition based on the principles of *Energy Democracy*, an emerging global social movement demanding that energy transitions be as *economically equitable* and *socially just* as they are *environmentally sustainable* (Fairchild and Weinrub 2017). For each of these dimensions—equity, justice, and sustainability[1]—I explicate specific criteria of assessment based on scholarship from policy studies, political theory, and environmental studies respectively. Then, I use Amartya Sen's *Capabilities Approach* (2009) to operationalize these criteria into three continua between existing and desired institutional alternatives and arrangements. The final section presents the three-axis framework alongside eight archetypal models of energy sector governance and demonstrates its utility by assessing two representative GECs from practice. I conclude with closing remarks on the potential of the framework to serve as a tool to encourage self-reflexive assessments of GEC capacities toward more radical enactments.

Literature Review

The advent of civil society in the traditional state-sponsored and market-oriented energy sector has led transition scholars to pay closer attention to the countervailing civic force presented by community-led energy transitions (Smith et al. 2016; Stirling 2014; van der Schoor and Scholtens 2015). Identifying a "qualitative difference" between mainstream-market-based business reforms in the energy sector,[2] and grassroots energy transitions involving committed citizen activists and organizers, Gill Seyfang and Adrian Smith describe "community-level activities as *innovative niches*" (2007, 585; emphasis in original) that "exist within the social economy" (591). This, in turn, leads to concerns around governance of grassroots innovations, chiefly toward preservation of their socially inclined orientation and to safeguard against cooptation by incumbent market players.

Insofar as grassroots energy transitions create new sites of innovative activity in the energy sector, their continued operation and subsequent expansion within the social economy require simultaneous destabilization of the dominant market-oriented and fossil-fuel-based status quo that is, the incumbent *socio-technical regime* (Smith 2007). It is to his end that Smith et al. highlight the importance of civic action in the energy sector that adopts a more antagonistic position to the incumbent status quo, not only to provoke "debate about what energy is for in society and how citizens are involved" (2016, 425; see also Cuppen 2018) but also to unveil "concentrated power and fallacies of technocratic control as problems instead of solutions for current ecological and social imperatives" (Stirling 2014, iii).

However, proposals for unmooring grassroots from the incumbent energy paradigm through mobilization of radical transformational politics encounter critical challenges on two accounts. First, grassroots efforts to actualize a "truly progressive social transformation" (Stirling 2014, 22) in the energy sector encounter pronounced regime pressures, often buttressed by technocratic and deterministic claims for utility control (Smith et al. 2016; Sovacool 2009; Stirling 2008, 2014). Second, grassroots communities indicate variable support for radical politics on account of differential sociocultural and political leanings across different countries (Frantzeskaki, Avelino, and Loorbach 2013) and in the case of United States even across states and localities (Byrne et al. 2007; Outka and Feiock 2012; Yi and Feiock 2014). The latter challenge is expressly illustrated through empirical research on growing bipartisan support for renewables in the United States, where different states are adopting differential and contrasting policy and regulatory approaches in tune with dominant political ideologies (Brown and Hess 2016; Coley and Hess 2012; Hess, Mai, and Brown 2016).

What further presents as a challenge to mobilizing radical transformational politics in the US energy sector is the political contentiousness around environmental issues (Brulle 2008; Brulle and Jenkins 2008) and the intensifying polarization regarding climate change (Fisher, Waggle, and Leifeld 2016; Hess and Brown 2017; Leiserowitz 2006; McCright and Dunlap 2003). Since personal politics and ideologies strongly influence individual attitudes and choices toward the environment and belief in climate change (Gromet, Kunreuther, and Larrick 2013; Hess and Brown 2017), Jason Morris's description of grassroots energy initiatives as "contextually contingent projects" is apt, where "the politics and practices of locality feature prominently and problematically" (2013, 17; see also Hess 2008).

Ultimately, as GECs continue to emerge from self-organized civic action in both liberal and conservative states in the United States as well as across the world, they are affected by political and ideological proclivities of GEC members in addition to incumbent utility pressures and geographic policy and regulatory variabilities. It is against this background that I question the uncritical optimism surrounding GECs often perceived as a potent countervailing and emancipatory force against the prevailing status quo. Although scholarship is beginning to acknowledge this concern (Van Veelen 2018), there is still a critical gap in the literature as regards to assessing the nature of emerging GECs and their capacities to actualize a fundamental energy transition. To overcome this gap, I develop an analytical framework with which to assess GECs' capacity for actualizing a progressive social transformation—that is, a fundamental energy transition in the energy sector.

Dimensions of a Fundamental Energy Transition

How should we conceptualize a fundamental energy transition? It could broadly be described as a transition in the "hardware" of the existing energy system, using Gordon Walker and Noel Cass's shorthand, that necessarily accompanies a transformation in the "softwar" (2007).[3] This, in turn, begs the question of what this "software" transformation (i.e., change in the underlying logics governing the mainstream energy system) should entail. One response is provided by the growing global social movement for *Energy Democracy*. In a recent edited volume, key organizers and activists describe "the way forward for a revolutionary movement in energy" (Fairchild and Weinrub 2017, x) as one that "must address three major aspects of our energy system: its relationship to the environment, to social justice, and to a new economy" (7). These three aspects provide the key dimensions for which to strive in order to ensure a departure from the existing carbon-intensive, unjust, and inequitable status quo. In what follows, I take these aspects as dimensions of a fundamental energy transition and explicate specific criteria from the literature to derive three continua or axes as building blocks for the proposed framework.

Economic Equity

Equity is in and of itself the subject of many debates, concerned with distribution of resources where the ultimate recipients of resources, the resources themselves, and the process of distribution are implicated in varying interpretations and determinations of the concept (Stone 1988). To the extent that I employ this concept to adjudicate GECs' capacity to fairly distribute resources within the society or community in which it arises, I draw on the distinction between *self-interest* and *public interest* by Deborah Stone, who notes that "citizens . . . have two sides: a private, rather self-interested side and a more public-spirited side, and we might think of the public interest as those things desired by the public-spirited side of citizens" (1988, 21). What this concept entails for assessing economic equity imparted by a GEC is attention not only to how its members organize, in other words its social organization, but also to where they stand vis-à-vis individualistic self-interest versus collectivistic public interest. This latter aspect is crucial as Stone instructively points out: "Where one stands on issues of distribution is determined not so much by the specifics of any particular issue (say, tax policy or student financial aid) as by a more general world view. This world view includes assumptions about the meaning of community and the nature of property, assumptions that transcend particular issues" (1998, 53).

To this end, the specific social organization a particular GEC adopts and its consequent impact for economic equity, are, at least in part, influenced by the values and interests of the GEC's members. Insofar as GECs operate within the social economy, one could argue that the values and interests of its members are squarely affected by their public-spirited side. However, there is reason to call this assumption into question in light of the varying political and ideological proclivities, as mentioned earlier, that GEC members may espouse: individualistic ideological preferences versus collectivistic preferences. This overlap is further exacerbated in light of the growing partnership and engagement between the market economy and the social economy (Hatzl et al. 2016; Hess 2013; see also Mayer et al. 1997). Empirical evidence points to nonprofit organizations and social enterprises increasingly adopting market instruments for revenue generation[4] (Dacin, Dacin, and Tracey 2011; Dey and Steyaert 2012) and for-profit firms increasingly pursuing social and environmental goals in response to consumer demands (Benlinsky 2014).

A useful approach to discern GECs' capacity for imparting economic equity, particularly in context of growing overlap between profit-motivated market-oriented individualism and solidarity-motivated social collectivism, is to focus on the relative influence of GECs' social organization and members' worldviews on *value capture* or *value creation*. Filipe Santos expresses the distinction between the two as follows: "Value creation from an activity happens when the *aggregate utility of society's members increases* after accounting for the opportunity cost of all the resources used in that activity. Value capture from an activity happens when *the focal actor is able to appropriate a portion of the value created* by the activity after accounting for the cost of resources that he / she mobilized" (2012, 337; emphasis added).

Insofar as inferences to this end focus on GEC members' inherent rationalities and dispositions and pay attention to the societal consequences of their social organization, for instance, exclusion based on GEC membership, a critical point is made by Mark Sagoff—

> We may be concerned as citizens, or as members of a moral or political community, with all sorts of values—sentimental, historical, ideological, cultural, and ethical—that conflict with the interests we reveal as consumers, buying shoes or choosing tomatoes. The conflict *within* individuals, rather than between them, may be a very common conflict. *The individual as self-interested consumer opposes himself as a moral agent and a concerned citizen.* (1988, 60–1; emphasis added)

The existence of these two contrasting positions *within* an individual actor, which manifest in her values and interests, implies their mutual opposability and coextensivity. In context of GECs emerging at the cusp of the market economy and the social economy, it is the espousal and, critically, *enactment* of market-oriented profit motives or community-oriented solidarity motives that, I argue, determine the reach of in/equitability through grassroots energy efforts.

Social Justice

While the dimensions of social justice and economic equity are conceptually inseparable, they can be analytically delineated in reference to GECs by arguing that attaining the latter doesn't entail advancing the former. This is to say that while it may be possible for a GEC to socially organize in a manner that leads to equitable outcomes for its members, it is *socially just* only if ownership and decision-making power in the cooperative is afforded to the wider population, encompassing diversity of citizens across income, education, race, and so on. This dynamic implies the question of scalability of GECs in order to reach a wider populace, an outcome contingent upon resource availability—financial and institutional, as well as social.

However, the retreat of the state under neoliberal globalization and subsequent incidence of market players in the social economy have diminished the reciprocity between citizens and the state in favor of market-oriented social cohesion. Describing this as "redistributive neoliberalism" David Hess (2011) highlights an accumulative intent on part of the incumbents. Debunking the possibility of market proponents to accomplish socially productive outcomes while adhering to their accumulative intent, Susan Fainstein clarifies the "contradiction between urban entrepreneurialism and social justice" (2001, 884). Due to pervasiveness of competitive profit-maximizing logics of the market, GECs' capacity to achieve socially just outcomes, therefore, rests on consistently progressive policy support (Seyfang, Park, and Smith 2013). In making this point more generally in the context of optimism around decentralized governance and overemphasis on participation,[5] without just distribution, Fainstein points out that "competition as the context of urban development . . . evades the role that the national state must play in assuring the economic well-being of individuals and the mitigating of competitiveness.

We thus *need to return to the concept of the enabling state* rather than simply the entrepreneurial state" (2001, 888; emphasis added).

In a similar vein, but in particular reference to the US energy sector, which vacillates between neoliberalism or state rollback and social liberalism or state rollout, Hess suggests the possibility of "opportunities for redistributive politics" during the rollout phase: "When one steps back and looks at the industry as a whole, there is a mixed regime of hegemonic social liberalism and hegemonic neoliberalism that provides opportunities for capital accumulation by the energy elites. But within the spaces opened up by restructuring, some *interesting opportunities have been created, perhaps inadvertently, for new forms of redistributive politics to emerge*" (Hess 2011, 1073; emphasis added).

Therefore, inasmuch as local community action by GECs parallels and cautiously engages with market-oriented competitive logics to ensure value creation, it continues to require progressive policy and regulatory support to affect socially just outcomes. In this context, discerning the capacity of GECs to advance social justice is contingent upon their policy and regulatory environment governing the corresponding electricity market environment.

Environmental Sustainability

Environmental sustainability is often operationalized in terms of avoided carbon emissions in national and international climate agreements. For GECs, this concern implicates the choice of technology—for instance, between solar photovoltaic systems versus natural gas—as well as GEC members' energy consumption behavior. To take carbon emissions reduction to assess GECs' capacity for environmental sustainability is also to acknowledge the interrelationship between carbon dioxide abatement and the harmful public health impacts of ozone and particulate matter air pollution, validated in the literature as being causally linked (Ebi and McGregor 2008; Jacobson 2008; see also Melamed, Schmale, and Schneidemesser 2014).

A critical, yet oft-ignored, aspect of environmental sustainability is the rapidly expanding system of global supply chains and extractive mineral enterprises in the renewables industry. The trade-off between point-of-use decarbonization and commodification of renewables akin to fossil fuels obscures their "fossilization" (Raman 2013). In this context, the call for reframing environmental sustainability by wresting control of "energy-as-commodity" in order to reanimate it as "energy-as-commons" is inexorably linked to creation of circular and sustainable local economies (Martinez 2017). However, if primacy is given only to this dimension of energy transitions, as is usually the case, without attention to the other two, that is, global supply chains and extractive mineral enterprises, we risk ignoring the key argument being made here, that energy transitions which do not advance economic equity and social justice are not fundamentally transformative of the incumbent energy system.

Taken together, the three dimensions underpin a *fundamental* energy transition—one that is economically equitable, socially just, and environmentally sustainable. In substantiating these aspects for assessing the capacity of GECs for actualizing a fundamental energy transition, three key assessment criteria come to the fore: the *inherent rationalities* or *dispositions* of individual GEC members, the facilitative influence of the *state rollout* in enabling a progressive policy and regulatory environment, and the need to acknowledge global consequences of local technology choices through assessment of *carbon content*.

The Capabilities Approach and Operationalizing the Assessment Criteria

Admittedly, the proposal to assess GECs based on criteria identified in the previous section runs into obvious ethical dilemmas. To adjudicate a GEC as more or less equitable, just, and sustainable, in relation to another, is to impose a normative conceptualization of these dimensions on the ethical position and moral sensibilities of GEC organizers and members. This dilemma gets further exacerbated in sociopolitical contexts punctuated by intensifying partisan polarization on already contentious policy issues (Pew Research Center 2014). Adopting a top-down and idealist position would, furthermore, ignore the fact that while some GECs "do have ambitions to expand and grow, others are simply providing local solutions to local needs as an end in itself, and have no desire to expand in the way that policymakers might hope" (Seyfang, Park, and Smith 2013, 988). Therefore, rather than impose what equity, justice, and sustainability should or ought to look like at the grassroots, I draw on Sen's "idea of justice" (2009) to focus on what people and communities do, in their everyday lives, and how it impacts their ability to make a change.

Sen's is a capabilities approach to justice and is counterposed to the more transcendental or ideal notion of justice dominant in Western philosophies,[6] such as Rawls's theory of justice (Sen 2009). Whereas Rawls's "veil of ignorance" thought experiment encouraged normativity, Sen's approach is predicated on a firmly rejecting normative conceptions of justice in favor of a comparative approach. This, Sen argues, is to unmoor assessments of justice and, by extension of equity and sustainability in the present case, from assumptions about "perfect institutions." By deferring decisions or judgments about what specifies equity, justice, and sustainability, Sen's approach embraces "a realist reading of behavioral norms and regularities" (81) toward a "reasoned public dialog" about actions and institutions in real-world contexts. This contextually specific approach to adjudicating assessments not only aligns with transitions scholarship emphasizing lived experiences and day-to-day practices of energy consumers (Shove and Walker 2007, 2010) but also encourages participative and reflexive stance for advancing toward a deliberated alternative energy future. Moreover, the approach "usefully provides conceptual coherence across contexts, whether regional or global, whilst having the flexibility to be contextually specified" (Day, Walker, and Simcock 2016, 262).

It should be noted, however, that Sen's approach remains controversial and criticized precisely for rejecting higher-order ideal notions of justice (Hinsch 2011; Valentini 2011). Specifically, the emphasis on "reasoned public dialog" at the expense of identifying and setting up "perfectly just institutions" against which to adjudicate assessments suffers from a failure to consider the impact of impasse-inducing "deep disagreements" in the course of deliberative public reasoning. The fact that different individuals and communities have different worldviews and hence varying conceptions of equity, justice, and sustainability must already be amply obvious. Therefore, using the capabilities perspective to develop the analytical framework—to comparatively assess different GECs—should not detract from acknowledging the need to ground these assessments normatively. It is to this task of normatively grounding the aforementioned assessment criteria that I now turn my attention.

The criteria—namely, rationalities of GEC members, situated policy, and regulatory environment and the impact of their actions on decarbonization—are operationalized through three axes—of rationality, regulation, and carbon content, respectively. To this end, each axis is conceived as a continuum between features that broadly define the existing mainstream energy system—that is, market-oriented, deregulated, and carbon-intensive—and those that are envisioned to predicate the alternate energy future—that is, community-oriented, (progressively) regulated, and low-carbon, respectively. Each of these constitute the building blocks of the proposed analytical framework and are expounded on below.

Rationality Continuum

This continuum (figure 4.1) is used to comparatively assess the capacity of GECs to impart economic equity. In transitioning toward an alternate energy future, the shift from profit maximization in the market economy is desired to give way to collective solidarity through a robust social economy. The continuum is deployed to assess the extent to which GECs depart from the former to transition toward the latter in terms of their relative emphasis on value creation as opposed to value capture. Notably, insofar as the two extremes are set up as existing and desired institutions, respectively, there is nothing perfect or ideal about them. In other words, the ends of the continuum are not considered as absolute in the sense of existing independently. This, in turn, predicates assessments of equity on what individuals and communities do insofar as their relative alignment, expressed via their values and interests and their enactments, toward self-interested consumerism or concerned citizenship (Sagoff 1988).

Ascertaining where a GEC is located, in a relative sense, on the continuum between the Market and the Polis,[7] requires referring to the values and interests of the GEC members as well as the social organization they adopt as they self-organize. To this end, I propose narrative analysis of in-depth and reflective interviews conducted with GEC organizers and members, content analysis of internal programmatic documents and communication, and participant observation of day-to-day processes of self-organization as appropriate methods.

Regulatory Continuum

This continuum (figure 4.2) is used to comparatively assess the ability of GECs to advance social justice. In contrast to the aforementioned rationality continuum, the regulatory continuum assesses the policy and regulatory environment within which the GEC arises and which may or may not be

FIGURE 4.1. Rationality continuum.

FIGURE 4.2. Regulatory continuum.

within GEC members' direct political influence. In other words, this continuum is deployed to assess the policy contexts within which a GEC carries out its activities, either under conditions of constraint due to state rollback (deregulation) or enablement (progressive regulation).

Here too, it must be clear that the extremes are not absolutes since deregulated and regulated markets in the energy sector coexist conterminously. Therefore, to locate a particular GEC initiative somewhere on this continuum entails reference to the surrounding political and regulatory contexts and the ability they confer upon the GEC to scale their enactments. Here, the entanglement of social justice, assessed in terms of state support, with economic equity must be amply clear. Insofar as state support to facilitate GEC activities is afforded through progressive policies and regulations, socially just outcomes are advanced only when GEC members' values and interests and their enactments actualize economic equity. To this end, locating a GEC on the regulatory continuum requires discursive analysis of policies and plans, whether local, state or national, to ascertain the degree as well as nature of state support available to GECs. As a consequence of divergent policies across local, state and national level (Byrne et al. 2007), close attention to the situated policy and implications for the GEC is necessary.

Carbon Continuum

The Carbon Continuum, or Carbon-Intensity Continuum (figure 4.3), depends on technology (and hence energy source) choices made within the GEC initiative and assesses the consequent extent of decarbonization. Here too, the extremes are not independently tenable. Even at its peak fossil fuel use, the energy sector depended, marginally, on renewable resources, such as hydropower plants, and after the Public Utilities Regulatory Policies Act (PURPA) of 1978 on cogeneration systems based not only on natural gas but also solar and biomass technologies (Hirsh and Sovacool 2006; Morris 2013). Furthermore, it is typical for local energy projects, pursued by GECs, to draw on multiple renewable and nonrenewable resources in their pursuit of self-sufficiency, indicating that absolute resource-related autonomy through renewables and from the dominant carbon economy is unachievable and may actually be undesirable (Rae and Bradley 2012).

This observation begs the question: what position on this continuum would then justifiably be deemed as a departure from the carbon-intensive status quo? Based on the *trias energetica* concept from the field of energy engineering, I contend that energy system sustainability can be enhanced by "1° reducing the demand for energy, 2° applying renewable energy sources wherever possible, and 3° filling in the remaining need as efficiently and cleanly as possible with fossil fuels" (Vandevyvere and Stremke 2012, 1314). This approach to decarbonization of energy systems achieves two main objectives for ongoing energy transitions, particularly at the urban scale: first, by prioritizing energy efficiency and renewable energy, it paves the way for rapid adoption, scale-up, and integration of "green" practices and technologies into the urban fabric; and second, it subordinates the use of fossil fuel resources, without risking the reliability of energy systems in transition by avoiding prematurely discontinuing fossil fuel use altogether.

Therefore, in addition to ascertaining the CO_2 reductions availed through GECs' activities, the continuum requires contending—to the extent possible due the small-scale scope of GECs—with environmental implications of global supply chains and commodification of renewables (Mulvaney 2013;

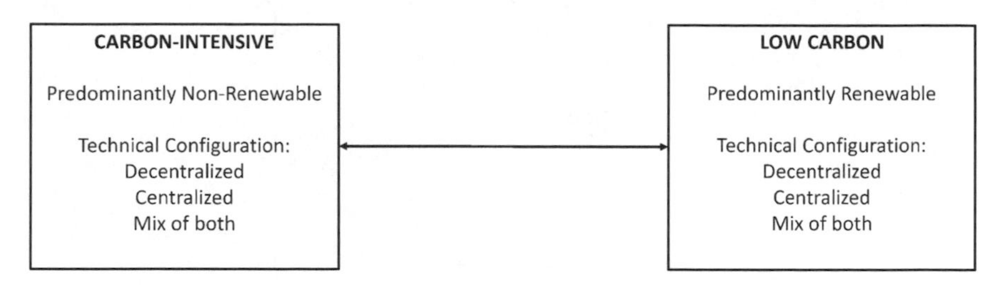

FIGURE 4.3. Carbon continuum.

Ottinger 2011; Raman 2013). For instance, this analysis could mean ascertaining whether the project used domestically manufactured technology and locally sourced resources.

Figure 4.4 shows the proposed three-axis analytical framework, developed by orienting the three continua discussed in the previous section in three dimensions. The Cartesian juxtaposition of the three axes helps explicate eight models conveyed by each of the eight quadrants. In comparatively assessing GECs and locating them on the three continua, the framework positions them in the state-space of any one of the eight quadrants. Whereas all the points in the three-dimensional space represent all possible combinations of social organization, nature, and degree of governance and extent of decarbonization, the eight models represent particular archetypal models evident in the existing energy sector (see table 4.1). As such, the 3D space delineated by each quadrant is continuous and not discrete, within which a GEC may be positioned.

To be clear, while the archetypes are exemplified in table 4.1 based on existing organization and governance and technological types within the mainstream energy system, the framework can just as feasibly be used to position GECs emerging from it. Let's take a closer look at two archetypes, 4 and 7, and justify their categorization by examining their social organization, role of state support, and decarbonizing potential. Both archetypes are predicated on standard well-known approaches to organization and governance of local energy provision: Community Choice Aggregation (CCA) in 4 and Solarize model in 7. In the CCA model, a local official or quasi-official entity—either the local government or government-run or contracted agency, aggregates energy demand from the community to procure supply based on consumer preferences. Here, the local-state-related entity performs

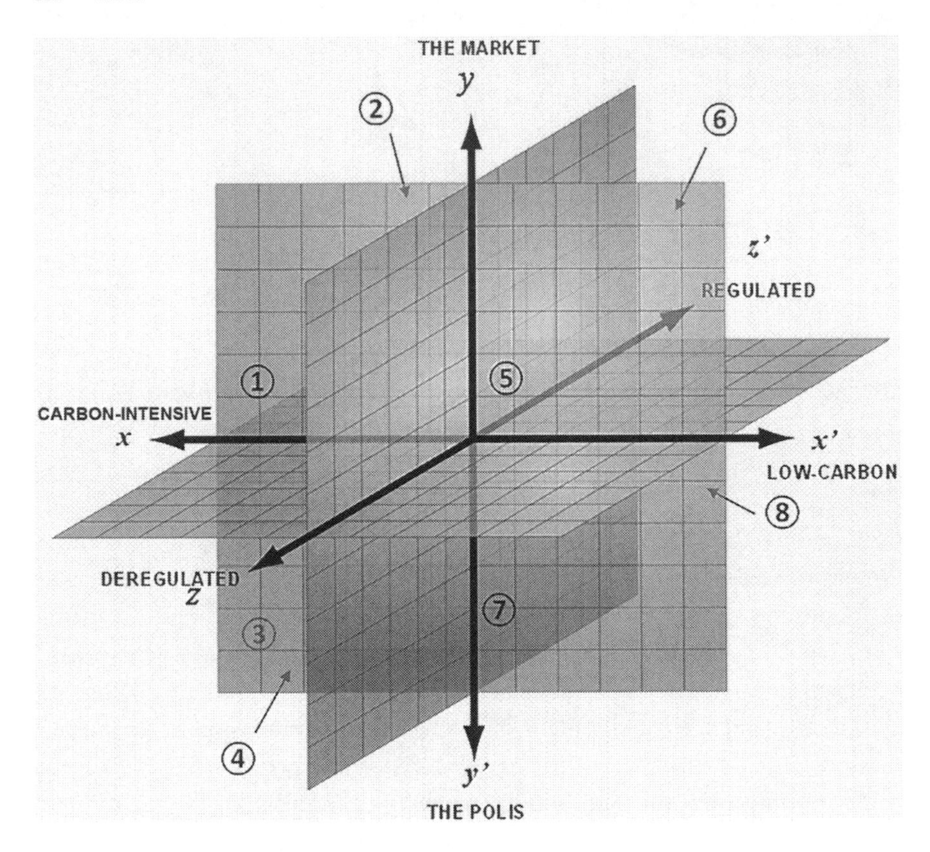

FIGURE 4.4. Three-Axis Analytical Framework.

a brokering function in tune with where the community prefers its energy to come from. This combination means that depending on local resource availability, feasibility, and community preference, a CCA may procure energy from carbon-intensive resources such as natural gas or low-carbon resources such as hydro or solar (CCP 2017). Typically, there tends to a mix of multiple sources, including renewable and nonrenewable, implying that locating a GEC on the carbon continuum entails measuring carbon emissions reduction achieved. In Solarize scenario, on the other hand, the state assumes a secondary role in creating a more-or-less conducive policy environment for residential solar installations. This environment exists because communities in this archetype self-organize to socialize costs and benefits of interacting with the market that is, private solar installers (Rubado 2010).

TABLE 4.1. Model archetypes and representative examples

	Quadrant	*Model Typology and Description*	*Example*
1	XYZ	Carbon-intensive, market-led, neoliberalism	Oil-based energy economy in Texas, with 75% deregulated retail markets (Electric Choice 2017)
2	XYZ'	Carbon-intensive, market-led, social liberalism	Shale gas extraction by private industry under stringent states regulations, e.g., in MD, MT, and TX (Richardson et al. 2013)
3	XY'Z	Carbon-intensive, community-led, neoliberalism	Partnerships between gas companies and communities for shale gas extraction in states with fewer stringent regulations e.g., IN, KY, and VA (Richardson et al. 2013)
4	XY'Z'	Carbon-intensive, community-led, social liberalism	State or local regulations promoting programs such as Community Choice Aggregation using publicly or privately extracted shale gas, e.g., Sonoma County, CA (Walton 2014)
5	X'YZ	Low-carbon, market-led, neoliberalism	Utility-scale (10 MW or more) renewable energy projects, e.g., Ivanpah solar project or solar module leasing by third-party firms, e.g., Solar City (Darling and Marcy 2016; SEIA 2017)
6	X'YZ'	Low-carbon, market-led, social liberalism	Large- to medium-scale renewable energy development by municipal governments through public-private partnerships, e.g., in Georgetown, TX (City of Georgetown TX 2016)
7	X'Y'Z	Low-carbon, community-led, neoliberalism	Distributed solar adoption through private installation company and citizen engagement, e.g., Solarize projects (Aylett 2013; Rubado 2010)
8	X'Y'Z'	Low-carbon, community-led, social liberalism	Distributed solar adoption through community-based cooperatives aided through state and local policies e.g., Co-op Power, MA (Co-op Power 2015)

Already, it is possible to see how the choice between natural gas versus solar energy puts the two models on opposite sides of the carbon continuum. Furthermore, insofar as the state acts on behalf of the community in a CCA, its ability to mitigate socioeconomic externalities—for instance, due to inequitable distribution of ensuing benefits—is predicated not only on ideologically sensitive preferences of the community members but also on progressive (re)distributive welfare actions performed by the state. In contrast, Solarize model tends to privilege homeowners with financial capacity to invest in solar modules, essentially excluding a segment of population, most of whom would typically be communities of color. Therefore, a CCA ensuring equitable and nondiscriminatory social organization would be more equitable than Solarize approach, which discriminates on the basis of homeownership and financial capacity. Depending on whether the local policy context of the CCA ensures redistributive outcomes through low-income assistance, it is relatively more or less just. However, it tends to be typically more just than Solarize, where local government remains in the background. Undeniably, these assertions are highly case and context dependent and need to be backed by empirical evidence. But what I have attempted to show, through illustration of archetypes 4 and 7, is the utility of the proposed analytical framework to offer a more nuanced understanding of how existing energy system and emerging GEC approaches may affect economic equity, social justice, and environmental sustainability.

This analysis brings me to the limitations and caveats of the analytical framework. First, in its current form, the proposed framework lacks metrics associated with the three continua and therefore serves as a metaphorical device to reflexively think about different approaches to energy system organization, governance, and technology choices. Second, given the possibility to intuit where a particular GEC lies on the three axes, potential for confirmation bias is high. This bias needs to be corrected by way of substantiating GEC assessment through empirical evidence. Third, the bounding of the extremes between institutional features of the existing and alternate (desired) energy system curtails envisioning an energy future that is probably more just and equitable than the "community-led, progressively regulated, and low-carbon" future I have elected. This possibility points to replacing the latter extremes by features of a more radical energy future such as the autarkic models of society envisioned by anarcho-syndicalists.[8] Last, given that the three axes afford assessment of GECs between nonabsolute extremes, the assessment cannot be performed unless it is comparative that is, between two or more GECs at a time. While not a limitation in and of itself, using the framework in practice does entails extensive empirical work on multiple cases.

Conclusion

It is highly unlikely that the rise of alternative citizen-led approaches to renewable energy provision within the existing energy system, in the United States as well as in other countries, faces no attempts for their mainstreaming. As a countervailing force within the mainstream energy sector, GECs are touted to advance more equitable, just, and sustainable outcomes. In the context of deepening political polarization in society and patchwork of policies and regulations punctuating urban and regional landscapes across the world, this chapter offers a simple three-axis framework with which to reflexively comprehend where we are and where we need to be in regards to ongoing energy transitions. But in offering a vision of the future toward which to advance, the framework is designed to avoid imposing normative conceptions of what progress should or ought to look like. Instead, by taking GECs as its unit of assessment, the framework provides a tool for GEC organizers, activists, local policy makers, and planners to self-reflect on their existing dispositions, preferences, and motivations, as well as evaluate their policy and regulatory environment and technology choices to deliberatively move toward more progressive social and sociotechnical transformation.

Notes

This research was conducted with a grant support from the Degenstein Foundation through their gift to Geisinger Health System.

1. Henceforth, I simply use the word *sustainability* instead of *environmental sustainability*.

2. An example is diversification of fossil-fuel-based companies and rural cooperatives into renewables.

3. This is also referred to in the literature as *social architecture* (Hoffman and High-Pippert 2005; Miller and Richter 2014) or '*institutional infrastructure* (Egyedi and Mehos 2012; Hargreaves et al. 2013).

4. This is especially the case as state support dwindled and is presently continually retrenched.

5. This is also referred to in the literature as the "localist trap" (Catney et al. 2014; Hess 2008).

6. Sen takes Rawls's theory to justice—justice as fairness—as a foil and representative of Western transcendental institutionalism to advance his idea of justice (2009).

7. I use the term *Polis*, following Stone, to denote the social economy that "must assume collective will and collective effort" (1988, 18).

8. This would require precisely explicating the features of such an autonomous society that are desired over the existing institutional alternatives (e.g., see Albert and Hahnel 1991).

References

Albert, Michael, and Robin Hahnel. 1991. *The Political Economy of Participatory Economics*. Princeton, NJ: Princeton University Press.

Aylett, Alex. 2013. "Networked Urban Climate Governance: Neighborhood-Scale Residential Solar Energy Systems and the Example of Solarize Portland." *Environment and Planning C: Government and Policy* 31 (5): 858–875. https://doi.org/10.1068/c11304.

Benlinsky, Michael. 2014. "The Rise and Struggles of Social Enterprises in 2013." *Stanford Social Innovation Review*, 2013–2015. http://www.ssireview.org/blog/entry/the_rise_and_struggles_of_social_enterprise_in_2013.

Brown, Kate Pride, and David J. Hess. 2016. "Pathways to Policy: Partisanship and Bipartisanship in Renewable Energy Legislation." *Environmental Politics* 25 (6): 971–990. https://doi.org/10.1080/09644016.2016.1203523.

Brulle, Robert J. 2008. "Politics and the Environment." In *The Handbook of Politics: State and Civil Society in Global Perspective*, edited by Kevin T. Leicht and J. Craig Jenkins, 385–406. New York: Springer.

Brulle, Robert J., and J. Craig Jenkins. 2008. "Decline or Transition? Discourse and Strategy in the U.S. Environmental Movement." Presented at the Annual Conference of the American Sociological Association, August, 2008, Philadelphia.

Bulkeley, Harriet, Vanesa Castán Broto, and Anne Maassen. 2014. "Low-Carbon Transitions and the Reconfiguration of Urban Infrastructure." *Urban Studies* 51 (7): 1471–1486. https://doi.org/10.1177/0042098013500089.

Byrne, John, Kristen Hughes, Wilson Rickerson, and Lado Kurdgelashvili. 2007. "American Policy Conflict in the Greenhouse: Divergent Trends in Federal, Regional, State, and Local Green Energy and Climate Change Policy." *Energy Policy* 35 (9): 4555–4573. https://doi.org/10.1016/j.enpol.2007.02.028.

Catney, Philip, Sherilyn MacGregor, Andrew Dobson, Sarah Marie Hall, Sarah Royston, Zoe Robinson, Mark Ormerod, and Simon Ross. 2014. "Big Society, Little Justice? Community Renewable Energy and the Politics of Localism." *Local Environment* 19 (7): 715–730. https://doi.org/10.1080/13549839.2013.792044.

CCP. 2017. "Community Choice Energy." *Center for Climate Protection*. Accessed April 13, 2018. https://theclimatecenter.org/our-work-2__trashed/community-choice/.

City of Georgetown TX. 2016. "Solar Power Deal Finalized with NRG Energy." https://georgetown.org/2016/11/18/solar-power-deal-finalized-with-nrg-energy/.

Coley, Jonathan S., and David J. Hess. 2012. "Green Energy Laws and Republican Legislators in the United States." *Energy Policy* 48 (August): 576–583. https://doi.org/10.1016/j.enpol.2012.05.062.

Co-op Power. 2015. "Co-op Power." Accessed July 10, 2020. http://www.cooppower.coop/.

Cuppen, Eefje. 2018. "The Value of Social Conflicts. Critiquing Invited Participation in Energy Projects." *Energy Research and Social Science* 38 (February): 28–32. https://doi.org/10.1016/j.erss.2018.01.016.

Dacin, M. Tina, Peter A. Dacin, and Paul Tracey. 2011. "Social Entrepreneurship: A Critique and Future Directions." *Organization Science* 22 (5): 1203–1213. https://doi.org/10.1287/orsc.12.3.391.10102.

Darling, David, and Cara Marcy. 2016. "About 30% of Distributed Solar Capacity Is Owned by Third Parties." *U.S. Energy Information Administration (EIA)*, December 7, 2016. Accessed July 30, 2017. https://www.eia.gov/todayinenergy/detail.php?id=29052.

Day, Rosie, Gordon Walker, and Neil Simcock. 2016. "Conceptualising Energy Use and Energy Poverty Using a Capabilities Framework." *Energy Policy* 93 (June): 255–264. https://doi.org/10.1016/j.enpol.2016.03.019.

Dey, Pascal, and Chris Steyaert. 2012. "Social Entrepreneurship: Critique and the Radical Enactment of the Social." *Social Enterprise Journal* 8 (2): 90–107. https://doi.org/10.1108/17508611211252828.

Ebi, Kristie L., and Glenn McGregor. 2008. "Climate Change, Tropospheric Ozone and Particulate Matter, and Health Impacts." *Environmental Health Perspectives* 116 (11): 1449–1455. https://doi.org/10.1289/ehp.11463.

Egyedi, Tineke M., and Donna C. Mehos, eds. 2012. *Inverse Infrastructures: Disrupting Networks from Below*. Northampton, MA: Edward Elgar.

Electric Choice. 2017. "Map of Deregulated Energy States (Updated 2017)." Accessed January 25, 2018. https://www.electricchoice.com/map-deregulated-energy-markets/.

Fainstein, Susan S. 2001. "Competitiveness, Cohesion, and Governance: Their Implications for Social Justice." *International Journal of Urban and Regional Research* 25 (4): 884–888. https://doi.org/10.1111/1468-2427.00349.

Fairchild, Denise, and Al Weinrub, eds. 2017. *Energy Democracy: Advancing Equity in Clean Energy Solutions*. Washington, DC: Island Press.

Fisher, Dana R., Joseph Waggle, and Philip Leifeld. 2016. "Where Does Polarization Come From? Locating Polarization within the U.S. Climate Change Debate." *American Behavioral Scientist* 57 (August): 70–92.

Frantzeskaki, Niki, Flor Avelino, and Derk Loorbach. 2013. "Outliers or Frontrunners? Exploring the (Self-) Governance of Community-Owned Sustainable Energy in Scotland and the Netherlands." In *Renewable Energy Governance*, edited by Evanthie Michalena and Jeremy Maxwell Hills, 101–116. New York: Springer.

Geels, Frank W. 2014. "Regime Resistance against Low-Carbon Transitions: Introducing Politics and Power into the Multi-level Perspective." *Theory, Culture and Society* 31 (5): 21–40. https://doi.org/10.1177/0263276414531627.

Gromet, Dena M., Howard Kunreuther, and Richard P. Larrick. 2013. "Political Ideology Affects Energy-Efficiency Attitudes and Choices." *Proceedings of the*

National Academy of Sciences 110 (23): 9314–9319. https://doi.org/10.1073/pnas.1218453110.

Hargreaves, Tom, Sabine Hielscher, Gill Seyfang, and Adrian Smith. 2013. "Grassroots Innovations in Community Energy: The Role of Intermediaries in Niche Development." *Global Environmental Change* 23 (5): 868–880. https://doi.org/10.1016/j.gloenvcha.2013.02.008.

Hatzl, Stefanie, Sebastian Seebauer, Eva Fleiß, and Alfred Posch. 2016. "Market-based vs. Grassroots Citizen Participation Initiatives in Photovoltaics: A Qualitative Comparison of Niche Development." *Futures* 78–79 (April–May): 57–70. https://doi.org/10.1016/j.futures.2016.03.022.

Hess, David J. 2008. "Localism and the Environment." *Sociology Compass* 2 (2): 625–638. https://doi.org/10.1111/j.1751-9020.2007.00082.x.

Hess, David J. 2011. "Electricity Transformed: Neoliberalism and Local Energy in the United States." *Antipode* 43 (4): 1056–1077. doi:10.1111/j.1467-8330.2010.00842.x.

Hess, David J. 2013. "Industrial Fields and Countervailing Power: The Transformation of Distributed Solar Energy in the United States." *Global Environmental Change* 23 (5): 847–855. https://doi.org/10.1016/j.gloenvcha.2013.01.002.

Hess, David J. 2015. "The Politics of Niche-Regime Conflicts: Distributed Solar Energy in the United States." *Environmental Innovation and Societal Transitions* 19: 42–50. https://doi.org/10.1007/s10787-012-0152-6.

Hess, David J., and Kate Pride Brown. 2017. "Green Tea: Clean-Energy Conservatism as a Countermovement." *Environmental Sociology* 3 (1): 64–75. https://doi.org/10.1080/23251042.2016.1227417.

Hess, David J., Quan D. Mai, and Kate Pride Brown. 2016. "Red States, Green Laws: Ideology and Renewable Energy Legislation in the United States." *Energy Research and Social Science* 11 (January): 19–28. https://doi.org/10.1016/j.erss.2015.08.007.

Hinsch, Wilfried. 2011. "Ideal Justice and Rational Dissent. A Critique of Amartya Sen's the Idea of Justice." *Analyse Und Kritik* 33 (2): 371–386.

Hirsh, Richard F., and Benjamin K. Sovacool. 2006. "Technological Systems and Momentum Change: American Electric Utilities, Restructuring, and Distributed Generation." *Journal of Technology Studies* 32, no. 2 (April): 72–85. http://scholar.lib.vt.edu/ejournals/JOTS/v32/v32n2/hirsh.html.

Hoffman, Steven M., and Angela High-Pippert. 2005. "Community Energy: A Social Architecture for an Alternative Energy Future." *Bulletin of Science, Technology and Society* 25 (5): 387–401. https://doi.org/10.1177/0270467605278880.

Hufen, J. A. M., and J. F. M. Koppenjan. 2015. "Local Renewable Energy Cooperatives: Revolution in Disguise?" *Energy, Sustainability and Society* 5 (1): 18. https://doi.org/10.1186/s13705-015-0046-8.

ILSR. 2016. "Introducing the Community Power Map." *Institute for Local Self-Reliance*. Accessed January 15, 2017. https://ilsr.org/introducing-the-community-power-map/.

Jacobson, Mark Z. 2008. "On the Causal Link between Carbon Dioxide and Air Pollution Mortality." *Geophysical Research Letters* 35 (3): 1–5. https://doi.org/10.1029/2007GL031101.

Klein, Sharon J. W., and Stephanie Coffey. 2016. "Building a Sustainable Energy Future, One Community at a Time." *Renewable and Sustainable Energy Reviews* 60: 867–880. https://doi.org/10.1016/j.rser.2016.01.129.

Leiserowitz, Anthony. 2006. "Climate Change Risk Perception and Policy Preferences: The Role of Affect, Imagery, and Values." *Climatic Change* 77 (1–2): 45–72. https://doi.org/10.1007/s10584-006-9059-9.

Martinez, Cecilia. 2017. "From Commodification to the Commons: Charting the Pathway for Energy Democracy." In *Energy Democracy: Advancing Equity in Clean Energy Solutions*, edited by Denise Fairchild and Al Weinrub, 21–36. Washington, DC: Island Press.

Mayer, Rudd, Eric Blank, Randy Udall, and John Nielsen. 1997. "Promoting Renewable Energy in a Market Environment: A Community-Based Approach for Aggregating Green Demand." Boulder, CO: Land and Water Fund of the Rockies/US Department of Energy Report.

McCright, Aaron M., and Riley E. Dunlap. 2003. "Defeating Kyoto: The Conservative Movement's Impact on U.S. Climate Change Policy." *Social Problems* 50 (3): 348–373. https://doi.org/10.1525/sp.2003.50.3.348.

Melamed, Megan L., Julia Schmale, and Erika von Schneidemesser. 2014. "Sustainable Policy: Key Considerations for Air Quality and Climate Change." *Current Opinion in Environmental Sustainability* 23 (December): 85–91. https://doi.org/10.1016/j.cosust.2016.12.003.

Miller, Clark A., and Jennifer Richter. 2014. "Social Planning for Energy Transitions." *Current Sustainable/Renewable Energy Reports* (September): 77–84. https://doi.org/10.1007/s40518-014-0010-9.

Morris, Jason. 2013. "The Evolving Localism (and Neoliberalism) of Urban Renewable Energy Projects." *Culture, Agriculture, Food and Environment* 35 (1): 16–29. https://doi.org/10.1111/cuag.12002.

Mulvaney, Dustin. 2013. "Opening the Black Box of Solar Energy Technologies: Exploring Tensions between Innovation and Environmental Justice." *Science as Culture* 22 (December): 230–237. https://doi.org/10.1080/09505431.2013.786995.

Ottinger, Gwen. 2011. "Environmentally Just Technology." *Environmental Justice* 4 (1): 81–85. https://doi.org/10.1089/env.2010.0039.

Outka, Uma, and Richard Feiock. 2012. "Local Promise for Climate Mitigation: An Empirical Assessment." *Environmental Law and Policy Review* 36 (3): 635–670. http://scholarship.law.wm.edu/cgi/viewcontent.cgi?article=1548&context=wmelpr.

Pahl, Greg. 2012. *Power from the People: How to Organize, Finance, and Launch Local Energy Projects*. White River Junction, VT: Chelsea Green Publishing.

Pew Research Center. 2014. "Political Polarization in the American Public." *Pew Research Center*, June 12. http://www.people-press.org/2014/06/12/political-polarization-in-the-american-public/.

Rae, Callum, and Fiona Bradley. 2012. "Energy Autonomy in Sustainable Communities: A Review of Key Issues." *Renewable and Sustainable Energy Reviews* 16 (9): 6497–6506. https://doi.org/10.1016/j.rser.2012.08.002.

Raman, Sujatha. 2013. "Fossilizing Renewable Energies." *Science as Culture* 22 (December): 172–180. https://doi.org/10.1080/09505431.2013.786998.

REN21. 2017. *Renewables 2017: Global Status Report*. https://www.ren21.net/wp-content/uploads/2019/05/GSR2017_Full-Report_English.pdf.

Richardson, Nathan, Madeline Gottlieb, Alan Krupnick, and Hannah Wiseman. 2013. *The State of State Shale Gas Regulation*. http://www.rff.org/files/sharepoint/WorkImages/Download/RFF-Rpt-StateofStateRegs_Report.pdf.

Rubado, Lizzie. 2010. "Solarize Portland: Community Empowerment through Collective Purchasing." *ACEEE Summer Study on Energy Efficiency in Buildings, Energy Trust of Oregon*. https://www.energytrust.org/wp-content/uploads/2016/12/101110_Rubado_SolarizePortland.pdf.

Sagoff, Mark. 1988. *The Economy of the Earth*. Cambridge: Cambridge University Press.

Santos, Filipe M. 2012. "A Positive Theory of Social Entrepreneurship." *Journal of Business Ethics* 111 (3): 335–351. https://doi.org/10.1007/s.

SEIA. 2017. "U.S. Solar Market Grows 95% in 2016, Smashes Records." *SEIA*. Accessed July 8, 2017. https://www.seia.org/news/us-solar-market-grows-95-2016-smashes-records.

Sen, Amartya. 2009. *The Idea of Justice*. Cambridge, MA: Harvard University Press.

Seyfang, Gill, Jung Jin Park, and Adrian Smith. 2013. "A Thousand Flowers Blooming? An Examination of Community Energy in the UK." *Energy Policy* 61 (October): 977–989. https://doi.org/10.1016/j.enpol.2013.06.030.

Seyfang, Gill, and Adrian Smith. 2007. "Grassroots Innovations for Sustainable Development: Towards a New Research and Policy Agenda." *Environmental Politics* 16 (4): 584–603. https://doi.org/10.1080/09644010701419121.

Shove, Elizabeth, and Gordon Walker. 2007. "Caution! Transition Ahead: Policies, Practice, and Sustainable Transition Management." *Environment and Planning A* 39 (4): 763–770. https://doi.org/10.1068/a39310.

Shove, Elizabeth, and Gordon Walker. 2010. "Governing Transitions in the Sustainability of Everyday Life." *Research Policy* 39 (4): 471–476. https://doi.org/10.1016/j.respol.2010.01.019.

Smith, Adrian. 2007. "Translating Sustainabilities between Green Niches and Sociotechnical Regimes." *Technology Analysis and Strategic Management* 19 (4): 427–450. https://doi.org/10.1080/09537320701403334.

Smith, Adrian, Tom Hargreaves, Sabine Hielscher, Mari Martiskainen, and Gill Seyfang. 2016. "Making the Most of Community Energies: Three Perspectives on

Grassroots Innovation." *Environment and Planning A* 48 (2): 407–432. https://doi.org/10.1177/0308518X15597908.

Sovacool, Benjamin K. 2009. "The Intermittency of Wind, Solar, and Renewable Electricity Generators: Technical Barrier or Rhetorical Excuse?" *Utilities Policy* 17 (3–4): 288–296. https://doi.org/10.1016/j.jup.2008.07.001.

Sovacool, Benjamin K. and Michael H. Dworkin. 2015. "Energy Justice: Conceptual Insights and Practical Applications." *Applied Energy* 142 (May): 435–444. https://doi.org/10.1016/j.apenergy.2015.01.002.

Stirling, Andy. 2008. "Opening Up' and 'Closing Down': Power, Participation, and Pluralism in the Social Appraisal of Technology." *Science, Technology, and Human Values* 33 (2): 262–294. https://doi.org/10.1177/0162243907311265.

Stirling, Andy. 2014. "Emancipating Transformations: From Controlling 'the Transition' to Culturing Plural Radical Progress." STEPS Working Paper 64. Brighton, UK: STEPS Centre.

Stone, Deborah A. 1988. *Policy Paradox: The Art of Political Decision Making*. New York: Norton.

Valentini, Laura. 2011. "A Paradigm Shift in Theorizing about Justice? A Critique of Sen." *Economics and Philosophy* 27 (3): 297–315.

van der Schoor, Tineke, and Bert Scholtens. 2015. "Power to the People: Local Community Initiatives and the Transition to Sustainable Energy." *Renewable and Sustainable Energy Reviews* 43 (March): 666–675. https://doi.org/10.1016/j.rser.2014.10.089.

Vandevyvere, Han, and Sven Stremke. 2012. "Urban Planning for a Renewable Energy Future: Methodological Challenges and Opportunities from a Design Perspective." *Sustainability* 4 (12): 1309–1328. https://doi.org/10.3390/su4061309.

Van Veelen, Bregje. 2018. "Negotiating Energy Democracy in Practice: Governance Processes in Community Energy Projects." *Environmental Politics* 27 (4): 644–665. https://doi.org/10.1080/09644016.2018.1427824.

Walker, Gordon, and Noel Cass. 2007. "Carbon Reduction, 'the Public' and Renewable Energy: Engaging with Socio-technical Configurations." *Area* 39 (4): 458–469. https://doi.org/10.1111/j.1475-4762.2007.00772.x.

Walton, Robert. 2014. "How Local Communities Are Taking Over Their Power Supply." *Utility Dive*, September 19, 2014. http://www.utilitydive.com/news/how-local-communities-are-taking-over-their-power-supply/310407/.

Yi, Hongtao, and Richard C. Feiock. 2014. "Renewable Energy Politics: Policy Typologies, Policy Tools, and State Deployment of Renewables." *Policy Studies Journal* 42 (3): 391–415. https://doi.org/10.1111/psj.12066.

Chapter 4 Summary

In context of increasing citizen activism in ongoing energy transitions—conceptualized as Grassroots Energy Communities (GECs)—this chapter introduces an analytical framework with which to grasp their nature. The chapter begins with a review of literature highlighting the need to ascertain the capacity of GECs to actualize a fundamental energy transition. The dimensions that characterize such a fundamental energy transition are then identified and articulated in terms of economic equity, social justice, and environmental sustainability. Drawing on Amartya Sen's Capabilities Approach, the chapter operationalizes these three dimensions as assessment criteria in terms of three distinct continua, namely, the Rationality Continuum, Regulatory Continuum, and Carbon Continuum, respectively. Taken as three axes, the three continua convey a three-dimensional space within which to locate different models of emerging GECs and to comprehend their departure from the market-oriented, state-regulated, and fossil-fuel-based status quo. Taking Community Choice Aggregation (CCA) and Solarize models as examples, the chapter illustrates the assessment of their capacity using the analytical framework to actualize a fundamental energy transition. Finally, the chapter concludes by underscoring the utility of the analytical framework for characterizing emerging models of the future energy system in pursuit of an equitable, just, and sustainable energy future.

KEY TAKEAWAYS

- Insofar as GECs are characterized as the venues from which future models of the energy system emerge, there continues to be a lack of consensus as regards the direction toward which their growth should be directed. Stated differently, there is lack of grasp over what constitutes a future energy system.
- Adopting the definition espoused by the growing global social movement for "Energy Democracy," a fundamental energy system, in this chapter, is defined as one that is economically equitable, socially just, and environmentally sustainable.
- Drawing on multidisciplinary scholarship from policy studies, political theory, and energy engineering, these dimensions of a fundamental energy transition—that is, equity, justice, and sustainability—are framed as a choice between (1) individualistic self-interest (value capture) versus collectivistic public interest (value creation), (2) entrepreneurial state and enabling state, and (3) fossil-fuel-based "energy-as-commodity" and renewable-based "energy-as-commons," respectively.

- Given the choices between disparate extremes, the building blocks of the analytical framework are three axes or continua between (1) market- and community-oriented rationality, (2) state rollout versus state rollback approaches to regulation, and (3) internationally traded fossil fuels versus locally resourced renewable resources.
- Juxtaposing the three continua or axes helps conceptualize the analytical framework as a three-dimensional space within which to locate different emerging models of energy systems, to ascertain their relative capacity to actualize a fundamental energy transition—one that is economically equitable, socially just, and environmentally sustainable.

Section Two

Methodological Approaches

5

Analysis of Research Methods Examining Shale Oil and Gas Development

FELIX N. FERNANDO, JESSICA D. ULRICH-SCHAD, AND ERIC C. LARSON

Introduction

Hydraulic fracturing and horizontal drilling (colloquially referred to as fracking) is arguably the most significant and contentious technological advancement in the nonrenewable energy industry during the last decade. The surge in shale oil and gas development in the United States, caused in part, by these technological advances, dissipated with the decline in prices since mid-2015. However, price recovery and current favorable political and regulatory outlook are instigating a resurgence in the shale oil and gas industry. Shale plays such as the Bakken, Haynesville, and Permian Basin experienced an increase in drilling activity in 2017. Therefore, future drilling and development activity can be expected to follow a miniboom/minibust cycle (Jacquet and Kay 2014) based on favorable market; technological, political, and pertinent regulatory dynamics such as technological advancements in fracking processes; price stability over breakeven levels; improved collection infrastructure; and supportive regulatory regimes. As

DOI: 10.5876/9781646420278.c005

the impacted communities and stakeholders evolve and adapt to the mini-booms and busts of the shale oil and gas industry, this cyclical industrial progression creates an intriguing and challenging landscape for social science research.

This chapter aims to elucidate the unique challenges, constraints, and opportunities associated with conducting social science research in the context of shale oil and gas development. Specifically, we (a) provide an in-depth review of quantitative and qualitative social science research focusing on issues pertaining to fracking, (b) outline the potential challenges and implications for researchers, and (c) highlight innovative ways the challenges are being addressed by social science scholars or could be addressed in future research. This chapter does not, however, describe the approaches used for transcribing, coding, preparing and cleaning, or analyzing data, nor does it address steps to ensure reliability and validity during data compilation and analysis. It must be noted that this chapter excludes a discussion of a stream of research that examines issues of income and natural resource curse using methods such as geospatial analysis, economic choice experiments, and econometric analysis.

The arguments and ideas presented and discussed in this chapter are based on (1) firsthand research experiences of the authors; (2) systematic review of sixty-eight original social science research articles published from 2005 to 2017 in the United States, Canada, Australia, and the United Kingdom; and (3) discussions with other prominent researchers in the field. The articles examined in the systematic review were identified by running searches using *hydraulic fracturing*, *fracking*, *shale oil and gas*, and other similar terms on multiple academic platforms such as ProQuest, EBSCOhost, JSTOR, and ScienceDirect. The articles were analyzed and organized by the research methods used.

This chapter is organized into three sections. The first section describes implications and considerations for conducting social science research in communities impacted by rapid shale oil and gas development and suggestions for longitudinal research. Integrating firsthand experiences with literature, the second and third sections collate different qualitative and quantitative methodological approaches, while outlining ways to address challenges in impacted community contexts.

Research Considerations and Implications during Miniboom/Minibust Cycles

Some of the major shale plays in the United States such as the Bakken and the Marcellus are located in areas that are considered largely rural and/or agricultural, while other shale plays such as the Barnett are located in largely urban areas. In addition, shale plays and the associated geographic footprint span multiple states. For example, development of the Bakken Shale directly impacts North Dakota, Montana, South Dakota, and some parts of Canada, while development of the Marcellus Shale directly impacts Pennsylvania, West Virginia, northeastern Ohio, and upstate New York. The vast geographic footprint, the urban/rural context, and the historical legacy of natural resource extraction require careful positioning and framing of the research context within which the social impacts and perceptions of stakeholders are examined and studied. In addition, the socially differential impacts, spatially differential impacts, and regionally differential impacts, based on the scale of research focus and analysis, present a complicated research landscape that should be considered and addressed through careful research design and methodology (Schafft, Brasier, and Hesse 2017).

At a regional level, the rapid workforce influx during the boom phase generates a swift increase in demand for community services and amenities. Urban areas that have the community infrastructure and capacity to absorb the demands of the workforce act as hubs from which the workers travel to drilling sites, activating the hub-and-spoke model (Brasier and Filteau 2015). Comparatively, rural communities in shale plays with no urban hubs experience significant strains irrespective of size, as the industry scrambles to secure resources necessary to support the workforce.

At the community level, the geospatial, temporal, and sociocultural dynamics of the community studied will significantly influence the process, frame, and the lens through which the social impacts of fracking will be perceived. The geospatial dynamics in terms of whether the community is located in the core/highly impacted area or in the periphery of a shale play will affect the nature and intensity of the impacts (Junod et al. 2018). Historical legacy of natural resource extraction booms and busts will also shape and influence community mobilizations and adaptations differently from a community experiencing rapid shale oil and gas development for the first time (Bugden, Evensen, and Stedman 2017). Kathryn Brasier et al. note that in the Marcellus Shale region, the changes are conditioned by the level of rurality of the county prior to development, geographic region, and preexisting trends in population and economic change (2015).

The miniboom phase has been largely described as a period of rapid influx of predominantly transient workforce, industrialization, shifts in land use, local inflation, overburdened infrastructure, social disruption, overwhelmed services, outmigration of longtime residents, and concerns of crime and safety. However, the boom phase also represents a period of new additions of services, community amenities, new younger residents, economic opportunity, religious and cultural diversity, and additions to community infrastructure. The community population dynamics also change during the boom phase. Notably, the characteristics of the longtime resident population might differ from the newcomer population. The longtime resident population of especially rural and agricultural communities have a higher percentage of older residents, while the economic opportunity attracts a younger populous to these communities, in addition to the transient workforce. The longtime residents might also differ based on quality-of-life preferences and place-based value conceptions (Fernando and Cooley 2016). The place-based values and contexts might also differ from one community to another and from one shale play to another.

At the individual or stakeholder group level, the uneven distribution of risks and benefits of shale oil and gas development has been well documented in literature (Brasier et al. 2011). Diverse stakeholder groups based on their positioning in the income/economic structure differentially experience and perceive the impacts of development (Fernando and Cooley 2015). For example, new residents living in mobile homes will face different risks and opportunities compared to longtime residents living in single-family homes (Brasier and Filteau 2015). Theodori reveals a paradox among the general population in the Barnett Shale region of Texas during the boom phase (2009). On the one hand, the general public typically dislikes the potentially problematic social and/or environmental issues from fracking. However, on the other hand, the general publics appreciate and view favorably the economic and/or service-related benefits that normally accompany such development. Fernando and Cooley outline similar perceptions among the general public in the Bakken Shale region of western North Dakota (2015). Numerous studies have examined the socioeconomic and community impacts during the miniboom phase.

The industrial activity and community impact dynamics associated with the miniboom/minibust cyclical progression are different compared to a singular boom/bust associated with conventional oil drilling or other large projects that generate similar impacts. Compared to the rapid exodus of workers and community members during a singular bust, the minibust phase largely represents a decline or a slowdown in mainly drilling and exploration operations and an outflow of transient workforce. However, workforce associated with production, collection, and maintenance can be expected to remain stable. The drilling technologies associated with fracking allow multiple wells to be sited on an existing well pad that lowers the incremental cost of new wells. As a result, some drilling activity can be expected even during the minibust phase.

The slowdown during the minibust phase could allow for self-correction of some of the negative impacts such as local inflation, strain on services, and infrastructure. During interviews conducted by Fernando with Stark County residents of the Bakken Shale area, participants described the minibust phase as a "period of adjustment" or as an "opportunity or a break to catch-up" (2016b) For example, a minibust presents an opportunity for housing market correction of unaffordable rents as well as a chance for the supply to catch up to the demand. The population also could stabilize as the largely transient workers associated with drilling and exploration move out of the community while production, collection, and maintenance workforce remain in the community. The new residents and families drawn by long-term production-oriented job opportunities could acclimatize during the minibust phase that could reverse or improve perceptions of social disruption during the boom phase. In addition, the slowdown enables the overburdened and overtaxed local infrastructure and services to plan and adjust. However, the demand and population decline during the minibust could present challenges for maintaining services, community amenities, and infrastructure additions, especially in rural communities. Very few studies have examined the socioeconomic and community impact dynamics during the minibust phase. Longitudinal social science research must continue to explore the community impacts, stakeholder perceptions, political mobilization, regulatory outcomes, and adaptive responses as miniboom and minibust cycles shape industrial progression.

The decline in industrial activity and pace could ease workloads, psychological stress, and overall tiredness among residents in impacted areas that may positively influence research participation, interest, and survey completion rates. The inmigration outmigration implications of longtime residents and newcomers present challenges for recruiting and maintaining the same sample of participants overtime. Longitudinal research must pay careful attention and acknowledge the sample changes over time. Finally, research must try to distinguish impacts associated with miniboom/busts from standard effects associated with any large-scale development or from other background economic trends in the study area.

As fracking is a fairly novel technology, lack of uniform regulations and/or policy regimes governing certain industry operations and lack of scientific knowledge on certain impacts have generated a broad public discourse. For example, while fracking has been perceived to have positive economic impacts, the impact on the natural environment and on overall quality of life remains a highly contentious debate (Anderson and Theodori 2009). It would be important to examine the changes in these discourses and regulatory regime changes over time.

Challenges and Implications for Qualitative Social Science Research

Qualitative studies are a vital tool for understanding the changes in community socioeconomic context and monitoring psychological, sociocultural, and environmental dynamics (Zilliox and Smith 2017). This section discusses how qualitative methods can be used to address implications and challenges in shale oil and gas development contexts.

Establishing Research Context and Framing of Research Issues

At the initial stages of a project, it is necessary to build a fundamental understanding of the context of the regional/community-level/historical legacy of the stakeholder group being studied. Analysis of news media articles, regulatory documents, and community-based texts can be used as a standalone approach or as a methodological primer to develop an understanding of the research context and the setting. Fernando outlines how analysis of the letters to editor of the *Williston Herald*, as a methodological primer, was used to build a fundamental understanding of perceived impacts of fracking in western North Dakota. As a methodological primer, media content analysis is useful in multiple ways (2016a):

A. Construct a fundamental or a preliminary understanding of the context, major impacts, and concerns.
B. Identify key informants and engaged citizens.
C. Highlight issues or clarify concerns that should be included in interview protocols or survey instruments.

Fracking can impose risks to some communities and benefits to others. How policy makers permit, regulate and monitor fracking can be influenced by differing perceptions of the risks and benefits. The media can play a critical role in portraying these perceptions (Blair et al. 2016). Adrianne Kroepsch used an analysis of opinion pages of three major newspapers in Colorado, policy-making documents, and neighborhood meeting discourses to examine the social construction of space in Colorado's northern Front Range (2016). Each of these sources provides important and unique insights. The print media provided a venue for broader arguments about development. The policy-making documents fielded more specific spatial recommendations, and the neighborhood meetings opened a window into spatial discourse grounded in the context (Kroepsch 2016).

Using media content analysis as a standalone methodological approach, Martha Powers et al. analyzed letters to the editor of the most widely circulated local newspaper in the most heavily drilled county in Pennsylvania (Bradford County) in order to understand how residents investigate risks associated with fracking (2015). The analysis of the letters demonstrated citizen discord and stress regarding four main issues: socioeconomic impacts, perceived threats to water, population growth and implications, and changes to the rural landscape, which are some of the commonly outlined themes in social impact literature. Content analysis approaches have the potential to contextualize the impacts and concerns of a place. However, the findings may not be broadly generalizable and the effects of nonparticipant bias limit broader application of the method.

Key informants can provide important knowledge and insights about community characteristics and context. Key informants include local elected officials; county and city government officials; local human and social service agency representatives; community and economic development organization leaders; industry representatives; local business owners (directly and indirectly affected by development); landowners (including those representing landowner coalitions); environmental activists; educators and Extension agents; and active local citizens (Anderson and Theodori 2009; Brasier et al. 2011). Publicly available listings, media content analysis, and county Extension agents can be used to identify key informants. Key informants can be engaged through interviews or focus groups. Key informants can also provide referrals to additional individuals useful for the research project (Ladd 2013). However, Brasier et al. (2011) also notes several limitations of key informants:

1. It might not be possible to recruit key informants from all of the organizations, institutions, or from different stakeholder groups potentially affected by fracking.
2. Key informants from industry and environmental organizations might be unresponsive or unwilling to participate. Contentious environmental concerns, proprietary nature of industry practices and techniques, and strenuous workloads might affect industry representatives' interest in participating. Industry representatives also might decline to answer questions on sensitive or highly debated issues associated with fracking.
3. Responses of elected officials or government officials such as economic development officials might reflect a set positional agenda.
4. Lack of organization and mobilization by impacted local groups will limit the ability to identify key informants from such groups, especially in the early stages of development.
5. Landowners who have signed nondisclosure agreements will be reluctant to participate or express their views freely.

Disaggregating or Contrasting the Experiences and Views of Different Community Subgroups or Geographic Scales

Fracking and related issues have generated a broad debate and conflicting viewpoints. Stakeholders with divergent viewpoints can respond differently to impacts and development. While open-ended questions enable participants to provide rich detailed responses that include examples from their experiences (Archbold 2015), qualitative interview data from different stakeholder groups impacted by fracking enable a broad comparative analysis of stakeholder attitudes and impact perceptions (Fernando and Cooley 2015).

Anderson and Theodori used key informant interview data collected in two Barnett Shale counties to investigate the reported positive and negative outcomes of fracking and to examine the similarities and differences in perceptions between respondents from each of the study counties (2009). Similarly, Brasier et al. used key informant interviews to understand impact perceptions and assess trust in industry and regulatory agencies in four counties in the Marcellus Shale area that experienced different levels of development and variations in natural resource extraction histories (2011). The findings of these studies highlight the usefulness and ability of key informants to provide a detailed and broad understanding of the social impacts that enable comparisons between different scales such as counties or core and periphery communities.

Q methodology (also referred to as Q sort) also can be used to investigate the microlevel discourses of stakeholders on socioeconomic, health, environmental, and other issues surrounding fracking processes. It allows researchers to identify important criteria and explicitly outline areas of consensus and conflict. Q method allows researchers to quantitatively map subjective attitudes and opinions on fracking, rendering them open to statistical analysis to enable researchers to identify a number of idealized accounts or discourses related to fracking. For example, Fry, Brannstrom, and Sakinejad, implemented Q method procedures using semistructured interviews to identify planning and development perspectives on gas well landscapes in multiple municipalities in the Dallas/Fort Worth area (2017). Geospatial data were used to identify gas well locations and compute drilling density within municipalities with drilling ordinances in order to create a rigorous sample frame for semistructured interviews. During the interviews, the participants were asked to conduct a Q sort of nine aerial photographs representing different community landscapes (Fry, Brannstrom, and Sakinejad 2017). Similarly, Matthew Cotton also used Q methodology to examine emergent perspectives on a range of environmental, health, and socioeconomic impacts associated with fracking in the United Kingdom (2015). Q method can be very helpful in unearthing perspectives without requiring participants to articulate them.

Focus groups provide an avenue to examine and collate different perspectives concurrently within an organized setting. However, if a focus group is made up of participants with differing viewpoints that can lead to a contentious debate, researchers should consider power dynamics (e.g., boss and employee), affiliations, or political differentials between participants.

Researchers must also maintain balance of discussion and encourage members to use others' ideas as prompts to stimulate their own perspectives. Because of the contentious nature of the debates surrounding fracking, it would be challenging to organize and stimulate an open discussion in a focus group comprising representatives from different stakeholder subgroups. Multiple focus groups, each consisting of the constitute membership from a particular stakeholder subgroup, should be used, if the aim of the study is to collate perspectives or impacts among different stakeholder groups.

Purposefully selecting the focus group members to represent an impacted subgroup, such as landowners impacted by fracking, would allow for a deeper exploration of the issues that are of specific importance to the targeted impacted group. Simona Perry used GIS and photo-voice to facilitate focus group discussions to examine the psychological and sociocultural determinants of community health in the Marcellus Shale (2013). The focus group participants were involved in the community-integrated GIS process during which they selected geographic places of special importance to them in the county, mapped their land, and identified their neighbors, all the while discussing their relationship to place and community. Focus group participants were also involved in a photo-voice process that involved taking photographs of things and places that exemplified their relationship with their land, the county, and the changes they were experiencing, then writing about those photographs and sharing them with others in the group.

Focus groups examining social impacts and perceptions of fracking can adopt a semistructured topic guide where open-ended questions spur discussion or a more deliberative approach. Deliberative focus groups promote discussion by first informing the participants about the issues and then facilitating discussion to examine the framing of the issues. For example, Laurence Williams et al. adopted a deliberative focus group approach to examine risks and benefits of fracking in the United Kingdom (2015). A topic guide and a series of concept boards (consisting of pictures, diagrams, newspaper headlines, quotes, etc.) were developed and refined during the pilot focus group. These materials were used to provide a coherent overview of the way fracking was being considered from various institutional perspectives for the subsequent focus groups.

Sample representation is an important consideration that must be addressed in studies that compare and contrast the impact perceptions of different stakeholder groups or different geographic scales. When participant groups are structured to explore challenges, issues, and concerns about fracking within a specific target group, the findings should not be generalized to the public.

Maintaining Confidentiality, Building Trust, and Addressing Research Subject Exhaustion

Perry notes that the successful recruitment of focus group participants took four months longer than anticipated (2013). Two things caused this delay: difficulties in gaining the trust of a diversity of rural landowners in the county and the inability to guarantee complete anonymity to potential focus group participants who had signed nondisclosure agreements or were in legal proceedings with a shale gas company. Qualitative interviews are useful to ensure confidentiality and anonymity for participants such as landowners who have signed nondisclosure agreements, but are willing to participate in research, and to accommodate those participants uncomfortable in a group setting (Perry 2013). Researchers must also try to establish and maintain connections and networks with community members that will build trust over time. Multiple interactions, continuous communication, and constructive feedback to stakeholders about findings will build trust and address fears of exploitation.

Scheduling conflicts and busy work schedules during times of rapid shale oil and gas development can affect participant recruitment. In addition, places such as Bradford County in the Marcellus and Williams County in the Bakken Shale region have seen extensive research, and residents can experience research fatigue. Collaborations between multiple institutions and research teams will enhance the quality of the overall research project and reduce the burdens on participants.

Challenges and Implications for Quantitative Social Science Research

Quantitative studies statistically assess a set of research questions or hypotheses that aim to explain impact perceptions, factors influencing impact perceptions, or the mechanics mediating psychological thought processes in the context of shale oil and gas development (Zilliox and Smith 2017). This section discusses how quantitative methods can be used to address implications and challenges in shale oil and gas development contexts. The section concludes with a discussion of mixed methods social science research in shale oil and gas contexts.

Developing Contextually Pertinent Survey Research Instruments

Social, environmental, or other issues pertaining to fracking can vary from one shale play to another or from one community to another. Therefore, the best practice is to develop contextually pertinent survey instruments that simultaneously allow for comparisons between plays. Studies examined in the review used a variety of sequential methodological approaches (using qualitative methods to shape survey instruments) to develop pertinent survey instruments. For example, Derrick Evensen and Rich Stedman conducted a content analysis of regional newspaper coverage in the Marcellus Shale region, and they conducted interviews with key informants to design the questions to be included in the survey (2017). Other studies used focus groups (McGranahan et al. 2017) and key informant interviews (Theodori 2009) to develop locally relevant questions for survey research. Focus groups are also useful for pilot testing of survey instruments to assess intelligibility of question wording and meaning (Evensen and Stedman 2017).

Examining Fracking-Related Issues in a Certain Geographic Region or within a Target Population

Table 5.1 summarizes the geographic focus, sample population, and effective response rates of surveys using different delivery methods. Mail surveys were used by numerous researchers to examine these issues in multiple shale plays. In the Marcellus Shale region, Brasier et al. used mail surveys to explore the nature of perceived risks associated with shale development in Pennsylvania and New York counties (2013). Similarly, Bugden, Evensen, and Stedman examined factors that influence perceptions and actions toward fracking using mail survey data from the "Twin Tiers" regions of New York and Pennsylvania (2017). Theodori used mail survey data from two counties located in the Barnett Shale region of Texas to empirically explore potentially problematic issues associated with fracking (2009). Comparatively, Fernando

TABLE 5.1. Comparison of research using survey approaches

Authors	*Survey Approach*	*Geographic Focus*	*Sample Population*	*Effective Response Rate*
Theodori 2009	Random mail survey	Two counties in the Barnett Shale region of Texas	1,600 randomly selected households; 67 surveys returned as undeliverable	39% (600 completed responses)
Brasier et al. 2013	Random mail survey	21 Pennsylvania and 8 New York counties in the Marcellus Shale	6,000 households; 521 surveys returned as undeliverable	35% (1,917 responses)
Jacquet and Stedman 2013	Random mail survey	Tioga and Bradford Counties in northern Pennsylvania	1,800 property owners; 49 surveys returned as undeliverable	59% (1,028 responses)
Fernando 2015	Delivery and pickup	Two hub towns (Williston and Dickinson) in the Bakken Shale region of North Dakota	700 residents in each community totaling 1,400 surveys	49% (682 responses)
Bugden, Evensen, and Stedman 2017	Random mail survey	34 municipalities in the "Twin Tiers" counties in the Marcellus Shale region	4,998 households; 629 surveys returned as undeliverable	28% (1,202 respondents)
McGranahan et al. 2017	Random mail survey	Six counties in the Bakken Shale region of North Dakota	1,000 randomly selected rural farmers and ranchers; 60 surveys returned as undeliverable	17% (177 completed responses were received)
Lachapelle and Montpetit 2014	Phone survey (land line and cell)	1,505 Quebec respondents, 415 Michigan residents, and 424 Pennsylvania residents	Random samples drawn from the Quebec, Michigan, and Pennsylvanian populations	26% in Quebec and 14% in the United States using a representative random digit dialing procedure
Theodori et al. 2014	Mixed mode (phone and mail)	21 counties in the central core and Tier 1 of the Marcellus Shale region in Pennsylvania	Phone survey: 2,000 random phone numbers entered into a phone bank; 496 were unusable. 1,504 usable phone survey sample; mail survey: 1,600 randomly selected households; 95 surveys returned as undeliverable	Phone survey: 27% (400 completed responses); mail survey: 27% (400 respondents; only the first 200 replies received from each of the well-density categories were included in the analysis)

continued on next page

TABLE 5.1—*continued*

Authors	*Survey Approach*	*Geographic Focus*	*Sample Population*	*Effective Response Rate*
Crowe, Ceresola, and Silva 2015	Mixed mode (electronic and mail)	Local government officials across six shale plays in the United States	753 local leaders	41% (308 completed responses)
Fedder et al. 2019	Mixed mode (electronic and mail)	Residents of two counties in the Bakken Shale play (one in Montana and one in North Dakota)	1,850 randomly selected residents	26% (391 completed responses)
Schafft et al. 2014	Electronic	309 school districts in the Pennsylvania's Marcellus Shale region	891 educators representing school district superintendents, high school principals, and high school directors of curriculum and instruction	42% (372 completed responses)
Andersson-Hudson et al. 2016	Electronic	National survey of United Kingdom residents	YouGovUK, which has approximately 400,000 members	3,822 completed responses
Stedman et al. 2016	Electronic	National survey of US residents	Nationally representative sample of individuals from Qualtrics's online panels	1,625 completed responses (57 people exited the survey prematurely)
Weible and Heikkila 2016	Electronic (email)	Policy actors from New York, Colorado, and Texas	The survey was emailed to 1,101 policy actors: 379 policy actors in New York, 398 in Colorado, and 324 in Texas	Overall response rate of 31% (344 responses); New York: 34% (129 completed responses); Colorado: 34% (137 completed responses); Texas: 24% (78 completed responses)

used a delivery and pickup approach to examine the quality-of-life perceptions of residents in two hub towns in the Bakken Shale play of North Dakota (2015). Compared to mail surveys, the delivery and pickup approach requires more time and resources, thus limiting geographic coverage area. However, this method can help to increase response rates, particularly among groups who prefer face-to-face interaction (Ulrich-Schad, Brock, and Prokopy 2017).

Examining Fracking-Related Issues over Broader Geographic Scales

Given the contentiousness of fracking, it is important to examine perceptions on fracking and regulatory dynamics over broader geographic scales. The advancements in electronic and phone survey approaches provide novel ways to focus on wider geographic scales or access harder-to-reach populations. For example, Erick, Lachapelle and Éric Montpetit used phone surveys to examine public opinion on fracking in the province of Quebec compared to Michigan and Pennsylvania using a representative random digit dialing procedure (2014). Similarly, Charles Davis and Jonathan Fisk used data from a national phone survey of American adults to examine public attitudes toward fracking and policies to identify factors that account for differing levels of support (2014).

Similar approaches can also be used to examine policy dynamics of different stakeholder groups over broad geographic scales. For example, Christopher Weible and Tanya Heikkila examine support of or opposition to fracking among policy actors (all levels of government, environmental and citizen organizations, oil and gas industry, and academics/consultants) across New York, Colorado, and Texas using data from an online survey (2016). Methodologically, this study provides a rare large-sample comparative analysis of policy actors in three states.

Approaching Hard-to-Reach Populations and Improving Response Rates

Electronic surveys with representative samples of residents of oil and gas development areas are more difficult given the complexity in obtaining complete sampling frames and the lack of internet usage and access among some subgroups of the study population, especially in rural areas with aging populations, where much of the development is taking place.

Most standard phone number lists only include landlines. Although random digit dial approaches using both landlines and cell numbers can be used to approach longtime residents, approaching new residents with out-of-state cell numbers can be difficult. However, phone surveys could be particularly useful when the target population might not be reachable using other delivery approaches. For example, it might not be possible to reach oil industry workers using mail- or electronic-based approaches. They are also one of the busiest stakeholder groups during high-activity periods. Using shorter instruments to reach oil industry workers through phone surveys might provide new insights about perceptions and attitudes among oil workers. However, such an effort would also require industry cooperation to generate phone number lists as random digit dialing approaches would be ineffective with out-of-state oil workers.

Busy work schedules and high-activity periods such as the summer can affect response rates. Stakeholder groups such as farmers and oil industry workers are busy during the summer months, and researchers must carefully consider the timing of survey delivery. Compared to single delivery modes, use of multiple delivery modes enhances access to the target population and provides the target population the flexibility to participate using the most convenient mode. Theodori et al. used a mixed mode (phone and mail) survey to examine the issues associated with public views on hydraulic fracturing and the management, disposal, and reuse of frac flowback wastewaters in Marcellus Shale region in Pennsylvania (2014). Jessica Crowe, Ryan Ceresola, and Tony Silva (2015) surveyed local government officials across six shale plays (Bakken, Eagle Ford, Marcellus, Permian, Monterrey, and New Albany) in the United States to examine local officials' positions on fracking using a mixed mode approach (electronic and mail). The electronic surveys (the link to the surveys) were delivered through email and regular mail for those officials whose email addresses were not available. Subsequently, a physical copy of the survey was mailed to those nonresponding officials. Similarly, Michael Fedder, Jessica D. Ulrich-Schad, and Julie Yingling (2019) used a mixed mode (a letter with the option of taking the survey on paper or online) survey to examine how residents of Bakken oil play counties in North Dakota and Montana perceived the environmental impacts of current and future oil and gas development relative to economic factors.

Multiple wave approaches also help to improve response rates. Dylan Bugden, Darrick Evensen, and Richard Stedman (2017) used a four-wave mailing approach (survey, reminder, second survey, and second reminder, with approximately 7–10 days between mailings). Other researchers used similar approaches as well (Jacquet and Stedman 2013: survey, two reminder letters, and another copy of the survey; Theodori 2009: survey, postcard reminder, and two follow-up surveys mailings). Out of the survey efforts examined, Crowe, Ceresola, and Silva adopted a strenuous multiwave (five waves) mixed mode (two modes) approach (2015).

Ensuring Representative Sampling

The extent of sample representation of the characteristics of the population affects the generalizability of the survey findings. For example, the sample of Brasier et al. (2013) was overrepresented by men, older residents, and educational attainment compared to the demographic characteristics of the general population of the surveyed counties. This imbalance suggests the need to be somewhat cautious in generalizing the survey results to the extent that these characteristics (gender, age, and educational attainment) influence perceptions of risk. One commonly used approach for addressing this issue is for researchers to create survey weights based upon available demographic data (e.g., race, sex, age) to make the sample statistics more representative of the study population (e.g., Fedder et al. 2019). For example, Adam Mayer used statewide survey data of Colorado residents to examine public perceptions of risk and benefits related to fracking (2016). The Colorado population has a very unequal spatial distribution—most of the population is concentrated in the metropolitan Denver area and neighboring cities such as Boulder and Colorado Springs. These areas have relatively little drilling. To ensure that the views of people who live in close proximity to fracking were represented, high-drilling rural areas were intentionally oversampled by ranking counties by the number of active wells and dividing this ranking into strata (Mayer 2016).

At larger geographic scales, the generalizability, validity, and applicability of the survey findings principally depend on sample representativeness of the considered population. In addition, use of panels to recruit participants are limited by the extent of representativeness of the characteristics of the panel members to the general population. Researchers must take measures to ensure that the characteristics of the selected sample approximates the general population considered, by using approaches such as quota-sampling (Evensen and Stedman 2017). For example, Stedman et al. administered a nationwide online survey in the United States using Qualtrics's online panels, where the geographic distribution of the respondents were quota-sampled to be consistent with the distribution of the national population (2016).

Importance and Limitations of Secondary Data

Quantitative analysis using census and other secondary data could be used to describe trends over time, to benchmark data collected using other methods, to capture information about historical trends, and to compare different geographic scales or community subgroups. Although secondary data such as economic data on income and wealth, cost of living, and housing quality and affordability and population data are sometimes available, it is important to note multiple complicating issues of data availability and usability:

A. There may be a lack of availability at the community level or in specific regions such as the Bakken Shale that is largely rural with low population density.
B. Census estimates of population data do not account for transient workers who commute for employment, or who live in workforce housing such as man camps (Ruddell 2011).
C. The overall population data does not provide inmigration and outmigration dynamics occurring within a community. The number and characteristics of residents leaving the community and the nature of newcomers into a community can provide valuable insights to community change over time (Fernando and Cooley 2015).
D. Organizations from different states use different data collection and reporting strategies that hinder cross-scale comparisons.

In addition, the rapid socioeconomic changes during the boom period require quick policy formulation and community adaptation strategies. The time lag associated with collection and compilation of some secondary data limits practical usability.

Methodological Triangulation and Mixed Method Approaches for Strong Social Science

Research Qualitative social science research examined in the systematic review used multiple qualitative methods to ensure validity, reliability, and applicability through methodological triangulation. Perry used ethnographic fieldwork to monitor the psychological and sociocultural determinants of community health in the Marcellus Shale (2013). The study is based on data collected through interviews, focus groups, and participant observations over a period of three years. The ethnographic data from Perry includes audio- and videorecordings of focus groups and interviews, photographs and writings from photo-voice, spatial data and maps from the GIS process, informational brochures and handouts from meetings, and field notes of participant observations and interviews, as well as historic photographs and documents from archival research (2013). Similarly, Anna Szolucha documents and analyzes the potential and current social impacts of shale gas developments in Lancashire, United Kingdom, using ethnographic fieldwork (2016). The study is based on data from attending the meetings and events of the local planning authorities, from public inquiry hearings, from local grassroots antifracking groups and national regulatory agencies, from extensive field notes, from a photo-voice exhibition, and from semistructured interviews. Triangulation validates qualitative data by cross-corroborating findings and themes from multiple sources. With the exception of Perry, the qualitative studies systematically analyzed were largely cross-sectional examinations (2013). Therefore, there is a clear need for longitudinal qualitative social science research to examine the social dimensions of fracking over time.

Use of mixed methods provides a more thorough and holistic exploration of the research questions or focus compared to use of either qualitative or quantitative methods. For example, Brasier et al. presents the findings of a mixed methods research project where quantitative and qualitative data gathered from four Pennsylvania counties were used to examine how various social indicators (including housing, health care, education, crime, and residents' perceptions of their communities) have changed as a result of Marcellus Shale development (2015). In-depth, qualitative data, gathered primarily through focus groups, were used to interpret trends in the quantitative data. While the quantitative data indicate that changes across many of the analyzed indicators are limited and often difficult to distinguish from broader regional and long-term trends, the qualitative data suggest substantial community change as well as trepidation about the future of the communities. Similarly, Skylar Zilliox and Jessica Smith used a mixed methods approach to examine how Memorandums of Understanding shaped public perceptions of the industry and the town government in a politically heterogeneous suburban Colorado town (2017). The study combined demographic analysis using census data, semistructured interviews with key players, in-person observations of and coding of Town Hall meetings, and public comments. These methods complement each other and offer a strategy for corroborating data. The interviews and public commentary data helped to develop a fuller picture of the motivations, concerns, and relationships identified in quantitative forms (Zilliox and Smith 2017). Matthew Fry, Christian Brannstrom, and Michael Sakinejad also used a variety of quantitative and qualitative data sources such as GIS-based well data; temporal characteristics of production sites and land cover change data; census data to characterize population growth over time; content analysis of city ordinances, comprehensive plans, city council meetings, and public comments; and semistructured interview data to identify planning and development perspectives on gas well landscapes and setback distances in multiple municipalities in the Dallas Fort Worth area (2017). Setback distances is a topic that has received little research attention but presents important policy implications. Use of mixed methods to examine research issues, especially those that have important policy implications, would enhance the usability and applicability of the findings.

Comparatively, Anna Anya Phelan and Sander Jacobs adopted a novel concurrent mixed methods approach, with both quantitative and qualitative data collected at the same time to empirically examine how quality of life in regional communities of the Surat Basin in Australia is being influenced by rapid development of coal seam gas projects (2016). The quantitative data were collected using a structured questionnaire in a cross-sectional mixed mode (online and paper) survey from 428 participants. The qualitative data collected included five open-ended questionnaire items completed by the same participants; twenty-four semistructured interviews with 41 participants, and direct observations. The qualitative findings provided a deeper story, enhancing the findings from the quantitative stage, and the strength of the merged-data analysis provided both statistical and narrative in the enhancement in validity and reliability of results.

Mixed methods approaches enable comprehensive and salient findings. Typically, quantitative methods are sequentially used after qualitative methods in mixed methods studies. However, qualitative methods can be used subsequent to quantitative methods to examine conflicting or contradictory findings. For example, if a survey reveals a conflicting or a contradictory finding compared to the literature, qualitative interviews or focus groups can be used to further examine the survey findings.

Conclusion

The discussion and analysis of studies presented in this chapter enables researchers to address opportunities and constraints of using various social science research methods to examine issues related to fracking. The synthesis presented in the chapter indicates that while a broad and diverse range of methods have been employed to examine the social impacts of fracking, opportunities remain to improve research efficiency and effectiveness. As the impacted communities and stakeholders evolve and adapt to the minibooms and minibusts of the shale oil and gas industry, the information presented would help researchers to navigate an intriguing and challenging landscape for social science research.

Acknowledgments

The authors thank Kathy Brasier and Gene Theodori for sharing their constructive input, ideas, and perspectives during the initial stages of drafting this chapter.

References

Anderson, Brooklynn J., and Gene L. Theodori. 2009. "'Local Leaders' Perceptions of Energy Development in the Barnett Shale." *Southern Rural Sociology* 24 (1): 113–129.

Andersson-Hudson, Jessica, William Knight, Mathew Humphrey, and Sarah O'Hara. 2016. "Exploring Support for Shale Gas Extraction in the United Kingdom." *Energy Policy* 98 (November): 582–589.

Archbold, Carol A. 2015. "Established-Outsider Relations, Crime Problems, and Policies in Oil Boomtowns in Western North Dakota." *Criminology, Criminal Justice Law, and Society* 16: 19–40.

Blair, Benjamin D., Christopher M. Weible, Tanya Heikkila, and Larkin McCormack. 2016. "Certainty and Uncertainty in Framing the Risks and Benefits of Hydraulic Fracturing in the Colorado News Media." *Risk, Hazards and Crisis in Public Policy* 6 (3): 290–307.

Brasier, Kathryn J., Lisa Davis, Leland Glenna, Timothy W. Kelsey, Diane K. McLaughlin, Kai Schafft, and Kristin Babbie, et al. 2015. "Communities Experiencing Shale Gas Development." In *Economics of Unconventional Shale Gas Development*, edited by William E. Hefley and Yongsheng Wang, 149–178. Cham, Switzerland: Springer International Publishing.

Brasier, Kathryn J., and Matthew R. Filteau. 2015. "Community Impacts of Shale-Based Energy Development: A Summary and Research Agenda." In *The Human and Environmental Impact of Fracking*, edited by Madelon L. Finkel, 95–114. Santa Barbara: Praeger.

Brasier, Kathryn J., Matthew R. Filteau, Diane K. McLaughlin, Jeffrey Jacquet, Richard C. Stedman, Timothy W. Kelsey, and Stephan J. Goetz. 2011. "'Residents' Perceptions of Community and Environmental Impacts from Development of Natural Gas in the Marcellus Shale: A Comparison of Pennsylvania and New York Cases." *Journal of Social* Sciences 26 (1): 32–61.

Brasier, Kathryn J., Diane K. McLaughlin, Danielle Rhubart, Richard C. Stedman, Matthew R. Filteau, and Jeffrey Jacquet. 2013. "Risk Perceptions of Natural Gas Development in the Marcellus Shale." *Environmental Practice* 15 (2): 108–122.

Bugden, Dylan, Darrick Evensen, and Richard Stedman. 2017. "A Drill by Any Other Name: Social Representations, Framing and Legacies of Natural Resource Extraction in the Fracking Industry." *Energy Research and Social Science* 29 (July): 62–71.

Cotton, Matthew. 2015. "Stakeholder Perspectives on Shale Gas Fracking: A Q-Method Study of Environmental Discourses." *Environment and Planning A* 47 (August): 1944–1962.

Crowe, Jessica, Ryan Ceresola, and Tony Silva. 2015. "The Influence of Value Orientations, Personal Beliefs, and Knowledge about Resource Extraction on Local Leaders' Positions on Shale Development." *Rural Sociology* 80 (4): 397–430.

Davis, Charles, and Jonathan M. Fisk. 2014. "Energy Abundance or Environmental Worries? Analyzing Public Support for Fracking in the United States." *Review of Policy Research* 31 (1): 1–16.

Evensen, Darrick, and Rich Stedman. 2017. "Beliefs about Impacts Matter Little for Attitudes on Shale Gas Development." *Energy Policy* 109 (October): 10–21.

Fedder, Michael, Jessica D. Ulrich-Schad, and Julie Yingling. 2019. "'You Shouldn't Worry Walking a Block and a Half to Your Car': Exploring Perceptions of Crime and Anomie in the Bakken Oil Play." *International Journal of Rural Criminology* 4 (2): 193–216.

Fernando, Felix N. 2015. "Quality of Life Perceptions among Stakeholder Groups in Western North Dakota." Presented at the Great Plains Sociological Association Annual Conference, October 15–16, 2015, Fargo.

Fernando, Felix N. 2016a. "Media Content Analysis as a Methodological Primer for Quality of Life (QoL) Studies: Case of Western North Dakota." *Northern Plains Ethics Journal* 4 (1): 23–49.

Fernando, Felix N. 2016b. "Mini-boom / Mini-bust Model of Modern Shale Oil Development: Challenges and Policy Implications." Presented at the International Symposium on Society and Resource Management, June 22–26, 2016, Houghton, MI.

Fernando, Felix N., and Dennis R. Cooley. 2015. "An Oil Boom's Effect on Quality of Life (QoL): Lessons from Western North Dakota." *Applied Research in Quality of Life* 11 (4): 1083–1115.

Fernando, Felix N., and Dennis R. Cooley. 2016. "Socioeconomic System of the Oil Boom and Rural Community Development in Western North Dakota." *Rural Sociology* 81 (3): 407–444.

Fry, Matthew, Christian Brannstrom, and Michael Sakinejad. 2017. "Suburbanization and Shale Gas Wells: Patterns, Planning Perspectives, and Reverse Setback Policies." *Landscape and Urban Planning* 168 (December): 9–21.

Jacquet, Jeffrey, and David L. Kay. 2014. "The Unconventional Boomtown: Updating the Impact Model to Fit New Spatial and Temporal Scales." *Journal of Rural and Community Development* 9 (1): 1–23.

Jacquet, Jeffrey B., and Richard C. Stedman. 2013. "Perceived Impacts from Wind Farm and Natural Gas Development in Northern Pennsylvania." *Rural Sociology* 78 (4): 450–472.

Junod, Anne N., Jeffrey B. Jacquet, Felix Fernando, and Lynette Flage. 2018. "Life in the Goldilocks Zone: Perceptions of Place Disruption on the Periphery of the Bakken Shale." *Society and Natural Resources* 31 (2): 200–217.

Kroepsch, Adrianne. 2016. "New Rig on the Block: Spatial Policy Discourse and the New Suburban Geography of Energy Production on Colorado's Front Range." *Environmental Communication* 10 (3): 337–351.

Lachapelle, Erick, and Éric Montpetit. 2014. "Public Opinion on Hydraulic Fracturing in the Province of Quebec: A Comparison with Michigan and Pennsylvania." *Issues in Energy and Environmental Policy* 17 (October): 1–21.

Ladd, Anthony E. 2013. "Stakeholder Perceptions of Socioenvironmental Impacts from Unconventional Natural Gas Development and Hydraulic Fracturing in the Haynesville Shale." *Journal of Rural Social Sciences* 28 (2): 56–89.

Mayer, Adam. 2016. "Risk and Benefits in a Fracking Boom: Evidence from Colorado." *The Extractive Industries and Society* 3 (July): 744–753.

McGranahan, Devan Allen, Felix N. Fernando, and Meghan L. E. Kirkwood. 2017. "Reflections on a Boom: Perceptions of Energy Development Impacts in the Bakken Oil Patch Inform Environmental Science and Policy Priorities." *Science of the Total Environment* 599–600 (December): 1993–2018.

Perry, Simona L. 2013. "Using Ethnography to Monitor the Community Health Implications of Onshore Unconventional Oil and Gas Developments: Examples from Pennsylvania's Marcellus Shale." *New Solutions* 23 (1): 33–53.

Phelan, Anna Anya, and Sander Jacobs. 2016. "Facing the True Cost of Fracking; Social Externalities and the Role of Integrated Valuation." *Ecosystem Services* 22 (B): 348–358.

Powers, Martha, Poune Saberi, Richard Pepino, Emily Strupp, Eva Bugos, and Carolyn C. Cannuscio. 2015. "Popular Epidemiology and 'Fracking': Citizens' Concerns Regarding the Economic, Environmental, Health and Social Impacts of Unconventional Natural Gas Drilling Operations." *Journal of Community Health* 40 (3): 534–541.

Ruddell, Rick. 2011. "Boomtown Policing: Responding to the Dark Side of Resource Development." *Policing* 5 (4): 328–342.

Schafft, Kai, Kathy Brasier, and Arielle Hesse. 2017. "A Multidimensional Reconceptualization of Boomtown Development and Its Impacts: Thinking Regionally, Spatially, and Contextually." Presented at the Annual Meeting of the Rural Sociological Society, July 27–30, 2017, Columbus, OH.

Schafft, Kai A., Leland L. Glenna, Brandn Green, and Yetkin Borlu. 2014. "Local Impacts of Unconventional Gas Development within Pennsylvania's Marcellus Shale Region: Gauging Boomtown Development through the Perspectives of Educational Administrators." *Society and Natural Resources* 27 (4): 389–404.

Stedman, Richard C., Darrick Evensen, Sarah O'Hara, and Mathew Humphrey. 2016. "Comparing the Relationship between Knowledge and Support for Hydraulic Fracturing between Residents of the United States and the United Kingdom." *Energy Research and Social Science* 20 (October): 142–148.

Szolucha Anna. 2016. *The Human Dimension of Shale Gas Developments in Lancashire, UK: Towards a Social Impact Assessment*. https://www.academia.edu/29073366/The_Human_Dimension_of_Shale_Gas_Developments_in_Lancashire_UK_Towards_a_social_impacts_assessment.

Theodori, Gene L. 2009. "Paradoxical Perceptions of Problems Associated with Unconventional Natural Gas Development." *Southern Rural Sociology* 24 (3): 97–117.

Theodori, Gene L., A. E. Luloff, Fern K. Willits, and David B. Burnett. 2014. "Hydraulic Fracturing and the Management, Disposal, and Reuse of Frac Flowback Waters: Views from the Public in the Marcellus Shale." *Energy Research and Social Science* 2 (June): 66–74.

Ulrich-Schad, Jessica D., Caroline Brock, and Linda S. Prokopy. 2017. "A Comparison of Awareness, Attitudes, and Usage of Water Quality Conservation Practices between Amish and Non-Amish Farmers." *Society and Natural Resources* 30 (12): 1476–1490.

Weible, Christopher M., and Tanya Heikkila. 2016. "Comparing the Politics of Hydraulic Fracturing in New York, Colorado, and Texas." *Review of Policy Research* 33 (3): 232–250.

Williams, Laurence, Phil Macnaghten, Richard Davies, and Sarah Curtis. 2015. "Framing 'Fracking': Exploring Public Perceptions of Hydraulic Fracturing in the United Kingdom." *Public Understanding of Science* 26 (1): 89–104.

Zilliox, Skylar, and Jessica M. Smith. 2017. "Memorandums of Understanding and Public Trust in Local Government for Colorado's Unconventional Energy Industry." *Energy Policy* 107 (August): 72–81.

Chapter 5 Summary

This chapter aims to elucidate the unique challenges, constraints, and opportunities associated with conducting social science research in the context of shale oil and gas development. Specifically, we (a) provide an in-depth review of quantitative and qualitative social science research focusing on issues pertaining to fracking, (b) outline the potential challenges and implications for researchers, and (c) highlight innovative ways the challenges are being addressed by social science scholars or could be addressed in future research. The arguments and ideas presented and discussed in this chapter are based on (1) firsthand research experiences of the authors; (2) systematic review of sixty-eight original social science research articles published from 2005 to 2017 in the United States, Canada, Australia, and the United Kingdom; and (3) discussions with other prominent researchers in the field. This chapter is organized into three sections. The first section describes implications and considerations for conducting social science research in communities impacted by rapid shale oil and gas development and suggestions for longitudinal research. Integrating firsthand experiences with literature, the second and third sections collate different qualitative and quantitative methodological approaches, while outlining ways to address challenges in impacted community contexts.

KEY TAKEAWAYS

- The industrial activity and community impact dynamics associated with the miniboom/minibust cyclical progression are different compared to a singular boom/bust associated with conventional oil drilling or other large projects that generate similar impacts.
- Very few studies have examined the socioeconomic and community impact dynamics during the minibust phase. Longitudinal social science research must continue to explore the community impacts, stakeholder perceptions, political mobilization, regulatory outcomes, and adaptive responses as miniboom and minibust cycles shape industrial progression.
- The synthesis presented in the chapter indicates that while a broad and diverse range of methods have been employed to examine the social impacts of fracking, opportunities remain to improve research efficiency and effectiveness.

- Qualitative studies are vital for understanding the changes in community socioeconomic context and monitoring psychological, sociocultural, and environmental dynamics. A variety of qualitative research methods such as interviews, focus groups, content analysis, and Q sort are being used to examine implications and challenges in shale oil and gas development contexts.
- Quantitative studies are vital for statistically assessing a set of research questions or hypotheses that aim to explain impact perceptions, factors influencing impact perceptions, or the mechanics mediating psychological thought processes in the context of shale oil and gas development. A variety of quantitative research methods such as multiple mode surveys using different delivery approaches and census / secondary data are being used to examine implications and challenges in shale oil and gas development contexts.

6

Identifying Energy Discourses across Scales in Canada with Q Methodology and Survey Research

JOHN R. PARKINS AND KATE SHERREN

Introduction

Social impacts from energy production encompass wide-ranging material domains such as employment and infrastructure (Cameron and Van Der Zwaan 2015). Researchers also identify impacts on culture and cognition, including critical changes in attitudes and emotions (Davidson 2017). A related area of interest involves the recursive relationship between technologies and public discourses and the ways in which these discourses can structure our perceptions and expectations of technological change. One example comes from the *Calgary Herald*, a major daily paper in the Province of Alberta, Canada. In an opinion column from a well-known environmental organization, the Pembina Institute, the author states, "The phase-out of coal-fired power is another climate policy with federal reinforcement. It is popular with a majority of Albertans who want to protect their families from the harmful impacts of burning coal. CEOs of major coal generators are saying that coal phase-out makes economic sense, and are accelerating

DOI: 10.5876/9781646420278.c006

timelines for conversions and closures compared to legislation" (Hastings-Simon and Dyer 2017).

In response to this opinion column, an online comment liked by many readers states that the Pembina Institute fails to mention that the carbon tax is deeply unpopular, that the tax will mean higher taxes at a time when the province is hurting economically, and that the province's climate policy is not gaining "social license" from its critics. This comment suggests that the closure of coal generators is clearly wrongheaded, much like other climate policies implemented by the provincial government.

Within energy-dependent jurisdictions, such as Alberta, these types of interactions in public and private life are ubiquitous, affecting our ideas, feelings, and preferences for technological change. There is nothing new about this phenomenon. The emergence of renewable energy technologies almost always intermingles new ideas, expressions, and narratives about how these technologies are accepted, integrated, or rejected within our homes, cities, and surrounding landscapes. As renewable energy technologies evolve, how we think and talk about energy landscapes is also changing. Energy discourses shape our thinking, our reactions, and our sense of what is desirable or undesirable in the surrounding landscape. Understanding discourses as the subconscious organization of collectively held values and mental models (Lakoff 2008), this chapter advances methodological innovations to identify subtle variations in discourses on energy production within and between regions of Canada. Two questions provide guidance for this study. Given that discourses on energy development are often context specific, are there ways of gaining local insight while also maintaining comparability across larger populations? What are the implications of better understanding regional- and national-level discourses on energy development for transition to low-carbon energy futures?

To answer these questions, we present results from mixed methods, involving Q methodology and a national survey of Canadian citizens. Since Q methodology is an established method for identifying discourse (Watts and Stenner 2012), our empirical work begins with the identification of discourses on energy development based on in-person Q sorts with individuals who are generally informed about energy issues in three Canadian provinces: Alberta, Ontario, and New Brunswick. Results from this work are presented elsewhere (Parkins et al. 2015). In this chapter, we explore how these nationally derived discourses play out at a regional scale in New Brunswick and compare the *informed discourses* with *layperson discourses* in national and regional population surveys. Those surveys included eighteen of the most factor-defining Q set statements about energy development (i.e., discursive fragments), to assess their salience within the general population. By using this mixed methods approach combining surveys and Q methodology—not commonly used but with some precedence (Danielson 2009)—we seek to maintain a focus on the local context of energy discourses, while achieving a comparative dimension of the analysis at the national scale. Before we provide details on methodology and results, the following section offers reasons for focusing on discourse with attention to culture and imagination.

Contending with Culture and Discourse

An emerging topic of interest within energy social sciences involves energy cultures (Strauss, Rupp, and Love 2013). In linking energy and culture, Janet Stewart (2014) invokes Wilhelm Ostwald's classic definition of culture "as the way in which humans influence nature's raw energies for their own purposes" (334). This definition is consistent with sociological understandings of culture as a set of skills and habits that allows groups to persist through time (Swidler 1986). Examining specific cultures of thought and practice, Janet Stephenson, Debbie Hopkins, and Adam Doering propose that energy cultures in the field of transportation consist of three core areas of influence: materials, practices, and norms (2015). Materials represent the physical technologies and infrastructures around us. Practices include the actions associated with energy, such as the act of getting to work or taking the bus. Finally, norms "are personal and social expectations about how life should be lived, including such things as expectations about speed of travel and aspirations for car ownership" (Stephenson, Hopkins, and Doering 2015, 357). These core areas are a point of focus in the social sciences of energy development, with research on public attitudes toward energy technologies (Jacquet 2012), the social practices of energy consumptions (Labanca 2017), and norms that influence social behavior (Horne and Huddart Kennedy 2017).

Ideas about norms are particularly relevant to this chapter, especially as they relate to personal expectations, aspirations, and a general sense of what is allowable and acceptable within energy landscapes. Our energy imaginaries represent a common understanding of what is allowable or not, making "possible common practices and a widely shared sense of legitimacy" (Taylor 2007, 172). The everydayness of energy systems is at times reflected in the invisibility of infrastructure, such as transmission lines that are ubiquitous in a way that renders them unnoticed (Hirsh and Sovacool 2013). The acceptability of our energy lifestyles (Shove 2017) includes the unquestioning desire for air travel and the expectation of warmth and comfort in our homes that is clearly but unconsciously linked to energy consumption. Thomas Loder describes this phenomenon of energy imaginaries as an aspect of environmentality, where the focus of research moves from issues of governance and decision making to issues of culture and how we come to formulate our preferences (2016). Consideration of culture, therefore, involves attention to politics, economics, and taken-for-granted value positions that condition the ways in which people think and act. Culture, as expressed through political and economic discourses, has a powerful impact on personal preferences (Loder 2016; Robbins 2007). Culture is intimately entangled in our thought processes, by opening certain paths of opportunity and at the same time constraining other choices, whereby "subjects are formed not solely through coercion, but through complex, ideologically-mediated encounters with both social and physical environments" (Loder 2016, 737). Q methodologists, making a distinction between individual preferences and collective views, also express these sentiments. Through Q methodology "the focus shifts away from personal meaning and knowledge structures toward their social counter-parts; the shared viewpoints, bodies of knowledge or discourses that represent the substantive, cumulative and publicly accessible product of innumerable human selection" (Watts and Stenner 2012, 42).

To reinforce the ways in which culture and discourse condition our responses to energy alternatives, critical theorists such as Andreas Malm, in his book on the origins of global warming, reminds us that the fossil economy was indeed created. Yet like a long-standing building, the structures of a fossil economy become entrenched in the environment. "Eventually it appears indistinguishable from life itself; business-as-usual" (2016, 13). Not only do our energy landscapes become normalized and our lives routinized in relation to these energy systems, our identities, our distinctiveness, and our sense of belonging are inextricably linked to these norms, expectations, and aspirations. Quoting Malm again: "You are a subject of the fossil economy, and since you repeat the act frequently . . . you cannot imagine *not* flying: [Fossil Capital] has constituted the subject, who cannot see himself outside of it and who rarely reflects upon, let alone articulates, the ideological affiliation. It is just there, in the veins of material life" (365).

In this analysis, Malm illustrates how culture itself does the heavy lifting to maintain status quo and establish enduring relations of power and acquiescence toward dominant political and economic interests. Malm's work is crucial to scholarship on energy cultures in part because it tracks the origins of these norms through the lens of historical and material change. In similar ways, the identification of discourses is a way to step back from the materials, practices, and perceptions of energy in society and to ask questions about the sources of inspirations, the enduring ideas and narratives that we root deeply in our consciousness and carry forward in everyday life. Where did these norms come from? What are the core ideas behind our sense of right and wrong, and how do these ideas differ across groups? Gaining insight into these issues is, in part, a process of learning about energy cultures and how cultural distinctives are held in tension with each other. These points of tension are important to understand because they represent a cleavage, a rift within the social fabric of existing and future energy landscapes that hold the potential for movement.

Consistent with John Dryzek and Jeffrey Berejikian (1993, 51), in this study we identify discourses as embodied sets of "capabilities which enable the assemblage of words, phrases and sentences into meaningful 'text' intelligible to readers or listeners." These texts may contain the same or similar words but may also be associated with different meaning systems and interpretations of energy technologies and energy landscapes. Such textual fragments, reflecting a small piece of the discourse on energy development, are a frequent part of social science studies, offering insight into the building blocks of ideas and meanings that jockey for dominance within our energy imaginaries.

To set the stage for the mixed method analysis presented in this chapter, in the remainder of this section, we review several examples of discursive framing around energy issues, with attention to distinct research methods. Examining the Marcellus shale gas development in Pennsylvania, Michael Finewood and Laura Stroup use secondary sources such as documents from government and industry to document the discursive strategies that form around support and opposition to hydraulic fracturing technologies (2012). They argue that "multi-scalar, pro-fracking narratives serve to obfuscate the drilling process and normalize impacts on the hydro-social cycle. This largely occurs through a discursive framing of natural gas as a green fossil fuel" (73). Accordingly, opposing discourses around risks to water are discounted in this context as an issue of costs versus benefits, emphasizing the need for local residents and communities to sacrifice local environments for the greater good. Similarly, in Poland, Aleksandra Lis and Piotr Stankiewicz, using observations, media analysis, and in-depth interviews, identify framing strategies in response to shale gas development (2017). These frames include dominant discourses about shale gas as a novel economic and strategic resource, with oppositional voices identifying shale gas as a threat. Across a number of these studies on hydraulic fracturing, we observe the association between discourses of hydraulic fracturing coupled with energy security, transition fuels, and climate change, allowing for potent ideological battlegrounds against the clearly negative impacts of hydraulic fracturing on local environments (Jackson et al. 2014).

In addition to studies that identify discourses through narratives embedded within media, interviews, and other available written materials, Hilary Boudet et al. deploy large-scale surveys to determine public perceptions of hydraulic fracturing (2014). Although survey researchers do not often frame their work in terms of discourses, survey participants are invited to respond to statements (i.e., discourse fragments) that offer insight into the ideas that people support and reject. In their study on public perceptions of fracking, Boudet and colleagues draw attention to a respondent's worldview as a potential factor leading to support or opposition to hydraulic fracturing. Drawing on Anthony Leiserowitz (2006), worldviews involve a general social, political, and cultural approach to the world that guides an individual's response to a situation. As such, worldviews can represent aspects of culture and energy imaginaries defined above. In their analysis of survey data, measurement of worldviews involves two predefined categories from the work of Mary Douglas and Aaron Wildavsky (1983). An individualist worldview involves preferences for autonomy and freedom. An egalitarian worldview involves preferences for justice and fairness. Based on their findings, opposition to fracking is associated with women, those who hold egalitarian worldviews, those who read newspapers more regularly, and those who associate fracking with environmental impacts. Support for fracking is associated with older respondents, those with a bachelor's degree or higher, political conservatism, and those who associate fracking with positive economic growth. An individualist worldview was not a predictor of support.

Finally, Q methodology is also a popular way of building insights from research on discourses. A number of scholars utilize this method in relation to energy discourses (Cotton 2015; Jepson, Brannstrom, and Persons 2012; Wolsink and Breukers 2010). For instance, using twenty-one respondents who participated in their Q sort, Jepson, Brannstrom, and Persons explored discourses associated with wind power development in Texas (2012). In their analysis, they identified five distinct discourses on wind development ranging from *wind welcomers*, with strong support for wind based on economic impacts, to *community advocates*, with modest support for wind and concern for negative impacts and marginalized groups. The Jepson study is significant in part because it provides insight into the finely tuned and locally derived discourses that emerge around energy development. Within the five energy discourses identified in this region of west Texas, all respondents maintained a strong sense of environmental skepticism. Unlike other parts of the world where climate concern and associated climate policy provide an underlying motivation for wind energy development, in this study respondents "express an unwavering commitment to possessive individualism" (Jepson, Brannstrom, and Persons 2012, 861). An explanation for this commitment is provided by the authors in terms of the cultural politics of Texas, including the links between environmentalism and liberalism, and the idea of energy production as a way of life.

We learn from this literature review that an analysis of discourses can offer insight into the ideologically mediated encounters with energy landscapes and the ways in which our ideas, expectations and aspirations can lock in a set of assumptions and preferences for future energy systems. Nevertheless, in as much as discourses represent deeply held and enduring social positions, they also represent points of contradiction and potential movement. As Malm's historical analysis indicates, culture and discourse can give us the tools to imagine something other than business-as-usual and perhaps some indication of how change can become imaginable (2016). Moreover, recent research on energy development highlights that perspectives on energy development can be highly context specific. As such, the remainder of this chapter offers a mixed method approach to the study of discourses across scales in Canada.

Combining Q Methodology and a National Survey

To identify discourses, as well as salient discourse fragments for later surveys, we draw on Q methodology, involving the sorting of statements on energy production in Canada. This method highlights points of divergence and convergence on energy production within the region, and our use of this method is consistent with other researchers in resource management settings (Clare, Krogman, and Caine 2013; Cotton 2015). To guide the initial development of these statements, and to ensure that we captured a "universe of statements" on energy development, we took inspiration from a set of sustainability discourses identified in John Barry, John Dryzek, and David Schlosberg (2005) with further insight from Terry Leahy, Vanessa Bowden, and Steven Threadgold (2010) and Sarah Strauss, Stephanie Rupp, and Thomas Love (2013). Eight thematic areas include the following: collapse, promethean, ecological modernization, agency and response, consumer sovereignty, place protection, distributed versus centralized, and global fairness. Using this initial set of themes, we identified a series of forty-eight statements pertaining to energy development to serve as our concourse. These statements came from the published literature, news media, and industry and environment-focused organizations, plus other sources to capture a "universe of ideas" about energy development. As the statements were intended for use across regions in Canada, they intentionally avoided technology-specific statements that might be irrelevant in particular places.

Our concourse of forty-eight statements covered all eight themes and were sorted on a sort-board resembling a pyramid structure (table 6.1), where more statements are allowed within the neutral region, and fewer statements are allowed at the tails of the distribution (i.e., strongly disagree or strongly agree). This process of sorting generates a forced choice set unique to each research participant. In each province (Alberta, $n = 16$, Ontario, $n = 21$, and New Brunswick, $n = 21$) members of our research team recruited a cross-section of local residents who were informed on local energy issues to be Q sort participants. For example in Ontario, research participants were selected who had knowledge of wind and local nuclear power development

TABLE 6.1. Number of statements and scores for each statement

No. of statements	1	2	4	6	7	8	7	6	4	2	1
Score	−5	−4	−3	−2	−1	0	+1	+2	+3	+4	+5
Instructions	Disagree most strongly					Neutral					Agree most strongly

(Hempel 2017). These three regions formed the basis of a "national" Q sort that reflects the aggregate of discourses that are present within these regions. After participants sorted the statements, factor analysis identified statistically distinct discourses based on similar Q sorts, allowing for calculations of how each factor would archetypically score each statement along with overall averages of support or opposition for specific statements in the concourse. To enable comparison of those averages with increasingly common online Q method, we also used a polling firm, Corporate Research Associates, to recruit a random sample of New Brunswick residents to complete an online Q sort exercise using the same concourse. This process resulted in eighty-six participants.

With the Q method, we can identify discourses within a smaller group, but the strength of this method also reveals a potential weakness. From Q methodology alone, one cannot determine the prevalence of discourses within a larger group or area (i.e., results are not generalizable). To extend our work from these regional settings to the national scale, a subset of eighteen statements with high valence (i.e., support and opposition) were then included in national and New Brunswick surveys to gauge the extent to which the defining ideas from our Q sorts were equally "defining" to a national sample of Canadians.

A polling firm, Corporate Research Associates, secured a national sample through nationwide bilingual email solicitation to over 450,000 Canadians. Completed in fall 2014, the final survey comprised 3,000 respondents with quotas for age, gender, region, and mother tongue in Quebec and New Brunswick. Income, education, and urban/rural distribution data were also collected and tracked for alignment with Statistics Canada national proportions (see details in Comeau et al. 2015). A full version of the study questionnaire is available through the project Dataverse (2017) online portal. In addition to the national survey described above, we gathered a separate sample of New Brunswick residents (n = 500) via the same online panel, because the NB share of the national sample was only seventy-one. This additional sample offered an opportunity to explore views on energy development in the province in more detail, with attention to a proposed hydroelectric dam redevelopment (Sherren et al. 2017). For this chapter, the national survey and the New Brunswick survey allow comparison between the discourses that were initially identified in the in-person Q method studies and average support or opposition with individual statements from both in-person and online Q, at provincial and national scales.

Comparing Q Sort and Survey Results

Starting with results of the Q sort research in three provinces, five factors identified across the three regions were based on all fifty-eight Q sorts, but there were regional differences where the factor-defining sorts derived (table 6.2). New Brunswick was the source of just over half (8) of the 15 factor-defining sorts that generated factor 1, "climate change is a primary concern." Alberta dominated factors 2 ("maintain the energy economy"; 4 out of the 7 defining sorts) and 3 ("build on resilience of nature and local energy systems"; 3 out of 4). Ontario dominated factors 4 ("markets and corporations will lead"; 2 out of 4) and 5 ("renewables are the path forward"; 3 out of 3). We can thus generate average scores for just New Brunswick on each of the statements based on Q sorts and expect some differences from the national results. Looking specifically at results in table 6.2, analysis indicates some consistencies, such as strong support for the statement "Companies must take responsibility and pay for the pollution they produce" as well as the rejection of claims such as "Most Canadians are well aware of the environmental impacts of energy development." Across the five discourses that were identified in our research, these statements were almost uniformly supported or rejected. In other words, knowledgeable people within their regional settings held consistent views on these issues. For other statements, there was much less consensus. For example, Q sort participants who defined factor 5 ("renewables are the path forward") strongly supported the statement "We need to find ways to develop untapped resources for renewable energy," but this statement was only moderately supported by others. Similarly, the statement "Nature will be fine no matter what humans do; it is a robust, self-correcting system" was strongly supported by participants who defined factor 3, but was resoundingly rejected by all other Q sort participants (table 6.2). Analysis of these

TABLE 6.2. Q statements for factor scores,[a] sorted by average and identified for distinct factor as follows: (1) climate change is a primary concern, (2) maintain the energy economy, (3) build on resilience of nature and local energy systems, (4) markets and corporations will lead, (5) renewables are the path forward

Statements	*National Factor Average*	*Factor*[b]				
		1	2	3	4	5
Companies must take responsibility and pay for the pollution they produce.	4.0	4	4	4	5	3
Market forces, not incentives or taxes, drive development and conservation of energy.	2.2	−1	3	2	4	3
We need to find ways to develop untapped resources for renewable energy.	2.0	2	2	1	0	5
Canadians have a duty to be global leaders by reducing our own energy consumption.	1.6	2	4	3	−2	1
Climate change poses a grave and urgent threat to our planet.	1.6	5	0	0	−1	4
Current trends in energy consumption are clearly unsustainable and must be reduced immediately.	1.0	4	0	0	−1	2
Growth in energy production is key to Canada's economic progress.	0.6	−2	5	−1	2	−1
Small and distributed energy sources are more resilient than centralized production.	0.4	1	−2	3	1	−1
Consuming too much energy is immoral.	0.0	0	−1	−2	−1	4
The local community should decide on the energy systems that are best for them.	0.0	−1	−3	2	1	1
All forms of energy should be more expensive.	−1.4	0	−2	−1	−4	0
Canada's commitment to democracy makes it an ethical supplier of energy.	−1.6	−3	0	−4	0	−1
The current fad for "going green" will accomplish nothing.	−1.8	−2	−3	−1	1	−4
Nature will be fine no matter what humans do; it is a robust, self-correcting system.	−2.0	−5	−5	5	−2	−3
A high level of energy consumption is part of the good life.	−2.2	−3	−2	−3	−2	−1
Canada's greenhouse gas emissions are justified because they are tiny compared to other countries.	−2.4	−4	−3	−2	0	−3
Only the world's poorest people will suffer from climate change.	−3.2	−3	−4	−5	−5	1
Most Canadians are well aware of the environmental impacts of energy development.	−3.2	−3	−2	−3	−3	−5

continued on next page

TABLE 6.2—*continued*

Statements	*National Factor Average*	*Factor*[b] 1	2	3	4	5
Number of Q sorts by province.	Total	Defining sorts, by factor				
Number of NB factor–defining sorts.	21	8	2	1	1	0
Number of ON factor–defining sorts.	21	4	1	0	2	3
Number of AB factor–defining sorts.	16	3	4	3	1	0

[a] Factor scores indicate how each statement was ranked (+5 to −5) within the "ideal" Q sort representing each discourse.

[b] Each factor represents a discourse composed of a group of participants who sorted the statements in a similar way.

results indicates that distinct discourses do exist within Canada. From this analysis alone, however, the prevalence of these discourses across Canada cannot be determined. The generalizability of these results to the broader population is thus limited.

To address these limitations in Q methodology, we provide results from the national and New Brunswick survey and how they relate to Q sort results. Within table 6.3, overall levels of agreement and disagreement with the eighteen statements are mostly consistent across the survey and Q sort results. For instance, on average, the most strongly supported statement in all datasets related to corporate responsibility: "Companies must take responsibility and pay for the pollution they produce." There are, however, several inconsistencies. On the national survey, with an average score of 2.8 (on a scale of 1 to 5), Canadians are somewhat neutral (neither showing support nor opposition) toward the claim that "most Canadians are well aware of the environmental impacts of energy development." This result contrasts with the Q method results showing the strongest disagreement with that particular statement (−3.2, on a scale from −5 to 5). Similarly, in response to the statement that "market forces, not incentives or taxes, drive development and conservation of energy," we observe a discrepancy between the national survey and the Q sort. On average, participants in the Q sort lent more support for this statement (2.2) than was reflected in the Likert-based national survey (3.2).

Leaving aside these small differences, on a whole, the similarities between the Q sort results and the national survey results are striking. The scatterplot in figure 6.1 illustrates this commonality with a regression line that fits

TABLE 6.3. Factor score averages and survey averages for Q sort statements, reported for the national sample and the New Brunswick sample

Statements	*National Factor Average*[a] n = 58	*National Survey Average*[b] n = 3000	*New Brunswick Factor Average*[a] n = 86	*New Brunswick Survey Average*[b] n = 500
Companies must take responsibility and pay for the pollution they produce.	4.0	4.3	3.4	4.3
We need to find ways to develop untapped resources for renewable energy.	2.0	4.1	2.6	4.1
Canadians have a duty to be global leaders by reducing our own energy consumption.	1.6	3.9	1.4	4.0
Climate change poses a grave and urgent threat to our planet.	1.6	3.9	2.0	3.9
Current trends in energy consumption are clearly unsustainable and must be reduced immediately.	1.0	3.6	1.2	3.6
Growth in energy production is key to Canada's economic progress.	0.6	3.5	1.2	3.5
Consuming too much energy is immoral.	0.0	3.4	−0.8	3.5
The local community should decide on the energy systems that are best for them.	0.0	3.3	0.2	3.4
Market forces, not incentives or taxes, drive development and conservation of energy.	2.2	3.2	0.4	3.3
Small and distributed energy sources are more resilient than centralized production.	0.4	3.1	−0.6	3.2
Canada's commitment to democracy makes it an ethical supplier of energy.	−1.6	2.9	−0.6	2.9
Most Canadians are well aware of the environmental impacts of energy development.	−3.2	2.8	−1.4	2.8
The current fad for "going green" will accomplish nothing.	−1.8	2.7	−2.8	2.5

continued on next page

TABLE 6.3—*continued*

Statements	*National Factor Average*[a] n = 58	*National Survey Average*[b] n = 3000	*New Brunswick Factor Average*[a] n = 86	*New Brunswick Survey Average*[b] n = 500
A high level of energy consumption is part of the good life.	−2.2	2.4	−2.6	2.3
Canada's greenhouse gas emissions are justified because they are tiny compared to other countries.	−2.4	2.4	−2.6	2.3
All forms of energy should be more expensive.	−1.4	2.3	−3.0	2.1
Nature will be fine no matter what humans do; it is a robust, self-correcting system.	−2.0	2.1	−4.0	2.0
Only the world's poorest people will suffer from climate change.	−3.2	2.1	−3.0	1.9

[a] Factor average indicates the average factor score for Q sort participants, ranked from −5 to +5, where 0 is neutral.

[b] Survey average indicates the average score for survey participants, on a scale from 1 (strongly disagree) to 5 (strongly agree), where 3 is neutral.

the national data with an R^2 of 0.8. These results from the national survey are also echoed in the provincial dataset from New Brunswick. We observe an even stronger fit between the two methods in New Brunswick, with an R^2 of 0.94. Like the national analysis, however, there are slight differences in the New Brunswick averages. For example, support for the statement "Canadians have a duty to be global leaders by reducing our own energy consumption" is slightly stronger in the New Brunswick survey compared to the New Brunswick Q sort.

Beyond the similarities and differences between methods, results also show consensus around key ideas for energy development in Canada. Statements that are most strongly supported by Canadians (table 6.3) include the desire to see companies take responsibility for the pollution they produce, the need to develop untapped renewable energy resources, and the expectation that Canada will be a global leader in reducing energy consumption. Conversely, the ideas that are most strongly opposed relate to the world's poor being alone in suffering from climate change, the idea that nature will be fine, no

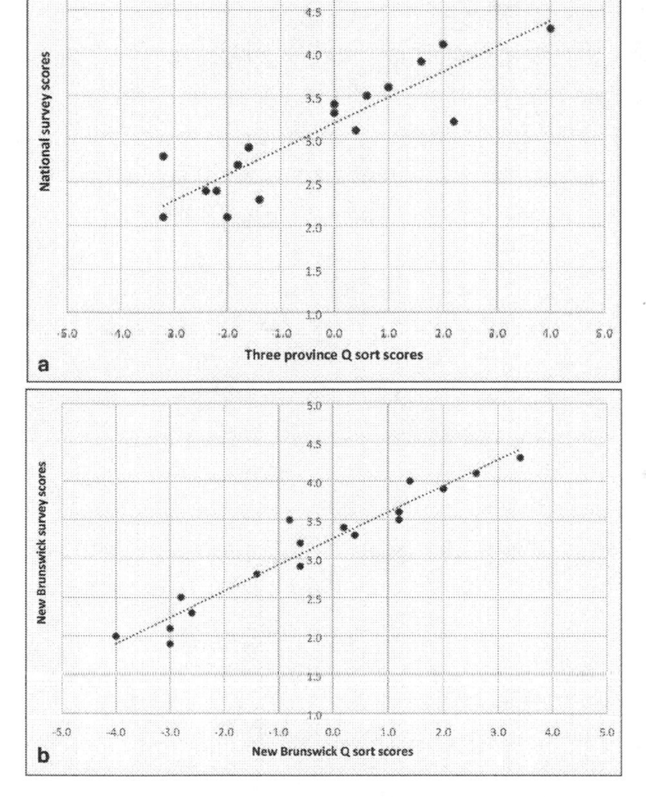

FIGURE 6.1. Average factor scores and survey scores for discourse statements on the National Survey, New Brunswick Survey (vertical scaled 1 to 5), and Q Sorts in three provinces and New Brunswick alone (horizontal scale −5 to +5).

matter what humans do, and the idea that all energy should be more expensive. These statements offer insight into the energy imaginaries of Canada. Broadly defined, they reflect the expectations and aspirations of Canadians about the future of energy development.

Implications for Discourse Analysis across Scales

As noted in the introduction, this chapter seeks to advance methodological innovations to identify variation in discourses across regions and implications for the emergence of low-carbon energy. Returning to our guiding questions from the introduction, we provide answers to these questions in the following two subsections.

Are There Ways of Gaining Local Insight While Also Maintaining Comparability across Larger Populations?

Our approach involves using Q methodology to identify statistically distinct discourses within a regional group of key informants combined with a probability sample to test the generalizability of discourses within a population. Using this mixed method approach, we maintain nuanced insight at the local scale while gaining some generalizability across scales of inquiry.

More specifically, results from this study suggest that a singular and overarching set of governing logics, a national discourse, and associated tensions within the discourse are present and observable across national and regional scales. The Q method concourse of statements was technology neutral, but with the backdrop of hydrocarbon development in Alberta, wind and nuclear development in Ontario, and hydroelectricity development in New Brunswick, we expected some of this geographic, cultural, and technological diversity to emerge in our analysis. What we observe in the data, however, is a high degree of alignment between regional discourses and national discourses. Key ideas about energy development are ranked almost exactly the same nationally as in New Brunswick. The national discourse on energy development in Canada tends to overwhelm all of the variation that might exist within a local setting.

From a methodological perspective, we find that the small sample of fifty-eight research participants whose Q sorts were used to identify the initial list of eighteen statements (i.e., high and low valence) offers a statistically robust reflection of the national context. With minimal exceptions, Q method insights are consistent with survey data. This outcome was found similarly by Kate Sherren, Logan Loik, and James A. Debner, whose small-*n* face-to-face Q sort pilot about coastal wetlands produced the same dominant factors as an $n = 183$ online Q sort, even though some statements were changed (2016). This is an important finding, in part, because one noted weakness of Q method studies is the lack of generalizability. Making a clear distinction between surveys and Q methodology, Stentor Danielson (2009) notes that conventional survey techniques have advantages over Q methodology. These advantages include generalizability to a population and explanation of results using other variables from the survey. Q methodology, on the other hand, has strengths in terms of generalizing the structures of a topic, "a research method capable of identifying the currently predominant social *viewpoints* and knowledge structures relative to a chosen subject matter" (Watts and Stenner 2012, 42; emphasis in original). Notwithstanding the noted lack of generalizability from Q method studies, the Q sort within New Brunswick with eighty-six participants accurately maps onto the results from our national and New Brunswick surveys (figure 6.1). This is a remarkable outcome, suggesting that Q method results are more generalizable than typically believed.

A final point on methodology, the cognitive work required to conduct a Q sort with a forced choice set of forty-eight statements appears to be similar to the cognitive work required for Likert-type scales. As illustrated in table 6.1, the standard procedure for conducting Q sorts requires that participants sort statements whereby a higher proportion of statements are located in the neutral categories (i.e., −1 to +1), and a smaller proportion are located in the extreme ends of the distribution (i.e., −5 or +5). Careful work by participants, sorting statements in relation to each other, and selecting a final Q sort on the surface appears to be quite different from a Likert scale response, where any number of statements within a list is supported or opposed. The similarities in responses by our participants in the survey and the Q sort study, however, give us reason to wonder how these exercises provide similar results. Our results demonstrate that people approach the statements in similar ways when sorting or doing Likert scales. Results also demonstrate that sample size for the elicitation of discourses and the generalizability of discourses may not be as important as one might think.

What Are the Implications for Transition to Low-Carbon Energy Futures?

Based on Malm's notion of hegemonic ideologies as all encompassing and overwhelming of other ways of thinking about and knowing the world, there are several ways of interpreting the results of this study; in particular, the uniformity of energy discourses across region and scale within Canada. On one hand, we know from case studies of energy discourses that ideas about energy development can be context specific, reflecting local culture and the emergence of energy technologies that are contending for dominance within a region. Support for wind energy in west Texas is a case in point, with highly context-specific discourses emerging around this form of renewable energy (Jepson, Brannstrom, and Persons 2012). The Texas case suggests that dominant discourses are fluid, adaptable and changing according to local context. On the other hand, these local variations in discourses may be subjected to a more overarching set of governing logics, ways of thinking and knowing about the world that are consistent with dominant ideas about capital, markets, corporate actors, and collective responsibilities.

Within the Canadian context, what are the governing logics associated with energy development? In part, these logics reflect how we see ourselves as Canadians, our sense of common practice, and our sense of legitimacy as a nation. With relatively strong and enduring national media such as the Canadian Broadcasting Corporation, and national daily papers such as the *Globe and Mail* and the *National Post*, our sense of nationhood is deeply connected to our history as a "staples state" (Brownsey and Howlett 2008) and our growing obsession with energy production in all forms. These ideas about abundance and inexhaustible natural resources continue to feed our collective culture in a way that some analysts identify as our "founding myths" (Klein 2016). Therefore, a national consensus on energy discourses should not be surprising. We Canadians see a role for energy companies to take leadership and responsibility for sustainable energy development, we see tapping new energy resources is a priority, and we show support for reducing energy consumption reflects collective responsibility and leadership, but we also know that strong policies to address climate change will hurt more than just the world's poor; we will all feel the impacts. These are some elements of a dominant discourse on energy development in Canada, overshadowing discourses that might exist within a local or regional context. The extent to which this analysis is accurate suggests that energy cultures in Canada reflect established and enduring relations of power and acquiescence toward dominant political and economic interests.

Conclusion

A point of focus in this chapter involves mixed methods to identify and understand linkages between discourses across scales. Given the strengths of Q methodology in gathering structured viewpoints on discourses within a group, and strengths of surveys to generalize research across larger populations, it makes sense to combine these methods as a tool for cross-scale analysis. Although this study resulted in similar outcomes between regional and national scales, other studies with different topics, different Q sort statements, and different geographies may reveal other insights. These methods may give opportunities to test the claims of scholars such as Malm (2016) that discourses around energy development are almost entirely trapped by ideologies of neoliberalism and global logics of fossil capital. For those who are not familiar with discourse analysis, we propose in this chapter that discourse analysis is more than variable analysis of attitudes and preferences for energy technologies. If the fossil economy has indeed "constituted the subject, who cannot see himself outside of it" (Malm 2016, 365), then any study of energy impacts will require some attention to the impacts of discourses on our collective abilities to imagine emergent low-carbon energy futures. We also see value in this work on discourses as a way to identify common ground between often competing interests but also as a way to surface new ideas and new hopes and aspirations for energy development that can lead to low-carbon futures.

References

Barry, John, John S. Dryzek, and David Schlosberg, eds. 2005. *Debating the Earth: The Environmental Politics Reader*. Oxford: Oxford University Press.

Boudet, Hilary, Christopher Clarke, Dylan Bugden, Edward Maibach, Connie Roser-Renouf, and Anthony Leiserowitz. 2014. "'Fracking' Controversy and Communication: Using National Survey Data to Understand Public Perceptions of Hydraulic Fracturing." *Energy Policy* 65 (February): 57–67.

Brownsey, Keith, and Michael Howlett, eds. 2008. *Canada's Resource Economy in Transition: The Past, Present, and Future of Canadian Staples Industries*. Toronto: Emond Montgomery Publications.

Cameron, Lachlan, and Bob Van Der Zwaan. 2015. "Employment Factors for Wind and Solar Energy Technologies: A Literature Review." *Renewable and Sustainable Energy Reviews* 45 (C): 160–172.

Clare, Shari, Naomi Krogman, and Ken J. Caine. 2013. "The 'Balance Discourse': A Case Study of Power and Wetland Management." *Geoforum* 49 (October): 40–49.

Comeau, Louise A., John R. Parkins, Richard C. Stedman, and Thomas M. Beckley. 2015. *Citizen Perspectives on Energy Issues in Canada: A National Survey of Energy Literacy and Energy Citizenship. Department of Resource Economics and Environmental Sociology, Project Report #15–01*. Edmonton: University of Alberta.

Cotton, Matthew. 2015. "Stakeholder Perspectives on Shale Gas Fracking: A Q-method Study of Environmental Discourses." *Environment and Planning A* 47 (9): 1944–1962.

Danielson, Stentor. 2009. "Q Method and Surveys: Three Ways to Combine Q and R." *Field Methods* 21 (3): 219–237.

Dataverse. 2017. *Energy Transitions Project Portal*. Alberta: University of Alberta. https://doi.org/10.7939/DVN/10302.

Davidson, Debra J. 2017. "Evaluating the Effects of Living with Contamination from the Lens of Trauma: A Case Study of Fracking Development in Alberta, Canada." *Environmental Sociology* 4 (2): 196–209.

Douglas, Mary, and Aaron Wildavsky. 1983. *Risk and Culture: An Essay on the Selection of Technological and Environmental Dangers*. Berkeley: University of California Press.

Dryzek, John S., and Jeffrey Berejikian. 1993. "Reconstructive Democratic Theory." *American Political Science Review* 87 (1): 48–60.

Finewood, Michael H., and Laura J. Stroup. 2012. "Fracking and the Neoliberalization of the Hydro-social Cycle in Pennsylvania's Marcellus Shale." *Journal of Contemporary Water Research and Education* 147 (1): 72–79.

Hastings-Simon, Sara and Simon Dyer. 2017. "Pembina: Reality Check. We Can't Turn Back Time on Alberta's Climate Policy." *Calgary Herald*, August 5, 2017.

Hempel, Anna. 2017. *Planning for Change in Rural Ontario: Using Visual Q-methodology to Explore Landscape Preference*. PhD diss., University of Guelph. http://hdl.handle.net/10214/10311.

Hirsh, Richard F., and Benjamin K. Sovacool. 2013. "Wind Turbines and Invisible Technology: Unarticulated Reasons for Local Opposition to Wind Energy." *Technology and Culture* 54 (4): 705–734.

Horne, Christine, and Emily Huddart Kennedy. 2017. "The Power of Social Norms for Reducing and Shifting Electricity Use." *Energy Policy* 107 (August): 43–52.

Jackson, Robert B., Avner Vengosh, J. William Carey, Richard J. Davies, Thomas H. Darrah, Francis O'Sullivan, and Gabrielle Pétron. 2014. "The Environmental Costs and Benefits of Fracking." *Annual Review of Environment and Resources* 39 (October): 327–362.

Jacquet, Jeffrey B. 2012. "Landowner Attitudes toward Natural Gas and Wind Farm Development in Northern Pennsylvania." *Energy Policy* 50 (November): 677–688.

Jepson, Wendy, Christian Brannstrom, and Nicole Persons. 2012. "'We Don't Take the Pledge': Environmentality and Environmental Skepticism at the Epicenter of US Wind Energy Development." *Geoforum* 43 (4): 851–863.

Klein, Naomi. 2016. "Canada's Founding Myths Hold Us Back from Addressing Climate Change." *Globe and Mail*, September 23, 2016.

Labanca, Nicola, ed. 2017. *Complex Systems and Social Practices in Energy Transitions: Framing Energy Sustainability in the Time of Renewables*. New York: Springer.

Lakoff, George. 2008. *The Political Mind: Why You Can't Understand 21st-Century Politics with an 18th-Century Brain*. New York: Penguin.

Leahy, Terry, Vanessa Bowden, and Steven Threadgold. 2010. "Stumbling towards Collapse: Coming to Terms with the Climate Crisis." *Environmental Politics* 19 (6): 851–868.

Leiserowitz, Anthony. 2006. "Climate Change Risk Perception and Policy Preferences: The Role of Affect, Imagery, and Values." *Climatic Change* 77 (1): 45–72.

Lis, Aleksandra, and Piotr Stankiewicz. 2017. "Framing Shale Gas for Policy-Making in Poland." *Journal of Environmental Policy and Planning* 19 (1): 53–71.

Loder, Thomas. 2016. "Spaces of Consent and the Making of Fracking Subjects in North Dakota: A View from Two Corporate Community Forums." *The Extractive Industries and Society* 3 (3): 736–743.

Malm, Andreas. 2016. *Fossil Capital: The Rise of Steam Power and the Roots of Global Warming*. London: Verso Books.

Parkins, John R., Christy Hempel, Thomas M. Beckley, Richard C. Stedman, and Kate Sherren. 2015. "Identifying Energy Discourses in Canada with Q Methodology: Moving beyond the Environment versus Economy Debates." *Environmental Sociology* 1 (4): 304–314.

Robbins, Paul. 2007. *Lawn People: How Grasses, Weeds, and Chemicals Make Us Who We Are*. Philadelphia: Temple University Press.

Sherren, Kate, Thomas Beckley, Simon Greenland-Smith, and Louise Comeau. 2017. "How Provincial and Local Discourses Aligned against the Prospect of Dam Removal in New Brunswick, Canada." *Water Alternatives* 10 (3): 697–723.

Sherren, Kate, Logan Loik, and James A. Debner. 2016. "Climate Adaptation in 'New World' Cultural Landscapes: The Case of Bay of Fundy Agricultural Dykelands (Nova Scotia, Canada)." *Land Use Policy* 51 (February): 267–280.

Shove, Elizabeth. 2017. "Energy and Social Practice: From Abstractions to Dynamic Processes." In *Complex Systems and Social Practices in Energy Transitions*, edited by Nicola Labanca, 207–220. New York: Springer.

Stephenson, Janet, Debbie Hopkins, and Adam Doering. 2015. "Conceptualizing Transport Transitions: Energy Cultures as an Organizing Framework." *Wiley Interdisciplinary Reviews: Energy and Environment* 4 (4): 354–364.

Stewart, Janet. 2014. "Sociology, Culture, and Energy: The Case of Wilhelm Ostwald's 'Sociological Energetics'—A Translation and Exposition of a Classic Text." *Cultural Sociology* 8 (3): 333–350.

Strauss, Sarah, Stephanie Rupp, and Thomas Love, eds. 2013. *Cultures of Energy: Power, Practices, Technologies*. New York: Routledge.

Swidler, Ann. 1986. "Culture in Action: Symbols and Strategies." *American Sociological Review* 51 (April): 273–286.

Taylor, Charles. 2007. *A Secular Age*. Cambridge, MA: Harvard University Press.

Watts, Simon, and Paul Stenner. 2012. *Doing Q Methodological Research: Theory, Method and Interpretation*. Thousand Oaks: Sage.

Wolsink, Maarten, and Sylvia Breukers. 2010. "Contrasting the Core Beliefs regarding the Effective Implementation of Wind Power: An International Study of Stakeholder Perspectives." *Journal of Environmental Planning and Management* 55 (5): 535–558.

Chapter 6 Summary

As renewable energy technologies evolve, how we think and talk about energy landscapes is also changing. Energy discourses shape our thinking, our reactions, and our sense of what is desirable or undesirable in the surrounding landscape. Understanding discourses as the subconscious organization of collectively held values and mental models (Lakoff 2008), this chapter advances methodological innovations to identify subtle variations in discourses on energy production within and between regions of Canada. Two questions provide guidance for this study. Given that discourses on energy development are often context specific, are there ways of gaining local insight while also maintaining comparability across larger populations? What are the implications of better understanding regional- and national-level discourses on energy development for transition to low-carbon energy futures?

To answer these questions, we present results from mixed methods involving Q methodology and a national survey of Canadian citizens. Ideas about norms are particularly relevant to this chapter, especially as they relate to personal expectations, aspirations, and a general sense of what is allowable and acceptable within energy landscapes. Furthermore, an analysis of discourses can offer insight into the ideologically mediated encounters with energy landscapes and the ways in which our ideas, expectations, and aspirations can lock in a set of assumptions and preferences for future energy systems. As Malm's (2016) historical analysis reminds us, however, culture and discourse offer tools to imagine something new, something other than business-as-usual, and perhaps some indication of how change can become imaginable. Building on these concepts, our results suggest that a singular and overarching set of governing logics, a national discourse, and associated tensions within the discourse are present and observable across national and regional scales. What we observe in the data (in spite of significant Canadian regional differences in energy technologies and policies) is a high degree of alignment between regional discourses and national discourses. The national discourse on energy development in Canada tends to overwhelm all of the variation that might exist within a local setting. We conclude with a discussion about overarching governing logics and ways of thinking and knowing about the world that are consistent with dominant ideas about capital, markets, corporate actors, and collective responsibilities.

KEY TAKEAWAYS

- Mixed methods include a national survey of energy discourses and Q methodology conducted in three Canadian provinces (Alberta, Ontario, and New Brunswick).
- With possibilities for diverse points of view to emerge within regions where energy technologies and policies differ, our study finds limited variation in discursive material across these regions.
- We conclude that discourses around energy developments are almost entirely trapped by ideologies of neoliberalism and global logics of fossil capital.

7

A Capitals Approach to Biorefinery Siting Using an Integrative Model

SANNE A. M. RIJKHOFF, KELLI ROEMER, NATALIE MARTINKUS,
TAMARA J. LANINGA, AND SEASON HOARD

Introduction

Biorefinery site selection is critical to producing economically viable and environmentally sustainable biofuels. Facility siting requires significant financial investments and long-term strategic commitments. These decisions are primarily driven by economic factors, such as access to labor and markets, raw materials, and cost-reducing local incentives (Noon, Zhan, and Graham 2002). Economic criteria will always be central to site selection decisions. However, nontechnical considerations are also critical and can be among the largest obstacles to successful high-tech facility siting (Plate, Monroe, and Oxarart 2010; Rösch and Kaltschmitt 1999; White 2010).

Understanding local communities' influence on the outcomes of biorefinery siting is gaining traction, but it is difficult to measure and incorporate into the site selection process. In cases where bioenergy projects failed despite positive feasibility studies, Christine Rösch and Martin Kaltschmitt identified a number of issues: funding, financing, and insuring problems; unfavorable administrative conditions; organizational difficulties; limited

DOI: 10.5876/9781646420278.c007

knowledge and information; and lack of public acceptance (1999). Leann M. Tigges and Molly Noble found that securing community support can lower implementation costs (2012). Similarly, Arifa Sultana and Amit Kumar (2012) assumed that inclusion of strong socioenvironmental considerations in facility siting would enhance project social acceptability, and Miet Van Dael et al. (2012) used proxies to estimate support for a bioenergy project, including unemployment numbers, community acknowledgment of the Kyoto protocol, and unutilized industrial areas. However, community support is only one among several assets necessary to sustain biorefineries. The presence of social assets, in sufficient levels, can aid project development, implementation, and sustainability (Martinkus et al. 2017b; Rijkhoff et al. 2017). This presence is not only beneficial for siting from the developer's point of view, but the community benefits as well. Facility development can lead to increased job opportunities, tax revenues, and local infrastructure improvements (Cambero and Sowlati 2016). It is therefore critical to engage communities in the decision-making process, especially in controversial decisions such as energy development, to address potential public suspicion or distrust of the process. Collecting community input and involving community members in the decision process will do much to enhance project understanding and cooperation (Parks, Joireman, and Van Lange 2013). In this chapter, we discuss how decision support tools can enhance industry and public understanding, thus aiding site selection decisions.

Our interdisciplinary team has developed a decision support tool (DST) based on multicriteria decision analysis to aid high-tech facility siting decisions, including biorefineries (Martinkus et al. 2017b), where use of such tools is increasing (see Perimenis et al. 2011). The tool was developed with the Northwest Advanced Renewables Alliance (NARA), a group examining aviation biofuels and co-products supply chains using postharvest forest residues (e.g., slash).[1] Although the DST was developed for, and is applied to, siting biorefineries in the Pacific Northwest (PNW) region of the United States, the nationwide, county-level datasets used in the tool make it flexible enough to explore site selection options for a wide range of high-tech industries.

Our DST combines traditional economic siting criteria with underutilized social assets, particularly critical for renewable energy projects such as forest biorefineries (Cambero and Sowlati 2014). The social assets dataset, developed by Rijkhoff et al., includes social, cultural, and human capitals (2017).

Our multicriteria DST converts disparate qualitative and quantitative data into a consistent quantitative dataset to identify suitable biorefinery sites through weighting economic, physical, and social criteria by importance (Wang et al. 2009). The multicriteria DST is not meant to identify the right site, but narrow a large list of potential facilities or sites to a few top candidates for further investigation.

The development and implementation of the DST contribute to the discussion of how social and technical issues related to energy production interact. Our approach goes beyond the dimensions of technology and economics to include essential social and human elements. Our innovative DST involves quantitative indicators for economic and social criteria that can be used to assess facility siting options both regionally and nationally. Traditionally, inclusion of social assets in decision-making processes is obtainable only using costly research methods. As such, our DST provides a powerful tool for initial assessment for industry and policy makers. It can be applied to energy or other siting issues, adding a more holistic understanding of communities to meet their shared goals.

The objectives of this chapter are to (1) describe the economic and social criteria used in the multicriteria biorefinery siting DST; (2) validate the social asset dataset with case study analyses of four high-tech facility siting cases; and (3) demonstrate how the DST uses nationally available, county-level data to develop a rough ranking of, in our example, forest biorefinery facility siting options. We conclude the chapter with insights and recommendations for siting decisions based on multicriteria decision making, aided by DSTs.

Methods

The development of the DST represents an interdisciplinary effort, utilizing a sequential mixed-method integrated design to inform site selection decision-making processes. In this section, we briefly describe the methodology used in developing and validating components of the multicriteria DST.

Community Capitals Framework

We use the Community Capitals Framework (CCF) to guide the development of the siting criteria and indicators. This systems framework identifies

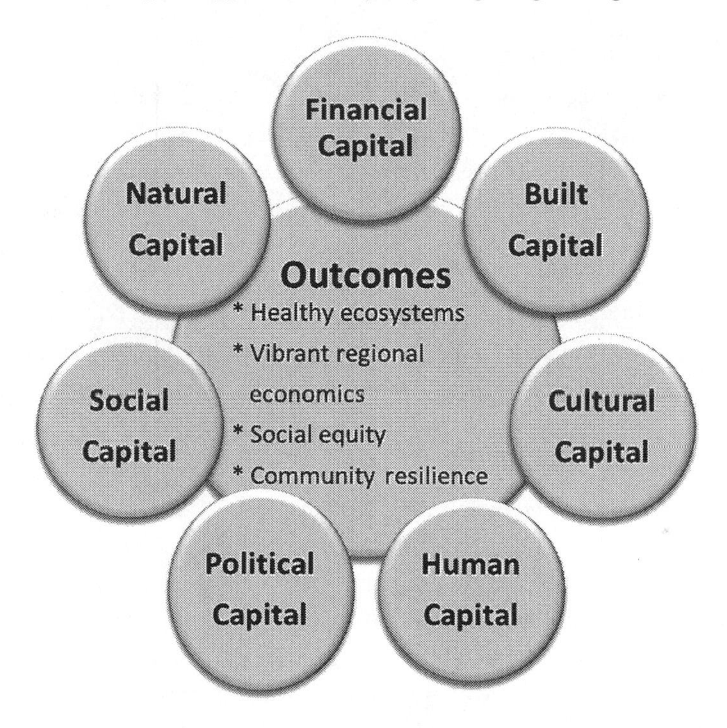

FIGURE 7.1. Community capitals framework. Source: Rijkhoff et al. 2017, based on Emery and Flora 2006, 21.

community assets as seven capitals: natural, built, social, cultural, human, financial, and political (figure 7.1) (Emery and Flora 2006). The CCF allows us to analyze community suitability more holistically for site selection. Six of the seven capitals are included in the DST; the exception is political capital, which is undeniably important but requires a more in-depth, site-specific analysis.[2]

To incorporate the CCF in the DST, two teams of researchers—representing economic, geospatial, political, and social sciences—conducted extensive literature reviews of multiple disciplines to identify potential quantitative indicators reflecting each capital. The teams also researched publicly and privately held secondary datasets that are available for all of the United States and at the county level, a sufficient scale to allow for regional comparison. After identifying indicators and datasets, the research teams created measures used in the DST. We describe the criteria and indicator development in "Developing the Indicators and Measures."

Case Studies of Social Assets

While extensive research and applications are available for quantitative indicators of natural, built, and financial capital in site selection, quantitative indicators of social assets are still relatively developmental and incomplete in site selection research. Further research is required to validate the social asset indicators and identify opportunities to improve the initial methodology for future DSTs. To examine social assets, we conducted case studies in four communities where high-tech biomaterial facilities were proposed or constructed through twenty-one semistructured phone interviews with stakeholders between January and March 2017. Participants were identified through purposive and snowball sampling methods to include a range of perspectives. They represent key actors involved in the facility implementation process, with knowledge beyond the general population.

Interview data were supplemented with secondary sources, including local newspaper articles, public meeting minutes, public outreach and information meeting videos, feedstock analysis plans, policy documents, and local internet sources for community history and background. All interview transcripts were coded thematically to ensure consistency throughout the cases. Two independent researchers coded a sample of the findings to ensure intercoder reliability. There was substantial agreement between raters, $K = 0.68$ ($p < .001$), 95% CI (0.52609, 0.82792).

DST and Application

The DST translates the CCF indicators into biorefinery siting criteria. Each criterion is assigned a weight and range of scale values. Weights define the relative importance of each criterion, and scale values provide a means for assessing existing facilities based on location-specific values relative to the range of values present. A higher scale value means a facility's location-specific value provides a lesser cost to operate the biorefinery than the other facilities. Each facility receives a score based on how well its assets provide for reduced costs; we favor repurposing existing facilities over greenfield development as a means to reduce capital expenditures (Martinkus and Wolcott 2017). The highest scoring facility would theoretically cost the least to repurpose and operate as a forest biorefinery. Both economic and social criteria are included in the DST as separate modules with separate weights; however,

TABLE 7.1. Decision support tool framework

	Economic Metric			*Social Metric*	
	Criterion 1 (C1)	*Criterion 2 (C2)*	*Criterion n (Cn)*	*Criterion 1 (C1)*	*Criterion n (Cn)*
SCALE, *s*					
5	*amax*	*bmin*	*cmin*	*amax*	*bmin*
4	*amax* − *B1*	*bmin* + *B2*	*cmin* + *Bn*	*amax* − *B1*	*bmin* + *Bn*
3	*amax* − 2*B1*	*bmin* + 2*B2*	*cmin* + 2*Bn*	*amax* − 2*B1*	*bmin* + 2*Bn*
2	*amax* − 3*B1*	*bmin* + 3*B2*	*cmin* + 3*Bn*	*amax* − 3*B1*	*bmin* + 3*Bn*
1	*amax* − 4*B1*	*bmin* + 4*B2*	*cmin* + 4*Bn*	*amax* − 4*B1*	*bmin* + 4*Bn*
Weight	*w1*	*w2*	*wn*	*w1*	*wn*

the scores from each are combined to create one final facility score. Table 7.1 illustrates the general form of the DST.

We apply the DST to locating a forest biorefinery in the PNW. The region is well suited to hosting such a facility, with existing physical infrastructure and ample biomass to meet the annual feedstock requirement. Ten active or recently decommissioned pulp mills are assessed for their repurpose potential as a forest biorefinery.

Developing the Indicators and Measures

DST development utilized mixed methods and two research teams to develop the tool's criteria guided by the CCF. The economic and geospatial team led identification of natural, built, and financial capitals, while the social asset team focused on measures for social, cultural, and human capital. Table 7.2 shows the multicriteria dataset used in the DST (Martinkus et al. 2017b).

Economic Criteria Development

Economic site selection criteria are identified from a biorefinery's technoeconomic analysis (TEA). The TEA delineates capital and operating costs for all processing units within the biorefinery and determines the minimum fuel selling price. Some cost components vary geospatially based on facility location (e.g., electricity rates, feedstock cost). By using these economic costs as siting criteria, we can identify sites that may operate at the least cost.

TABLE 7.2. Overview of indicators used in the decision support tool (DST)

Assets	*Community Capitals*	*Indicators*
Economic and Physical Criteria	*Natural capital* Source: Martinkus et al. (2017a)	Annual biomass availability
	Built capital Source: Esri (2017); Martinkus and Wolcott (2017)	Multimodal transportation (as part of biomass availability) Facility repurpose potential
	Financial capital Source: US Energy Information Administration (2014a)	Electricity/natural gas rate Labor costs (wages)
Social Assets	*Social capital* Source: Rupasingha, Goetz, and Freshwater (2006) (2009 data)	# Rent-seeking groups: political, labor, professional, and business organizations # Non-rent–Seeking Groups: civic organizations, bowling centers, golf clubs, fitness centers, sports organizations, and religious organizations # Nonprofit organizations % Voter turnout
	Cultural capital Source: WESTAF (2010)	# Arts-related organizations # Arts-related business # Occupational employment in the arts $ Revenues of arts-related goods and services
	Human capital Source: County Health Rankings (The Robert Wood Johnson Foundation 2013)	Health: % Low birthrate % Premature deaths % Obese (BMI > 30) % Self-reports of poor health condition (physically and mentally)

Note: All counts (#) and amounts ($) are calculated as a rate of the population per 10,000.

NATURAL CAPITAL

Natural capital is represented by the amount of feedstock available to a facility at a given cost. In our case, we assess the availability of postharvest forest residue as the feedstock for a forest biorefinery producing aviation biofuel. Forest residue volumes available at discrete locations throughout the PNW are determined as twenty-year averages (Martinkus et al. 2017a).

BUILT CAPITAL

Built capital is reflected in both the road transportation system and the existing facility infrastructure assessed for its repurpose potential. The forest residue volumes are routed along road networks utilizing least-cost routing algorithms to determine the least cost path between each biomass source point and industrial facility. The sum of all transportation costs are aggregated over all source points until annual biorefinery demand is met (Martinkus et al. 2017a). The total weighted average delivered feedstock cost to meet annual biorefinery demand is determined for each existing industrial facility and is used as a siting criterion.

The capital cost to construct a biorefinery may be reduced if existing facility infrastructure and assets can be repurposed. Existing facilities are assessed against the requirements of a greenfield biorefinery for their infrastructure compatibility to estimate the capital percent savings realized through repurposing (e.g., Martinkus and Wolcott 2017). This criterion prioritizes existing facilities that would theoretically cost the least to repurpose (Chambost, Mcnutt, and Stuart 2008).

FINANCIAL CAPITAL

Financial capital is examined here as the costs associated with biorefinery operation that vary geospatially, including energy and labor costs. Electricity and natural gas criteria are developed by aggregating energy data to the county level and averaging over the years 2010–2014 (US Energy Information Administration 2014a, 2014b). The labor criterion is developed through averaging weekly labor rates at the county level over the years 2012–2014 (US Bureau of Labor Statistics 2015).

Economic Criteria Weight Development

The economic criteria weights are developed using the biorefinery's TEA. Regional average energy rates and weighted average feedstock cost are inputted into the TEA, and regional average *annual* costs for the siting criteria are determined. The region examined is all counties in which candidate facilities reside. The repurpose potential criterion "cost" is developed by converting the total capital expenditure for constructing a greenfield biorefinery into an annualized expense, assuming a plant life (*n* years) and discount rate *r* using the Capital Recovery Equation. The average annual costs are summed, and the percentage of each criteria cost, out of total average cost, is determined. These percentages are the basis for the criteria weights. The percentages are normalized based on the range of scale values in the DST so facility scores can be calculated out of a total score of 100 for ease of understanding the results.

Social Asset Criteria Development

The DST's social asset criteria were developed by Rijkhoff et al. (2017) to give initial insights into nontechnical siting considerations. The social asset dataset provides quantitative proxy measures for social, cultural, and human capital (Martinkus et al. 2017b; Rijkhoff et al. 2017). Social asset quantification is difficult since each consists of multiple, often qualitative, indicators. However, while limited, the measures provide essential data for initial evaluation of candidate communities.

SOCIAL CAPITAL

Social capital reflects community connections, both among people and through organizations. It positively influences economic growth, promotes trust, and increases collective action through social networks that aid cooperation (Coleman 1988; Montgomery 2000; Putnam 2000; Rupasingha, Goetz, and Freshwater 2006). Quantitative indicators often used to measure social capital include the number of *rent-seeking* and *non-rent–seeking groups*,[3] nonprofit organizations in a community, and voter turnout. These indicators are incomplete representations of social capital; however, they are good proxies for measuring community-level social capital (Putnam 2000; Rupasingha, Goetz, and Freshwater 2006). The DST's social capital score was developed with the following indicators: all community organizations, nonprofit organizations, and voter turnout (Rijkhoff et al. 2017).

Cultural Capital

Cultural capital, also sometimes called creative capital, refers to community traditions and languages, people's perceptions of and interaction with the world around them, and acceptance of creativity and innovation (Emery and Flora 2006). Creativity is important for project success (Budd et al. 2008; Martinkus et al. 2014; and Martinkus et al. 2017b). Florida developed a comprehensive measurement of the creative using four indicators: the creative class (people with jobs that require creativity), innovation, high-tech industry, and diversity. Together, all weighted equally, these indicators form the Creative Vitality Index (CVI) (Florida 2002). Our cultural capital measure includes the number of arts-related organizations and businesses, the number of people employed in these organizations and businesses, and the revenues of arts-related goods and services (Rijkhoff et al. 2017).

Human Capital

Human capital addresses people's skills and abilities, which helps with assessing local workforce quality. Human capital is important to equitable and sustainable development solutions (Pretty and Ward 2001). The original social asset dataset used the following human capital indicators: community health, poverty levels, unemployment rate, and education levels (Rijkhoff et al. 2017). For the current DST, we utilized the community health indicators only to avoid potential overlap between the economic and social asset indicators.

Social Asset Scores

For the three social asset capitals, Rijkhoff et al. (2017) calculated a single capital score for each county, reflecting its performance for that capital. For social and human capital, individual indicators were multiplied by their factor loadings to create a single capital score. Since CVI is a nationwide index, the adapted CVI score for each county was utilized. Rijkhoff et al. (2017) used the US Census Region West to develop cutoff scores for each capital based on average regional performance. A county over the cutoff outperforms regionally and thus may be a stronger candidate for facility siting than those with scores below the cutoff value. Table 7.3 shows the cutoff scores for the Census Region West compared to those at the National level from Martinkus et al. (2017b).

The strength of the social assets dataset in the DST is the ability to narrow potential candidate sites by incorporating social criteria that are often

TABLE 7.3. Social, cultural, and human capital cutoff scores used in the decision support tool (DST)

Cutoff Scores		
	Census Region West	*National*
Social Capital	.0413	−.0043
Cultural Capital	.686	.491
Human Capital	−1.4247	.0838

Note: Asset ranges are provided in table 7A.1 of the appendix.

limited in siting decision frameworks. In fact, Rijkhoff et al. (2017) found in analysis of community-level projects that the social asset criteria were associated with project outcomes; cases above the cutoffs had positive results, while cases below had negative results. Social asset criteria inclusion does not replace the need for community engagement to ensure project success. Communities with high social capital can engage their networks to oppose a project; therefore, assessing community support and communicating with key actors are necessary before making final siting decisions.

Despite its benefits, the social asset criteria have limitations. To ensure national comparability, Rijkhoff et al. (2017) focused on obtaining county-level quantitative indicators, resulting in the exclusion of potential indicators that are unavailable at this level (e.g., trust). Additionally, robust indicators are unavailable at a lower level, such as city or town, which limits full consideration of sites in specific communities. Last, the social asset database does not include political capital; however, this indicator is currently being addressed for future inclusion.

To address these limitations, we conducted case studies to validate current indicators. These case studies examine the role of social, human, and cultural capital in the success or failure of community-level projects. The cases also assess indicators absent from the social asset dataset and suggest additional metrics for future DST iterations.

Ground-Truthing Case Studies

To assess the impacts of social assets in siting decisions, we selected four communities in the PNW where a high-tech biomaterial industry was proposed or constructed: Missoula, Montana; Lakeview, Oregon; Boardman,

Oregon; and Tacoma, Washington. Stakeholder interviews explore the role of social, cultural, and human capital in facility development and implementation. Case study findings are compared to Rijkhoff et al.'s (2017) initial social asset measures to validate and refine future measures and to explore their impact in project implementation. The case and stakeholder characteristics are listed in tables 7.4 and 7.5.

Conceptualizing Success Each case was examined for its level of success or failure based on a protocol adapted from a retrospective analysis of past complex policy projects (Rijkhoff et al. 2017).

SUCCESS

Smooth operation throughout its existence: encountered little to no community resistance; encountered no significant legal roadblocks; stayed economically viable throughout its operation.

PARTIAL SUCCESS

Currently operating and producing fuel or biomaterials, or successfully moving through the public permitting process with local support: may have encountered legal or economic roadblocks (e.g., major environmental lawsuits or violations, or economic viability struggles); operates normally today and can remain economically viable.

PARTIAL FAILURE

Currently operating but forced to diversify operations for economic viability, compromising original purpose: lawsuits or environmental violations have impeded ability to operate continuously; financial problems have forced only periodic operation, limiting production.

FAILURE

Never built: encountered local opposition or severe economic constraints; operations shutdown completely due to local resistance, environmental litigation, or economic ruin.

TABLE 7.4. Overview of case study communities

City	*County*	*State*	*Incoming Facility*	*Feedstock Source*	*Product*	*Status*	*Rating*
Missoula	Missoula	Montana	Blue Marble Biomaterials	Agriculture and Forest Biomass	Biochemical	Operating	Success
Lakeview	Lake	Oregon	Red Rock Biofuels	Forest Biomass	Aviation Jet Fuel	Not yet built	Partial success
Boardman	Morrow	Oregon	ZeaChem, Inc.	Agriculture and Forest Biomass	Ethyl Acetate/Ethanol	Operating	Partial failure
Tacoma	Pierce	Washington	Northwest Innovation Works	Liquid Natural Gas	Methanol	Never built	Failure

TABLE 7.5. Interview participants

Case	*Stakeholder Type*	*# Interviewees*
Missoula, MT	Nonprofit (1); government employee (1); university (1); private sector (1).	4
Lakeview, OR	Nonprofit (2); local/state government official (3); incoming facility (1).	6
Boardman, OR	Government employee (1); incoming facility (1); university (1); Local industry/private sector (2).	5
Tacoma, WA	Nonprofit (1); Local/state government official (1); government employee (2); university (1); local citizen (1).	6
	Total	21

Analysis

To assess the quantitative social asset indicators, a thematic analysis based on key literatures was performed; results were then compared to Rijkhoff et al.'s (2017) original county scores. Table 7.6 shows the conceptual coding framework used for interview analysis to examine the role of social, cultural, and human capital in facility siting decisions.

Missoula, Montana—Success

Blue Marble Biomaterials, operating since 2012, manufactures specialty chemicals from cellulosic biomass for food flavoring, fragrances, and cosmetics. It has been operating smoothly without community resistance. Key to its success are Missoula-based entrepreneurial and economic development organizations that help start-ups access grants, connect with stakeholders, and identify site locations. This formal infrastructure was important in relocating the facility from Seattle, Washington, where there was less stakeholder and economic support.

Rijkhoff et al. (2017) found favorable levels of social, cultural, and human capital for Missoula County (table 7.7). These scores are supported by interview data, which found the presence of several concepts of social capital, including bridging social capital and communication (table 7.7). Bridging social capital was particularly apparent, with a number of organizations and networks assisting start-ups and entrepreneurial ventures through the implementation process.

Cultural capital concepts focused on shared community values and sense of place. For example, Blue Marble uses cellulosic material from forest and agricultural waste to create biochemicals that replace their petroleum counterparts found in foods, fragrances, and other consumables. Stakeholders claimed the facility was seen as a "green" and "clean" industry and was readily accepted by the community.

Missoula stakeholders addressed availability of a skilled and educated workforce, components of human capital. All participants emphasized the University of Montana's role in connecting the company with a skilled workforce: "Having access to the University of Montana played a pretty key role for them. Whether it be recruiting students or doing research with a faculty, and that's been a big part of their reason for locating there" (interview with local nonprofit, February 2017).

TABLE 7.6. Interview coding themes

	Key Concept	*Descriptions*
Social Capital	Trust Bridging Bonding	Reflects the strength of connections among people and organizations within the community to make things happen (Emery and Flora 2006; Flora and Flora 2013). Social capital is often classified as either bonding social capital (ties that link individuals or groups with similar backgrounds) or as bridging social capital (connecting diverse groups within the community to each other and groups outside the community) (Emery and Flora 2006; Flora and Flora 2013; Putnam, Feldstein, and Cohen 2003).
	Modes of communication	Contact within, between, and across stakeholders and groups (Inkeles 2000).
Cultural Capital	Values History Legacy (industrial, environmental, social)	Consists of symbols and language and determines a community's distinctive character (Jacobs 2011). Cultural capital reflects the way people see the world and how they act within it, as well as their history, traditions, and language (Emery and Flora 2006; Flora and Flora 2013). Legacy is that which communities seek to pass on to the next generation.
	Sense of place	Describes our relationship with places, expressed in different dimensions of human life: emotions, biographies, imagination, stories, and personal experiences (Feld and Basso 1996).
Human Capital	Labor and workforce	Health of the potential workforces and ability of the community to be resourceful and access outside resources and bodies of knowledge (Emery and Flora 2006). The local labor force affects the community's success in attracting or supporting new business enterprises (Flora and Flora 2013).
	Skills (technical and intrapersonal) Knowledge (formal and informal) Experience	The ability of people based on their characteristics—such as formal and informal knowledge, technical and intrapersonal skills, experience, leadership, and talent—to develop and enhance their resources (Emery and Flora 2006; Gutierrez-Montes 2005).
	Capacity	Ability to access outside resources to contribute to local community and economic development (Emery and Flora 2006; Flora and Flora 2013).

Lakeview, Oregon—Partial Success

Red Rock Biofuels is the second biomass-related project to attempt to site in Lakeview, Oregon. The first, Iberdrola Renewables, failed because it could not obtain a power purchasing agreement from the local utility. However, this unsuccessful attempt demonstrated project feasibility, access to feedstock, physical infrastructure, and community support that appealed to Red

TABLE 7.7. Missoula, Montana: Interview findings compared to quantitative indicators

Assets	*Quantitative Social Indicators*		*Quantitative Levels*			*Qualitative Social Indicators*	*Present*[a]
Social Capital	Source: Rupasingha, Goetz, and Freshwater (2006) (2009 data)	# Rent-seeking groups: political, labor, professional and business organizations	Social Cap. 1997	Social Cap. 2005	Social Cap. 2009	Trust	—
		# Non-rent–seeking groups: civic organizations, bowling centers, golf clubs, fitness centers, sports organizations, and religious organizations				Bridging	√
						Bonding	—
		# Nonprofit organizations	1.20	2.07	1.84	Modes of communication	√
		% Voter turnout					
Cultural Capital	Source: WESTAF (2010)	# Arts-related organizations	CVI 2008	CVI 2009	CVI 2010	Values	√
		# Arts-related business				History	√
		# Occupational employment in the arts	.921	.956	.946	Legacy	√
		$ Revenues of arts-related goods and services				Sense of place	√
Human Capital	Source: County Health Rankings (The Robert Wood Johnson Foundation 2013)	% Low birth-weight	Health 2013		2.38	Labor and workforce	√
		% Premature deaths	Obesity 2013		−5.30	Skills	√
		% Obese (BMI > 30)	Poverty 2013		−.95	Knowledge	√
		% Self-reports of poor health condition (physically and mentally)	Education 2013		16.00	Experience	√
			Language 2013		−2.90	Capacity	—

Note: Shaded cells indicate scores that are better than the cutoff scores and, thus, levels that are more likely to be favorable for positive project implementation. Cutoff scores are based on averages for the respective years and variables for the region West (US census region) of over 446 counties and divided by themselves so that all scores are comparable (Rijkhoff et al. 2017). Due to missing data in the health scale for some of the counties, raw obesity scores were added as comparison (Rijkhoff et al. 2017).

[a] A positive qualitative social indicator is noted with a (√); if the indicator was not present during the interview process then (—).

Rock Biofuels. While this facility has not yet been built, the company has attained several approvals for an enterprise zone, pipeline capacity, an urban growth boundary amendment, and air quality permits. It is labeled a partial success, largely due to the substantial community support for the project.

Rijkhoff et al. (2017) found favorable levels of social capital and mixed levels of cultural and human capital for this case (table 7.8). Interview data corroborate the high levels of social capital but indicate some social assets played less of a role. For instance, social capital concepts—including trust, bridging, bonding, and communication—were present in every interview (table 7.8). Lake County has established trust and working relationships between stakeholders through the Lakeview Stewardship Group; they have worked together on contentious forest management issues for over a decade. Some environmental advocates are unsupportive of an industry focused on small-diameter wood utilization due to fear of environmental degradation. However, the executive director of a participating nonprofit believes that these groups are not opposing the project, because of significant levels of established trust: "I think what we've got is a lot of trust over the years that we've built with the collaborative here and to have those honest discussions without threats and those kinds of stuff, is doable" (interview with local nonprofit, January 2017). The project, moving forward without opposition, is actively supported by local leaders who engage their networks to increase project viability by hosting conference calls to keep multiple stakeholders informed.

Cultural capital themes in Lakeview reflect local history, legacy, and community values. Stakeholders indicated that the project fits with the community's goals of creating a new economy based on natural resource and renewable energy development. They also have a social legacy that is more supportive of biomass projects than neighboring counties.

Absent in the interviews was a discussion of the local workforce. Only one stakeholder mentioned that implementing projects of this size could present challenges for rural communities. Project-related skills, technical knowledge, and local stakeholder experience are contributing to Lakeview's ability to support Red Rock through the implementation process.

The qualitative findings support the quantitative indicators, which showed favorable levels of social capital in Lakeview. The quantitative metrics did not indicate favorable levels of cultural capital, with the exception of 2009, and human capital measures are mixed, with underperforming education

TABLE 7.8. Lakeview, Oregon: Interview findings compared to quantitative indicators

Assets	*Quantitative Social Indicators*		*Quantitative Levels*			*Qualitative Social Indicators*	*Present*[a]
Social Capital	Source: Rupasingha, Goetz, and Freshwater (2006) (2009 data)	# Rent-seeking groups: political, labor, professional, and business organizations	Social Cap. 1997	Social Cap. 2005	Social Cap. 2009	Trust	√
		# Non-rent–seeking groups: civic organizations, bowling centers, golf clubs, fitness centers, sports organizations, and religious organizations	3.00	2.64	2.81	Bridging	√
						Bonding	—
		# Nonprofit organizations				Modes of communication	√
		% Voter turnout					
Cultural Capital	Source: WESTAF (2010)	# Arts-related organizations	CVI 2008	CVI 2009	CVI 2010	Values	—
		# Arts-related business				History	√
		# Occupational employment in the arts	−.179	−.20	−.3171	Legacy	√
		$ Revenues of arts-related goods and services				Sense of place	—
Human Capital	Source: County Health Rankings (The Robert Wood Foundation 2013)	% Low birth-weight	Health 2013		−1.98	Labor and workforce	—
		% Premature deaths	Obesity 2013		−1.60	Skills	√
		% Obese (BMI > 30)	Poverty 2013		.87	Knowledge	√
		% Self-reports of poor health Condition (physically and mentally)	Education 2013		−1.00	Experience	√
			Language 2013		−4.00	Capacity	—

Note: Shaded cells indicate scores that are better than the cutoff scores, and thus, levels that are more likely to be favorable for positive project implementation. Cutoff scores are based on averages for the respective years and variables for the region West (US census region) of over 446 counties and divided by themselves so that all scores are comparable (Rijkhoff et al. 2017). Due to missing data in the health scale for some of the counties, raw obesity scores were added as comparison (Rijkhoff et al. 2017).

[a] A positive qualitative social indicator is noted with a (√); if the indicator was not present during the interview process then (—).

levels. However, cultural and human capital themes identified in stakeholder interviews suggest the presence of values, legacy, skills, and local experience that support the implementation process of the biofuel refinery (table 7.8).

Boardman, Oregon—Partial Failure

ZeaChem, Inc., is an ethanol and biochemical refinery located in the Port of Morrow's industrial park outside Boardman, Oregon. In 2012, ZeaChem completed construction of its demonstration biorefinery. Citing financial and technical challenges, ZeaChem has not reached commercial scale and is operating at limited capacity. Other setbacks include the loss of its primary feedstock source. ZeaChem is considered a partial failure because of its inability to scale up.

Rijkhoff et al. (2017) found mixed levels of human capital, and low levels of social and cultural (table 7.9) capital, which are partially supported in this case. The main social capital concepts were bonding, bridging, and communication. ZeaChem was said to be proactive with its communication and outreach efforts. Evidence of bonding social capital was present when a stakeholder described the community as insulated from outside influences (Roemer 2017). Early on, the company hired a local entrepreneur to coordinate communication between the company, community stakeholders, and outside networks and organizations. This early outreach contributed to building the refinery. However, this early engagement and support could not counter the technical and financial challenges that have kept ZeaChem operating at limited capacity.

The community's legacy and values stood out for cultural capital. The community is supportive of most economic development projects and especially projects that relate to food, agricultural, and timber industries. Stakeholders saw the project as a natural fit and easily supported the project. "We're an agricultural based economy, and so those industrial opportunities [are] value-added opportunities" (interview with local government employee, February 2017). They also said the project aligned with community values to have a "green industry." Boardman's industrial legacy and community values contributed to project acceptance and support.

Key human capital themes present were labor and workforce, skills, knowledge, and capacity. Participants cited challenges of getting technical experts from Denver, Colorado, and San Francisco, California, to relocate.

TABLE 7.9. Boardman, Oregon: Interview findings compared to quantitative indicators

Assets	*Quantitative Social Indicators*		*Quantitative Levels*			*Qualitative Social Indicators*	*Present*[a]
Social Capital	Source: Rupasingha, Goetz, and Freshwater (2006) (2009 data)	# Rent-seeking groups: political, labor, professional, and business organizations	Social Cap. 1997	Social Cap. 2005	Social Cap. 2009	Trust	—
		# Non-rent-seeking groups: civic organizations, bowling centers, golf clubs, fitness centers, sports organizations, and religious organizations				Bridging	√
		# Nonprofit organizations	.05	−.28	−.02	Bonding	√
		% Voter turnout				Modes of communication	√
Cultural Capital	Source: WESTAF (2010)	# Arts-related organizations	CVI 2008	CVI 2009	CVI 2010	Values	√
		# Arts-related business				History	—
		# Occupational employment in the arts	−.55	−.59	−.55	Legacy	√
		$ Revenues of arts-related goods and services				Sense of place	—
Human Capital	Source: County Health Rankings (The Robert Wood Foundation 2013)	% Low birth-weight	Health 2013	−.92		Labor and workforce	√
		% Premature deaths	Obesity 2013	−1.00		Skills	√
		% Obese (BMI > 30)	Poverty 2013	.67		Knowledge	√
		% Self-reports of poor health condition (physically and mentally)	Education 2013	23.60		Experience	√
			Language 2013	4.40		Capacity	√

Note: Shaded cells indicate scores that are better than the cutoff scores and, thus, levels that are more likely to be favorable for positive project implementation. Cutoff scores are based on averages for the respective years and variables for the region West (US census region) of over 446 counties and divided by themselves so that all scores are comparable (Rijkhoff et al. 2017). Due to missing data in the health scale for some of the counties, raw obesity scores were added as comparison (Rijkhoff et al. 2017).

[a] A positive qualitative social indicator is noted with a (√); if the indicator was not present during the interview process then (—).

Stakeholders contended that when the project reached commercial scale, Boardman and the surrounding area could meet its labor and workforce needs. However, ZeaChem remains operating at the demonstration level. A ZeaChem representative stated that in addition to technical challenges, coordinating technical experts' schedules to work at the demonstration site likely contributed to financial and technological delays. This opinion is supported by initial social asset metrics, as Rijkhoff et al. (2017) predicted education levels below the regional average.

The qualitative data demonstrate concepts of social and cultural capital; however, the quantitative indicators show unfavorable levels of social and cultural capital, with both underperforming compared to the regional average.

Tacoma, Washington—Failure

The port of Tacoma was the proposed location of the Northwest Innovation Works (NWIW) natural gas–to–methanol production plant. The facility would export methanol to produce olefins for use in plastics and other goods. Despite early political support from Washington's governor and other public officials, the plant faced strong local resistance. The company terminated the facility lease before the environmental review was conducted.

Rijkhoff et al. (2017) did not find favorable levels of social, cultural, or human capital with the exception of education (table 7.10). The analysis of social capital in Tacoma demonstrated negative indicators of social capital. Communication failure and breakdowns between stakeholder groups eroded public support. All interviewees indicated that NWIW failed to address the community's environmental and safety concerns. Additionally, the port notified the surrounding communities and others about meetings through traditional modes of outreach (e.g., press releases, newspaper articles, etc.); however, this information did not reach community members, who learned about the incoming facility through social media. Feeling inadequately informed and distrustful, people began organizing in opposition to the facility. Several participants noted this swell of public participation was unusual and the result of high levels of public distrust.

Key concepts of cultural capital were found to influence the opposition to the incoming methanol refinery in Tacoma. Perceived negative industrial, environmental, and social legacies led some community members to voice

TABLE 7.10. Tacoma, Washington: Interview findings compared to quantitative indicators

Assets	*Quantitative Social Indicators*		*Quantitative Levels*			*Qualitative Social Indicators*	*Present*[a]
Social Capital	Source: Rupasingha, Goetz, and Freshwater (2006) (2009 data)	# Rent-seeking groups: political, labor, professional, and business organizations	Social Cap. 1997	Social Cap. 2005	Social Cap. 2009	Trust	X
		# Non-rent–seeking groups: civic organizations, bowling centers, golf clubs, fitness centers, sports organizations, and religious organizations				Bridging	√
						Bonding	—
			−.90	−.70	−.75	Modes of communication	X
		# Nonprofit organizations					
		% Voter Turnout					
Cultural Capital	Source: WESTAF (2010)	# Arts-related organizations	CVI 2008	CVI 2009	CVI 2010	Values	X
		# Arts-related business				History	X
		# Occupational employment in the arts	−.76	−.06	−.36	Legacy	X
		$ Revenues of arts-related goods and services				Sense of place	X
Human Capital	Source: County Health Rankings (The Robert Wood Foundation 2013)	% Low birth-weight	Health 2013		.64	Labor and workforce	√
		% Premature deaths	Obesity 2013		3.80	Skills	—
		% Obese (BMI > 30)	Poverty 2013		−1.37	Knowledge	—
		% Self-reports of poor health condition (physically and mentally)	Education 2013		3.40	Experience	—
			Language 2013		−1.90	Capacity	—

Note: Shaded cells indicate scores that are better than the cutoff scores and, thus, levels that are more likely to be favorable for positive project implementation. Cutoff scores are based on averages for the respective years and variables for the region West (US census region) of over 446 counties and divided by themselves so that all scores are comparable (Rijkhoff et al. 2017). Due to missing data in the health scale for some of the counties, raw obesity scores were added as comparison (Rijkhoff et al. 2017).

[a] A positive qualitative social indicator is noted with a (√); if the indicator was not present during the interview process then (—); if the indicator was present in a negative manner, then (X).

concerns about environmental degradation and excessive water and energy consumption. Additionally, there was a social theme pervasive throughout the interviews that Tacoma gets Seattle's unwanted projects: "So, . . . while Tacoma apparently has become the petrochemical kitchen of the NW, Seattle gets the high-tech incubator jobs" (interview with local community member, February 2017).

The concepts of human capital focused primarily on the type and availability of the local workforce. When asked why they thought NWIW selected Tacoma, after first mentioning the natural and physical infrastructure, the interviewees described the workforce as ideal for this industry. This finding supports Rijkhoff et al.'s (2017) quantitative assessment indicating supportive levels of education in the county.

While this case demonstrates social organization forming against the project, influenced by negative industrial histories and legacies, neither the interview findings nor quantitative indicators (table 7.10) suggest favorable levels or themes of social, human, or cultural capital in support of this project implementation.

Case Study Conclusions

Data from the four cases show that social, cultural, and human capital can play a significant role in complex project siting decisions. In these cases, the social asset metrics are mostly supported by the interviews, lending support for their inclusion in our DST. However, the data also reveal both limitations of these measures and opportunities for improvement.

In the two cases of success/partial success, the quantitative indicators showing higher levels of social capital were supported by stakeholder interviews. Stakeholders noted the importance of bridging organizations in supporting projects through the implementation process, particularly well-connected nonprofit organizations. These groups facilitated communication between stakeholders and with other sectors of the community.

In addition, communication played a significant role in two ways: (1) how well and often multiple stakeholders communicated and (2) how well companies or public officials presented information to the public. In the successful case, stakeholders mentioned a major factor in Blue Marble's decision to relocate from Seattle to Missoula was the ease of access to, and communication with, vital local stakeholders. In contrast, NWIW's failure to address the local environmental and safety concerns in Tacoma contributed to strong opposition of the methanol refinery.

Previous work (Fey, Bregendahl, and Flora 2006; Klamer 2002) finds that cultural capital plays an important role in community economic development, including a community's capacity for creativity, innovation, and willingness to take risks and their shared legacy and values (Tigges and Noble 2012). The cultural capital measure used in the DST, the CVI, indicated favorable levels in the Missoula case, but case study analysis suggests these measures can be improved. For example, evidence of creativity, innovation, and risk taking was prominent in Lakeview. In both Lakeview and Boardman, the biofuel refineries were publicly supported because the projects aligned with a shared and valued legacy of timber and agricultural industries. In Tacoma, a legacy of environmental degradation, combined with NWIW's failure to address the community's fears, fueled opposition to the project. These aspects of cultural capital are not currently measured by the quantitative cultural capital indicator.

Whether the case study findings can validate human capital quantitative indicators is unclear. Both Tacoma and Missoula stated that the local workforce could support the incoming refinery. In Boardman, ZeaChem faced challenges coordinating outside experts' schedules for its demonstration facility, which stakeholders indicated impacted start-up and development. Lakeview stakeholders only mentioned the workforce in relation to having a small administrative staff, possibly indicating limited human capital.

There are limitations to the case studies. With only four communities, the size and scope of the assessments were restricted, and facilities examined were at different stages of development. For example, Lakeview is determined a partial success because, while the facility has not yet been built, the company has attained approvals for an enterprise zone, pipeline capacity, an urban growth boundary amendment, and air quality permits, and it is preparing for construction (Liedtke 2018). In contrast, ZeaChem is a partial failure because of financial and technical reasons, not insufficient social assets. As these cases are in various stages of development, future research is needed to determine if initial designations of success or failure still apply. However, at this time, the case studies indicate that the quantitative social asset indicators currently being used in the DST provide a more robust assessment for facility siting decisions compared to those that rely on traditional siting criteria (Martinkus et al. 2017b).

Decision Support Tool: Full Model

The DST's social asset inclusion creates a multicriteria tool that combines these assets with traditional economic site selection criteria (table 7.2). The current DST refines the social asset metrics of Rijkhoff et al. (2017) by adopting a scale measure of each capital rather than the cutoff scores, allowing for a more nuanced analysis of assets, especially for communities close to the cutoff.

As mentioned, each facility receives a score based on how well its assets compare to the siting criteria. Individual facility scores are calculated using the Weighted Sum Method (Wang et al. 2009), which represents the sum of individual criterion weights multiplied by location-specific scaled values and by an overall economic or social metric weight (equation 1).

$$F_j = \sum_{x=1}^{2} \theta_x \sum_{i=1}^{n} w_i \, s_{ji} \quad where \sum \theta_x = 1 \qquad Equation\ 1$$

where F_j is the score for facility j, w_i is the weight for criterion i, S_{ji} is the scaled value for criterion i at facility j, n is the total number of criteria, and theta x is the overall user-defined weight for metric x. Here, each metric is assigned a theta$x$ = 0.5, thus giving both metrics equal weight and importance. The overall weights can be adjusted (as long as they sum to 1) to provide greater importance to either the economic or social metric, thus potentially altering the final site selection.

Each criterion's range of facility-specific economic and social values is used to determine the range associated with each scale value. The criterion-based "bin" values (***Bi***) are determined by dividing the range of facility values ($a_{i,max}$, $a_{i,min}$) by the maximum scale value (***smax***) for each criterion i (equation 2, table 7.1).

$$B_i = \frac{a_{i,max} - a_{i,min}}{s_{max}} \qquad Equation\ 2$$

The maximum scale value is assigned to the minimum or maximum value in each criterion's range of values that denotes the most positive influence on facility siting, such as low electricity rate or high infrastructure compatibility. The subsequent scale values are calculated by either adding or subtracting B_i based on the positive or negative influence of the criterion (table 7.1).

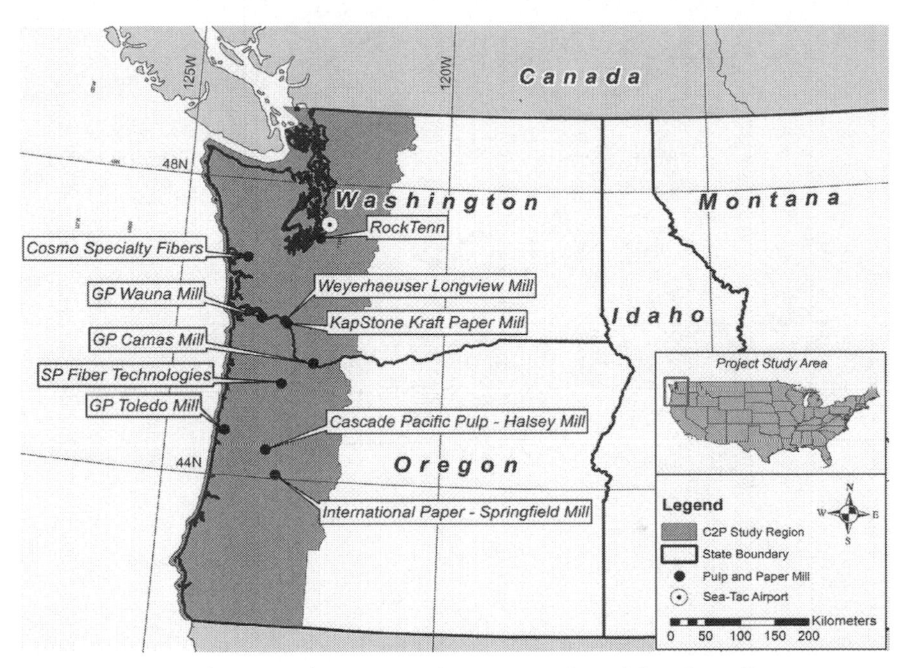

FIGURE 7.2. Cascade-to-Pacific (C2P) study region and candidate biorefinery sites.

Decision Support Tool Application

We applied the DST to locating a forest biorefinery in a region of the PNW we call the Cascade-to-Pacific (C2P), representing western Oregon and Washington (figure 7.2). Ten active or recently decommissioned pulp mills are assessed for their repurpose potential as a forest biorefinery serving the Seattle-Tacoma International Airport. This region has the existing physical infrastructure and ability to meet an annual feedstock requirement of about 830,000 bone-dry tons (BDT) of postharvest forest residues. Such a facility could produce approximately 36 million gallons of isoparaffinic kerosene (IPK), or aviation biofuel, from the feedstock through an enzymatic hydrolysis, fermentation, and catalytic conversion process (Zhu 2015). See table 7.11 for the decision matrix, and table 7.12 for the scaled values and overall facility scores.

Results

If the facilities are evaluated based on economic metrics alone (Economic Score column), Cosmo ranks highest due to infrastructure compatibility, low

TABLE 7.11. Decision matrix for C2P facility site assessment

	Economic Metric					*Social Metric*		
	Natural Capital	*Financial Capital*			*Built Capital*			
	Weighted Average Delivered Feed-stock Cost ($/BDT)	*Electricity ($/kWh)*	*Natural Gas ($/k.c.f.)*	*Average Wage ($/week)*	*Infrastructure: % Cost Reduction from Greenfield*	*Social Capital*	*Cultural Capital*	*Human Capital*
Scale								
5	62.7	0.047	7.5	571	40.6%	0.6	1.0	−2.9
4	65.3	0.051	7.9	627	39.2%	0.3	0.8	−2.0
3	67.8	0.055	8.3	683	37.8%	−0.4	0.7	−1.1
2	70.4	0.058	8.6	739	36.4%	−0.5	0.6	−0.2
1	72.9	0.062	9.0	795	34.9%	−0.9	0.4	0.8
weights	5.9	4.0	0.9	1.8	7.4	6.67	6.67	6.67

electricity rate, and relatively low costs for feedstock and fuel transport (table 7.12). These criteria are weighted the highest due to the large annual operational expenses they impose on the biorefinery. However, when considering the county's social assets, Cosmo ranks fourth due to low cultural and human capital (Social Score column). The low human capital score suggests that repurposing this pulp mill into a forest biorefinery may be hard due to workforce issues, suggesting the need to relocate labor or to provide ongoing training. In addition, the lower cultural capital scores could suggest a community with limited ability to adapt or creatively approach problems that might occur in the permitting process or other development stages.

When considering economic and social criteria equally, GP Wauna ranks highest (Facility Rank column). This facility has low infrastructure compatibility (the highest economic weight), yet it scores relatively high for all other economic metrics, and it ranks highest for all social metrics. To convert GP Wauna into a biorefinery, construction improvements may cost more, but the facility may also experience a faster permitting process due to the community's high social cohesion and potential ability to tackle difficult issues productively. This presumed capability translates into a faster start-up time, with construction costs recouped sooner, compared to a facility that may require less capital to repurpose but that encounters delays in permitting or due to public opposition.

TABLE 7.12. Scaled facility values and resulting scores C2P

	Economic Metric Weight, $\theta x = 0.5$							*Social Metric Weight, $\theta x = 0.5$*					*Final Facility Score and Rank*	
Facility	*Weighted Avg. Delivered Feedstock Cost (\$/BDT)*	*Electricity (\$/kWh)*	*Natural Gas (\$/k.c.f.)*	*Avg. Wage (\$/wk.)*	*% Cost Reduction from Greenfield*	*Economic Score*	*Weighted Score (Econ. Score x 0.5)*	*Social Capital*	*Cultural Capital*	*Human Capital*	*Social Score*	*Weighted Score (Social Score x 0.5)*	*Total Weighted Facility Score (= Wtd. Econ. Score + Wtd. Social Score)*	*Facility Rank*
Cascade Pacific	3	1	5	3	1	39.0	19.5	3	1	3	46.7	23.3	42.8	9
Cosmo	4	5	2	4	5	89.6	44.8	3	1	1	33.3	16.7	61.5	4
GP Camas	4	3	5	1	1	49.3	24.7	1	3	5	60.0	30.0	54.7	7
GP Wauna	4	4	5	4	1	58.7	29.4	5	5	5	100.0	50.0	79.4	1
GP Toledo	1	2	5	5	1	34.8	17.4	5	5	3	86.7	43.3	60.7	5
IP Springfield	3	2	5	3	1	43.0	21.5	3	5	4	80.0	40.0	61.5	3
Kapstone	5	5	2	1	1	60.5	30.3	2	1	1	26.7	13.3	43.6	8
RockTenn	4	1	1	1	1	37.7	18.9	1	3	3	46.7	23.3	42.2	10
SP Fiber	4	5	5	3	3	75.7	37.9	2	2	5	60.0	30.0	67.9	2
Weyerhaeuser	5	5	2	1	4	82.7	41.4	2	1	1	26.7	13.3	54.7	6

Note: Raw facility scores can be found in appendix table 7A.2.

As we have stated previously, the biorefinery siting DST is meant to refine a large list of potential facilities to a few top candidates for further investigation. Assessments are performed based on publicly available data, which enhances both its trust and transparency and which increases compatibility and comparability across the United States. Once a list of preferred candidate facilities is identified, in-depth community and facility analyses must be performed to identify the major concerns that would prevent or delay the facility's efficient repurposing.

Conclusion and Strategic Considerations

A multicriteria DST can be instrumental in effective facility siting due to its unique ability to combine multifaceted and divergent assets into a single framework for assessing candidate communities. The CCF provides a systematic framework for determining levels of community assets (or capitals) and the interaction between assets; however, the framework's qualitative nature limits its effectiveness for broad-based community comparison because of its data-rich needs. Our innovative, cross-disciplinary DST links the CCF to quantitative indicators to assess facility siting options both regionally and nationally. As such, this tool is powerful for initial assessment. Unfortunately, highly technical facility siting analysis are often conducted without critical social assets included, which limits the ability to predict successful implementation. Social assets not only provide assessment of the likelihood of community support, but they also address other critical aspects of project success that can aid project development, implementation, and sustainability. Our integrated DST was developed to identify forest biorefinery sites in the PNW, but it could be applied throughout the United States, for bioenergy or other industrial siting applications.

Case studies findings illustrate that the selected social asset indicators are effective proxies for assessing the presence of social, cultural, and human capital, despite using somewhat dated quantitative data (e.g., 2009, 2010, and 2013).[4] Although the case studies reveal the social asset indicators could be refined further, the current dataset is adequate for identifying host communities more likely to sustain a biorefinery. The case studies also indicate that the type of facility matters, particularly regarding whether it is a good fit with the community's culture. However, this is not a limitation of the DST because physical indicators, such as infrastructure and feedstock, would need to be adjusted to accommodate different types of facilities, and social assets are no different. Incorporating social assets into the DST does not eliminate the need for context-specific knowledge of candidate communities in order to enhance the likelihood of community acceptance of a siting decision. The tool does, however, provide an important initial assessment of these communities, resulting in a reduced number of sites to consider in greater detail.

The multicriteria DST has many strategic applications. High-tech facility siting is complex and often highly contentious. A critical role in the decision-making process for tools such as our DST is being a "boundary object" or a tool that can facilitate communication and learning between individuals and groups with disparate interests (McKnight and Zietsma 2007). Historically, DSTs have been used by industry leaders performing site selection to identify candidate locations that meet their needs and provide reduced capital and operational costs (Noon, Zahn, and Graham 2002). However, with growing interest in engaging stakeholders and the general public in environmental decisions, the DST can be extremely useful in creating a common understanding (Beierle 2002).

Our DST is also useful for economic developers to attract bioenergy, high-tech, or renewable energy developments to their region, especially given the growing interest in state- and local-level renewable energy production (Durkay 2017). The DST supports economically sustainable development by favoring existing facilities but is equally adept at assessing greenfield locations. With the incorporation of social assets, stakeholders can determine a community's readiness to accept a specific proposal and identify areas of limited social capacity. For instance, a community that underperforms in social capital may benefit from strategic engagement of economic developers to enhance citizen trust and support for particular industries (e.g., biorefineries or wind/solar farms). Lower levels of human capital may indicate a need to improve local workforce skills or bringing in key operational personnel. The DST analysis can indicate a communities' existing assets and highlight weaknesses to bolster or enhance the communities' appeal to industry investment.

While a DST can be a powerful tool for site selection, it is not without its weaknesses as the quality of the data used in the tool determines its accuracy. These tools require frequent updating, can be complex and difficult to understand, and require transparency about their uses and limitations. Weights should be evident and grounded in the literature. In our DST, we weighted the economic siting criteria using a quantitative assessment of the biorefinery TEA, while all social capitals were weighted equally. Performing sensitivity analyses, where weights and cost estimates are adjusted, can provide additional levels of understanding about the DST's outputs (Martinkus et al. 2017b). A DST also does not inform about risks involved in project development and implementation. Despite these limitations, the DST provides more information to industry, community leaders, and local stakeholders about candidate communities. This information can enhance the likelihood of implementation success through strategic engagement that reduces costs associated with community opposition, citizen distrust, and insufficient labor. The DST presented here can play an important role in high-level policy decisions and local permit approval. Moreover, DSTs can be crucial for nascent industries that must operate at reduced costs until market share can be achieved.

Appendix

TABLE 7A.1. Social, cultural, and human capital cutoff scores and ranges

Asset	*Indicators*		*Census Region West*	*National*
Social Capital		Average	.041	−.004
		Range	−3.06–7.88	−4.29–23.08
Creative Capital[a]		Average	.69	.49
Human Capital	Health	Average	−1.425	.084
		Range	−7.66–6.21	−7.66–12.50
	Obesity	Average	25.8%	30.3%
	Poverty	Average	.33	−.15
		Range	−5.65–7.82	−5.65–7.82
	Education	Average	58.0%	54.2%
	Language	Average	3.20%	1.80%

[a] For Creative Capital there is no range available due to the nature of the data. Please see Rijkhoff et al. (2017), for more details.

TABLE 7A.2. Original facility scores from decision support tool in the PNW

	Economic Metric					*Social Metric*		
	Natural Capital	*Financial Capital*			*Built Capital*			
Facility	*Weighted Average Delivered Feedstock Cost ($/BDT)*	*Electricity ($/kWh)*	*Natural Gas ($/k.c.f.)*	*Average Wage ($/week)*	*Infrastructure: % Cost Reduction from Greenfield*	*Social Capital*	*Cultural Capital*	*Human Capital*
Cascade Pacific	$68.59	.066	7.8	724	34%	−.46	.30	−.71
Cosmo	$65.51	.048	8.8	647	41%	−.30	.31	1.49
GP Camas	$67.60	.055	7.5	851	34%	−1.29	.60	−2.40
GP Wauna	$65.26	.053	7.8	629	34%	.64	.99	−2.61
GP Toledo	$75.44	.062	7.8	571	34%	.29	.90	−0.58
IP Springfield	$69.69	.062	7.8	712	34%	−.15	.96	−1.62
Kapstone	$62.97	.047	8.8	823	34%	−.66	.33	1.67
RockTenn	$67.35	.063	9.4	812	34%	−1.10	.66	−.91
SP Fiber	$66.49	.049	7.8	688	38%	−0.68	.51	−2.88
Weyerhaeuser	$62.71	.047	8.8	823	39%	−.66	.33	1.67

Acknowledgments

This work, as part of the Northwest Advanced Renewables Alliance (NARA) was funded by the Agriculture and Food Research Initiative Competitive Grant (2011-68005-30416), USDA National Institute of Food and Agriculture.

Notes

1. NARA, funded by the US Department of Agriculture's National Institute of Food and Agriculture (NIFA), facilitates the development of sustainable biojet fuel, high-value co-products made from lignin, supply chain coalitions, rural economic development, and bioenergy literacy in Washington, Oregon, Idaho, and Montana. For more information, see http://nararenewables.org/.

2. The most recent development iteration of the social asset dataset includes an initial measure of political capital. However, this capital is still being assessed. Preliminary findings are available per request.

3. Rent-seeking groups include political, labor, professional, and business organizations. Non-rent–seeking groups include civic organizations, bowling centers, golf clubs, fitness centers, sports organizations, and religious organizations.

4. An updated social asset database is currently in progress. Since it proved to be valid with the older data, we are confident that an updated version will perform just as well or even better. Information and data for the new version are available upon request.

References

Beierle, Thomas C. 2002. *Democracy in Practice: Public Participation in Environmental Decisions*. New York: Routledge.

Budd, William, Nicholas Lovrich Jr., John C. Pierce, and Barbara Chamberlain. 2008. "Cultural Sources of Variation in US Urban Sustainability Attributes." *Cities* 25 (5): 257–267.

Cambero, Claudia, and Taraneh Sowlati. 2014. "Assessment and Optimization of Forest Biomass Supply Chains from Economic, Social and Environmental Perspectives: A Review of Literature." *Renewable and Sustainable Energy Reviews* 36 (C): 62–73.

Cambero, Claudia, and Taraneh Sowlati. 2016. "Incorporating Social Benefits in Multi-objective Optimization of Forest-Based Bioenergy and Biofuel Supply Chains." *Applied Energy* 178 (September): 721–735.

Chambost, Virginie, Jim Mcnutt, and P. R. Stuart. 2008. "Guided Tour: Implementing the Forest Biorefinery (FBR) at Existing Pulp and Paper Mills." *Pulp and Paper Canada* 109 (7): 1–9.

Coleman, James S. 1988. "Social Capital in the Creation of Human Capital." *American Journal of Sociology* 94: 95–120.

Durkay, Jocelyn. 2017. "State Renewable Portfolio Standards and Goals." National Conference of State Legislatures. Accessed September 2017. http://www.ncsl.org/research/energy/renewable-portfolio-standards.aspx.

ESRI, 2017. ArcGIS. Redlands, CA: Environmental Systems Research Institute.

Emery, Mary, and Cornelia Flora. 2006. "Spiraling-Up: Mapping Community Transformation with Community Capitals Framework." *Community Development* 37 (1): 19–35.

Feld, Steven, and Keith H. Basso. 1996. "Introduction." In *Senses of Place*, edited by Steven Feld and Keith H. Basso, 1–12. Santa Fe, NM: School of American Research Press.

Fey, Susan, Corry Bregendahl, and Cornelia Flora. 2006. "The Measurement of Community Capitals through Research." *Online Journal of Rural Research and Policy* 1 (1): 1–28.

Flora, Cornelia B., and Flora, Jan L. 2013. *Rural Communities: Legacy and Change*. Boulder, CO: Westview Press.

Florida, Richard. 2002. "The Rise of the Creative Class." *Washington Monthly*, May 15–25.

Gutierrez-Montes, Isabel A. 2005. "Healthy Communities Equals Healthy Ecosystems? Evolution (and Breakdown) of a Participatory Ecological Research Project towards a Community Natural Resource Management Process." PhD diss., Iowa State University, Ames.

Inkeles, Alex. 2000. "Measuring Social Capital and Its Consequences." *Policy Sciences* 33 (3–4): 245–268.

Jacobs, Cheryl. 2011. "Measuring Success in Communities: The Community Capitals Framework." Extension Extra. Paper 517. South Dakota State Univerrsity, Brookings.

Klamer, Arjo. 2002. "Accounting for Social and Cultural Values." *De Economist* 150 (4): 453–473.

Liedtke, Kurt. 2018. "Red Rock Gets Go-Ahead for Lakeview: Renewable Biofuels Plant to Be Constructed in Lakeview." *Herald and News*, April 15. https://www.heraldandnews.com/news/local_news/red-rock-gets-go-ahead-for-lakeview/article_253409ba-63aa-5c42-becd-ea9ac2711805.html.

Martinkus, Natalie, Greg Latta, Todd Morgan, and Michael Wolcott. 2017a. "A Comparison of Methodologies for Estimating Delivered Forest Residue Volume and Cost to a Wood-Based Biorefinery." *Biomass and Bioenergy* 106 (November): 83–94.

Martinkus, Natalie, Sanne A.M. Rijkhoff, Season A. Hoard, Wenping Shi, Paul Smith, Michael Gaffney, and Michael Wolcott. 2017b. "Biorefinery Site Selection Using a Stepwise Biogeophysical and Social Analysis Approach." *Biomass and Bioenergy* 97 (February): 139–148.

Martinkus, Natalie, Wenping Shi, Nicholas Lovrich, John Pierce, Paul Smith, and Michael Wolcott. 2014. "Integrating Biogeophysical and Social Assets into Biomass-to-Biofuel Supply Chain Siting Decisions." *Biomass and Bioenergy* 66 (July): 410–418.

Martinkus, Natalie, and Michael Wolcott. 2017. "A Framework for Quantitatively Assessing the Repurpose Potential of Existing Industrial Facilities as a Biorefinery." *Biofuels, Bioproducts and Biorefining* 11 (2): 295–306.

McKnight, Brent, and Charlene Zietsma. 2007. "Local Understandings: Boundary Objectives in High Conflict Settings." Proceedings of OLKC 2007. https://warwick.ac.uk/fac/soc/wbs/conf/olkc/archive/olkc2/papers/mcknight_and_zietsma.pdf.

Montgomery, John D. 2000. "Social Capital as a Policy Resource." *Policy Sciences* 33 (3): 227–243.

Noon, Charles E., F. Benjamin Zhan, and Robin L. Graham. 2002. "GIS-Based Analysis of Marginal Price Variation with an Application in the Identification of Candidate Ethanol Conversion Plant Locations." *Networks and Spatial Economics* 2 (1): 79–93.

Parks, Craig D., Jeff Joireman, and Paul AM Van Lange. 2013. "Cooperation, Trust, and Antagonism: How Public Goods Are Promoted." *Psychological Science in the Public Interest* 14 (3): 119–165.

Perimenis, Anastasios, Hartley Walimwipi, Sergey Zinoviev, Franziska Müller-Langer, and Stanislav Miertus. 2011. "Development of a Decision Support Tool for the Assessment of Biofuels." *Energy Policy* 39 (3): 1782–1793.

Plate, Richard R., Martha C. Monroe, and Annie Oxarart. 2010. "Public Perceptions of Using Woody Biomass as a Renewable Energy Source." *Journal of Extension* 48 (3): 1–15.

Pretty, Jules, and Hugh Ward. 2001. "Social Capital and the Environment." *World Development* 29 (2): 209–227.

Putnam, Robert D. 2000. *Bowling Alone: The Collapse and Revival of American Community*. New York: Simon and Schuster.

Putnam, Robert D., Lewis M. Feldstein, and Don J. Cohen. 2003. *Better Together: Restoring the American Community*. New York: Simon and Schuster.

Rijkhoff, Sanne A. M., Season A. Hoard, Michael J. Gaffney, and Paul M. Smith. 2017. "Communities Ready for Takeoff: Integrating Social Assets for Biofuel Site-Selection Model." *Journal of Politics and Life Sciences* 36 (1): 14–26.

The Robert Wood Johnson Foundation and the University of Wisconsin Population Health Institute. 2013. County Health Rankings National Data. Accessed April 5, 2014. http://www.countyhealthrankings.org/.

Roemer, Kelli. 2017. "Exploring the Role of Social Assets in Refinery Implementation: Using Case Study Research to Ground-Truth CAAM." Master's thesis, University of Idaho, Moscow.

Rösch, Christine, and Martin Kaltschmitt. 1999. "Energy from Biomass: Do Non-technical Barriers Prevent an Increased Use?" *Biomass and Bioenergy* 16 (5): 347–356.
Rupasingha, Anil, Stephan J. Goetz, and David Freshwater. 2006. "The Production of Social Capital in US Counties." *Journal of Socio-economics* 35 (1): 83–101.
Sultana, Arifa, and Amit Kumar. 2012. "Optimal Siting and Size of a Bioenergy Facility Using Geographic Information System." *Applied Energy* 94 (July): 192–201.
Tigges, Leann M., and Molly Noble. 2012. "Getting to Yes or Bailing on No: The Site Selection Process of Ethanol Plants in Wisconsin." *Rural Sociology* 77 (4): 547–568.
US Bureau of Labor Statistics. 2015. *Quarterly Census of Employment and Wages*. Washington, DC.
US Energy Information Administration. 2014a. "Electric Power Sales, Revenue, and Energy Efficiency Form EIA-861." Detailed Data Files. Washington, DC.
US Energy Information Administration. 2014b. "Natural Gas Annual Respondent Query System." Washington, DC.
Van Dael, Miet, Steven Van Passel, Luc Pelkmans, Ruben Guisson, Gilbert Swinnen, and Eloi Schreurs. 2012. "Determining Potential Locations for Biomass Valorization Using a Macro Screening Approach." *Biomass and Bioenergy* 45 (September): 175–186.
Wang, Jiang-Jiang, You-Yin Jing, Chun-Fa Zhang, and Jun-Hong Zhao. 2009. "Review on Multi-criteria Decision Analysis Aid in Sustainable Energy Decision-Making." *Renewable and Sustainable Energy Reviews* 13 (9): 2263–2278.
Western States Arts Federation (WESTAF). 2010. "The Creative Vitality™ Index: An Overview." Accessed April 5, 2014. http://www.westaf.org/publications_and_research/cvi.
White, Eric M. 2010. "Woody Biomass for Bioenergy and Biofuels in the United States: A Briefing Paper." General Technical Report PNW-GTR-825. Portland: US Forest Service.
Zhu, Junyong. 2015. "Task C-Conversion: Pretreatment and Aviation Teams." Northwest Advance Renewables Alliance 3rd Cumulative Report. Pullman, WA.

Chapter 7 Summary

The chapter presents the development and application of a siting decision support tool (DST) for biorefineries that combines economic and social assets. The multicriteria tool is particularly useful to aid high-tech facility siting decisions. Economic siting criteria are represented by major biorefinery operational costs (e.g., feedstock and utilities) that vary geospatially. Through assessing location-specific costs at each facility, a site may be identified that provides reduced annual operational costs, resulting in a more cost-competitive fuel. Social criteria are characterized by the ability of communities to accept and sustain complex new industrial projects. We identify three key assets impacting project sustainability: social capital, creative capital, and human capital. We include several social asset metrics measured at the county level, such as community innovation and collective action capacity. The social asset database used in the DST represents the first quantitative model of these important nontechnical assets that is both comparative and adaptive for site selection in the United States.

We validate the strength of the selected social assets with four case studies of successful and unsuccessful biorefinery facilities sited in the US West. This case study analysis draws on interviews with key stakeholders to examine the role social assets played in the successful adoption and implementation of these industries. The findings support the selected social assets, showing they are effective proxies for assessing the presence of social, creative, and human capital. We also report on the results of applying the multicriteria DST to pulp mills in the Pacific Northwest for their repurpose potential as biorefineries. The DST provides a quantitative way to evaluate these existing facilities based on multiple location-specific siting criteria, providing a score to rank each facility based on its economic and social assets. We use the DST to evaluate ten existing mills for their fit and likely success as a repurposed biorefinery. Our analysis refines the initial list of ten sites to a few select locations. By doing so, the DST increases the likelihood of successful implementation of a repurposed biorefinery.

The DST illustrates an interdisciplinary approach to addressing economic and social barriers in bioenergy facility siting. The strength of the DST is its applicability and adaptability for use in both bio- and traditional energy plant siting decisions across the nation. By combining economic and social criteria, the DST provides industry, community, and government decision-makers with a ranked list of locations for siting high-tech plants that are more likely to have sustained economic success.

KEY TAKEAWAYS

- A multicriteria decision support tool (DST) is developed, tested and validated.
- The tool integrates quantitative measures of economic and social assets relevant to siting decisions.

- The decision support tool is instrumental in effective initial facility siting assessments.
- The tool contains social asset data for most counties in the United States, thus providing difficult-to-obtain social asset data at a national level.
- Case study findings support the use of selected social assets in assessing potential bioenergy sites.

Section Three
Case Studies and Applications

8

Cultural Counterpoints for Making Sense of Changing Agricultural and Energy Landscapes

A Pennsylvania Case Study

WESTON M. EATON, C. CLARE HINRICHS, AND MOREY BURNHAM

Introduction

Recent literature on social responses to renewable energy technology development suggests place meanings shape attitudes and behaviors toward such development locally (McLachlan 2009). Scholars working in this vein have drawn insight from the rich literature on place attachment and sense of place (Stedman 2003). Current research draws from social psychology, and, in particular, Social Representation Theory for insight into social responses to renewable energy development (Batel and Devine-Wright 2015; Devine-Wright et al. 2017). Social Representation Theory proposes that people construct meaning for new ideas and things, or what we can call social objects, such as novel energy technologies, by drawing from the store of meanings and symbols they deploy in their everyday lives. In this chapter, we contribute to this literature by highlighting how land meanings influence people's responses to renewable energy technologies, in a case where such energy development may change patterns of use and management of privately

DOI: 10.5876/9781646420278.c008

owned rural landscapes. Drawing from Kai T. Erikson (1976), we conceptualize land meanings as a continuum of cultural meanings salient for a group. We use examples drawn from a case study of landowner and farmer responses to the emerging bioenergy industry in the northeastern United States to demonstrate how land meanings shape how they make sense of new energy technologies.

Although we present this chapter from our vantage points in the rural and environmental social sciences, our broader research has been conducted within a larger interdisciplinary endeavor centered on partnership and dialogue with academic disciplines ranging from biological engineering to resource economics to plant genetics to supply chain management to policy and legal studies, as well as with bioenergy industry stakeholders and educators. The complex challenges of energy transitions have stimulated recognition that human factors, beyond economic assumptions about human behavior in respect to technology, merit much greater attention (Pellegrino and Musy 2017). Investigating cultural considerations among stakeholders in energy development represents an opportunity for "up-front engagement" rather than the "end of pipe" role that social scientists typically provide in relation to technological change (Lowe, Phillipson and Wilkinson 2013). Insights on the significance of culture and context for stakeholder interests and positions may temper and improve planning, design, and deployment of new energy projects.

Cultural Theory and Social Representations

Cultural sociologists examine how meanings, ideas, and other symbolic components of social life emerge, coalesce, and interrelate, with implications for societal stability and change. Classical social theorists such as Talcott Parsons and Max Weber viewed culture as composed of a group's central values and ideas that then shape behavior in direct, linear fashion. Contemporary cultural theorists have complicated this view. They assert that familiar, accessible symbols, ideas and practices provide resources that people draw upon, sometimes selectively, to make sense of and respond to challenges and opportunities in their lives (Swidler 1986).

Erikson's research on Appalachian culture complements contemporary critiques of classical theorists' notions of culture (1976). He argued that previous conceptualizations of culture failed to account for seeming contradictions in everyday social life. For Erikson, a group's culture is composed not only of core values but also of tensions or counterpoints—what he called "axes of variation"—across the expectations, purposes, obligations, responsibilities, and other sensibilities common to a group. Erikson developed his theory through studying change in the social life of an Appalachian community after a failed coal impoundment caused the disastrous Buffalo Creek flood. He outlined several axes of variation salient for Appalachian people, including tensions between self-resignation and assertion and between self-centeredness and group-centeredness, and a predominant tension "between a sense of independence at one pole and a need for dependency on the other" (91).

Erikson showed that in responding to the devastation wrought by the Buffalo Creek flood, common community characteristics and tendencies were transformed, resulting in what appeared as new cultural sensibilities. He argued that the disaster's immediate impact, including loss of life and annihilation of the community's physical environment and built infrastructure, hastened a shift along axes of variation toward the community's socially weaker, less resilient, and more isolating cultural traits. As a central finding, Erikson (1976, 84) asserted, "such shifts do not represent a drastic change of heart, not a total reversal of form, but a simple slide along one of the axes of variation characteristic of that social setting."

Erikson developed his theory to understand the aftermath of a sudden event with immediate biophysical and social impacts (1976). We contend that such conceptualizations of culture can be extended to understand social responses to slower moving, emerging land use and energy technology projects that will also have biophysical and social impacts. To do so, we focus on efforts in western Pennsylvania to introduce a new bioenergy crop industry where perennial nonfood crops are grown on privately owned land to provide biomass feedstocks for renewable bioproducts and energy (Burnham et al. 2017; Eaton et al. 2017). While natural or technological disasters trigger acute change, bioenergy development involves gradual changes in an area's economy and land use with uncertain and potentially uneven implications for residents (Rossi and Hinrichs 2011). Thomas Beamish's *Silent Spill* (2002) usefully illustrates how community responses to gradual technological or environmental change may differ from those associated with acute change. Investigating community responses to a decades-long underground oil spill in coastal California that caused no visible environmental devastation or dramatic social harm, he found that the slow, emergent character of the spill precipitated social accommodation rather than acute response or reaction.

As discussed below, changes in social and economic life related to an emerging bioenergy industry experienced by people living in rural western Pennsylvania have been slow—more like the tempo of the emergent issue Beamish (2002) uncovered in coastal California than the sudden crisis caused by the flooding of Buffalo Creek. Area bioenergy development fosters change through the gradual introduction and incorporation of novel or unfamiliar technological and knowledge objects, including new crops and land management or marketing practices that introduce new risks and possibilities (Eaton and Wright 2015). Yet while the implications of change via bioenergy development may differ from the environmental disasters either Beamish (2002) or Erikson (1976) studied, recognizing that a group's culture comprises multiple tensions or counterpoints, and that members of a group may draw on different ends of their cultural continuum in response to an uncertain future, may offer new insights regarding community responses to emerging land use transitions and energy development projects.

To examine how cultural endpoints connect with the varying meanings individuals and groups are related to a new bioenergy industry, we connect the axes of variation concept with Social Representation Theory, which examines the social-psychological processes by which people make sense of new social objects, or "turn the unfamiliar familiar" (Moscovici 1988). This theory emphasizes two central processes: (1) *anchoring*, which connects new social objects with established and locally salient meanings; and (2) *objectification*, which links novel ideas with a group's commonly understood symbols or metaphors.

Examining these social-psychological processes provides insight into how people make sense of renewable energy development and how different, even contradictory, interpretations can emerge for new objects both within and across individuals (Jovchelovitch 2007). Reviewing research on public responses to renewable energy development, Susana Batel and Patrick Devine-Wright argue that new energy technologies are "deployed in specific historical and societal contexts, in which different representations coexist" (2015, 314) and that changes associated with this process result "in the coexistence of competing and even contradictory meanings, not only within the same culture, society and groups, but within the same individual" (316). Overall, their work suggests that "institutional, cultural and contextual/relational dimensions" (317) of renewable energy technologies shape meaning construction processes and affect community response. They further call for researchers to ask, "'When?' 'How?' and 'For what purpose?' different meanings are used at societal, group, and individual levels" (317).

In bringing Erikson's "axes of variation" (1976) together with Social Representation Theory, we posit that anchoring and objectification processes are rooted in groups' cultural endpoints. Together these two approaches suggest that groups' latent meanings and symbols—the bedrock of new social representations—span tensions common to that group. Conceptualizing a group's culture in terms of axes of variation opens a line of inquiry into whether and how drawing from one cultural endpoint versus another shapes how people anchor and objectify new social objects including energy crops.

In the section that follows, we present an overview of the energy and agricultural history of Crawford County, our western Pennsylvania case study site, to provide a backdrop on longer-term local processes of social and economic change. This foundation informs our subsequent discussion of how bioenergy crop production represents a new, unfamiliar social object for local people.

Background

Landscape, Agriculture, and Extraction

Our research took place in Crawford County, Pennsylvania. The landscapes of northwestern Pennsylvania and eastern portions of Ohio (see figure 8.1) resulted from the retreat of massive ice sheets 17,000–22,000 years ago that once covered this area of the Great Lakes basin. While farmland in Ohio and in adjacent Pennsylvania counties not in the glacier's path tends to be flat with well-drained soil, Crawford County has rolling terrain intersected with broad, flat floodplains and wetlands. The depressions common throughout the area formed when buried ice that broke from retreating glaciers later melted, leaving lakes, swamps, and irregular landscapes accented by hills, gorges and gullies carved by drainages and springs. On today's working landscapes, portions of land remain either forested or as wetlands, meaning most agricultural holdings rarely exceed a few hundred contiguous acres of tillable land.

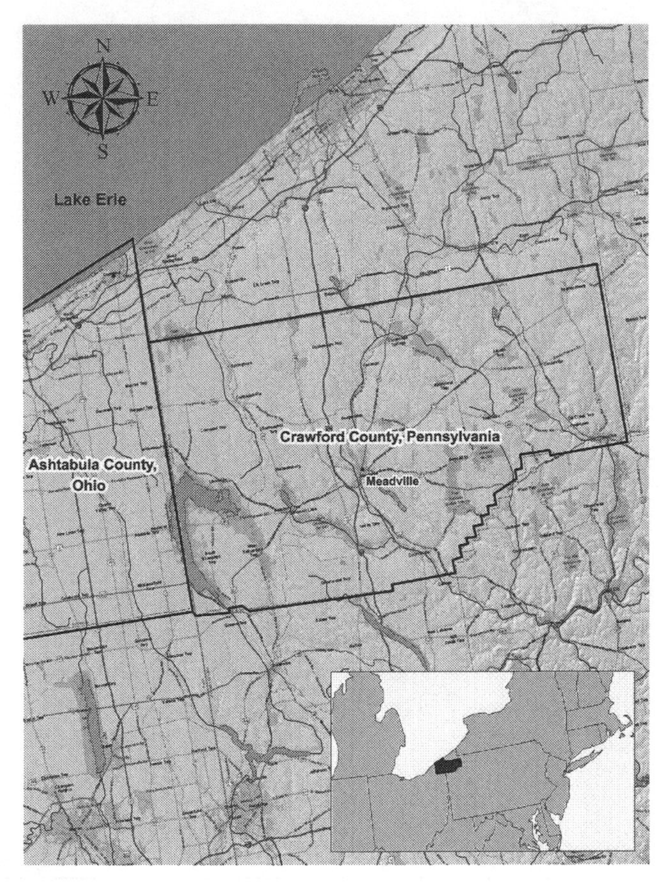

FIGURE 8.1. Crawford County, Pennsylvania. Created by James Spayd, Penn State Institutes of Energy and the Environment.

Retreating ice sheets are also responsible for the unique character of the area's soils (Crawford County Conservation District n.d.). Eroded glacial deposits provided the base for Venango-Frenchtown-Cambridge—referred to locally as "Frenchtown." This and other soil types in Crawford are highly erodible and poorly draining due to a lack of gravel. The combination of wet, hilly, poor draining and irregularly shaped fields have earned farmland in the area the term *marginal land*.

The county population grew dramatically after the Civil War, when Union Army veterans were awarded land or purchased land through the Homestead Act revision of 1870 and as the oil boom took off. The first successful drilling for oil in the United States occurred in 1859 in Titusville in southeastern Crawford County. Over the next forty years, development of the Oil Creek region intensified, driven by outside economic and political interests. Industrializing and polluting the western Pennsylvania landscape, the oil business in its heyday nonetheless served as an economic magnet for hopeful laborers, including many from Crawford County farms (Black 2000).

Similar to other parts of the country, farming in Crawford County transitioned away from highly diversified enterprises to greater specialization in dairy, beef, pork, and cash crops in the mid-twentieth century (Cochrane 1993). As late as the 1950s, farm markets in county urban centers such as Meadville remained packed with fresh produce, and the area's dairy industry still thrived. However, this phase of local agricultural life and economy was coming to a close. Technological innovations, including disposable packaging for milk, and the opening of national and global agricultural markets in the latter half of the twentieth century had a dramatic effect on the area's dairy industry—an industry already limited by the county's inability to scale because of small field sizes (Buttel, Larson, and Gillespie 1990). Likewise, corn and soybeans produced diminishing returns on investment in competition with the productive lands to the east and west and ongoing globalization of commodity markets.

Oil and gas extraction and speculation, and to some extent coal, have provided limited economic relief. As noted, Crawford County is home to the original oil boom of 1859–1870 and has since offered steady, though modest supplies of conventionally extracted oil and gas. Landscapes in Crawford are dotted with rusting wells and tanks installed during the previous century, with some still providing modest returns. The next oil and gas boom many had hoped for has, as of the early twenty-first century, yet to arrive. Although Crawford County sits promisingly above the Utica Shale, the Utica is thinner and deeper than the productive Marcellus, and therefore more difficult and expensive to drill.

For Crawford County residents who remained in agriculture, row crops and beef cattle have become alternatives to their predecessors' focus on dairy. Hay, straw, and forage crops; corn and soy; and increasingly beef cattle are the dominant commodities in Crawford County. Egg production—and, for a smaller number of farmers, pork, poultry, and a few very large dairy farms—also play a smaller but important role in the local agricultural economy. In 2012, average farm size in Crawford County was about 170 acres, and county farms had on average net cash farm income of just under $24,000 (USDA 2012). As elsewhere in Pennsylvania, fewer Crawford farmers now make their entire living from their farm operation, and many farming families rely on off-farm work.

These and other transitions in Crawford County's agricultural, extraction and manufacturing economies throughout the latter half of the twentieth and the early twenty-first centuries contributed to fluctuating rental and selling prices for farmland and other economic vulnerabilities, including decreases in residential property values and the tax base. Beyond driving residents to pursue nonfarm economic activities, this transition in the local economy had two other notable effects. First, due to disruptions in generations-long patterns of renting and owning farmland, new residents have bought land in the county, and others have moved to the area. Newer residents may have land use and management goals that differ from those espoused by longtime area farmers.

A second effect of the transition occurring in Crawford County is the growth of new land use, agricultural- and extraction-based industries over the past several decades. Commercial gravel and sand extraction now occurs on former agricultural land, and several large egg farms have come into production. Further, there has been a push to produce traditional crops for new markets. For example, many area farmers grow hay and straw not only for animal feed and bedding, but now also to supply western Pennsylvania's booming mushroom industry with material for compost. Yet while these newer industries entail purposes and practices long familiar to both older and newer residents, the area's growing bioenergy industry is one-step further removed from familiar farming practices.

Bioenergy as a New Social Object in Crawford County

Over the last decade, there has been a strong push in Crawford County to develop a bioenergy industry based on perennial grasses. For advocates to achieve their goals, landowners will need to grow perennial grasses, including switchgrass and miscanthus, to supply an emerging energy and bioproducts industry (Eaton et al. 2018). These perennial grasses resemble, but also differ from grasses such as fescue or timothy traditionally grown for hay. Residents here encounter the area's bioenergy industry through materials and events promoted by university research and Extension and government conservation organizations, as well as two companies that lease private land to grow energy crops. One of these firms, Ernst Conservation Seeds, leases and sometimes purchases land to grow a range of perennial crops—including milkweed, often viewed by locals as "weeds"—in order to harvest and sell seeds worldwide. Ernst's bioenergy division, however, has focused on establishing switchgrass—a native, perennial, warm-season clumping grass—for both seeds and biomass (Burnham et al. 2017). Ernst currently has 10,000 noncontiguous acres of switchgrass in production across Crawford County, which it envisions as providing pellets for home heating and other bioenergy applications. Due to what Ernst Seeds views as largely technical bottlenecks, however, the company's harvested switchgrass currently provides feedstock for grass pellets that it markets as absorbents for the oil, gas, and environmental management industries, rather than for producing bioenergy. In short, the energy pellets originally envisioned were difficult to properly pelletize and could not be readily used as an energy feedstock, while the high ash content was problematic for burning in residential-scale pellet stoves. Rather than a setback, however, proponents see the building up of a supply of biomass—no matter its current use—as necessary for the future "scaling up" of bioenergy across the region (Burnham et al. 2017).

Aloterra, a second firm operating just on the Ohio side of Pennsylvania's western border in Ashtabula County, also leases private, marginal land for growing bioenergy crops. Its business model varies from Ernst's in two central ways. First, rather than growing switchgrass, which is visually similar to other grasses grown as hay in the region, Aloterra's business model rests on growing and marketing the taller, visually distinctive grass miscanthus. While Aloterra selected miscanthus for its biophysical benefits (e.g., high-yield potential, ability to withstand poor soils), this particular crop has also been controversial, due to its rumored invasiveness. Organizations such as Penn State Extension have published fact sheets detailing the sterility of miscanthus, but suspicion that the plant could spread to neighboring fields remains evident in public discourse in Crawford County (Eaton et al. 2017). Second, rather than making absorbents, Aloterra has targeted end uses related to the renewable energy industry, including compostable table serviceware, and it has plans to develop a line of renewable building materials.

In short, while growing perennial grasses may not be entirely new for Crawford residents, leasing land to a private company rather than a neighboring farmer; living within fields of unusually tall, unfamiliar grasses grown on former dairy farms; and producing crops for energy and renewable materials rather than food purposes collectively make grass-based bioenergy crops new social objects. As we elaborate below, for some residents, bioenergy crops represent something positive for their land, family, and community and seem to "fit" with extant land use practices and cultural meanings in the area. However, others view bioenergy development as violating something sacred and traditional, and as reopening old troubles, or even opening new ones.

Research Methods

This study centers on seventeen interviews conducted with landowners in and around Crawford County. Interviewees represent a cross section of active and retired farmers who grow or grew hay and/or row crops on between 20 and 400 acres of land and about half of whom also had managed a dairy or small beef cattle operation. We also draw on a focus group held with local agricultural leaders, Natural Resource Conservation Service employees, Audubon Society employees, and a Penn State Extension professional, as well as an additional six interviews with local officials and employees from the area's two bioenergy companies.

We used three approaches to develop our list of potential interviewees. First, Penn State Extension professionals in the area provided contact information and entrée with area landowners. We also attended open houses, an animal auction, and other field day events organized by Penn State Extension and Ernst Conservation Seeds and used these opportunities to introduce our research and solicit interviews. Finally, we asked interviewees to suggest others in the area who were likely to have opinions different from their own on local bioenergy. We met interviewees either on their farm or in nearby public spaces. Interviews ranged from thirty-five minutes to two hours, typically lasting ninety minutes. We used a semistructured interview approach (Creswell 1994) in which we prompted interviewees to discuss the story of their land—how long they had owned their land, its biophysical properties (e.g., soil type, acreage), their goals for managing land, and what they hoped the future held for their land. We then asked interviewees about energy crops and their views on bioenergy development in Crawford County. Interviews were recorded and transcribed verbatim. Our analysis process involved identifying an initial set of codes for "land meanings" and "energy crops meanings" that were further sorted into the themes described below. To present our results, we use vignettes about the lives of individual landowners who were selected because their stories were particularly apt in reflecting our identified themes. We altered nonessential but potentially revealing details and used pseudonyms to protect the privacy of research participants.

Findings

Life in Crawford County is slowly shifting from rural land uses and farming practices that have been commonly shared, understood, and accepted toward new patterns and possibilities that now include bioenergy development. Similar to Erikson (1976), our interview findings show that people draw from different ends of a continuum of cultural resources to make sense of and take action in response to the changes associated with bioenergy. We illustrate two prominent counterpoints with a small selection of extended narrative accounts, rather than a larger number of shorter accounts, because fuller descriptions can show how the cultural counterpoints described below are intermingled with the meanings people here attribute to their land and land in their community. We then turn to the question of how cultural counterpoints drive the social-psychological processes suggested in Social Representation Theory by assessing the range of responses to ongoing bioenergy development evident in our interviews.

Cultural Counterpoints: Social Representations of Land in Western Pennsylvania

A common thread across our interviews with western Pennsylvania landowners is the strain they feel between, on the one hand, a sense of obligation to their family's previous generations and, on the other, a yearning for autonomy in the form of flexibility and control over their own land use decisions.

The phrase *moral obligation* aptly captures the sense of responsibility experienced by interviewees for continuing to manage their land in ways they believe prior generations of their family would expect them to. In particular, they felt meeting these expectations was the right or good thing to do, whereas failing to do so amounted to personal failure. This sense of duty to the past comes about through the entanglement of one's personal identity with the meanings given to the farmwork they and their family had undertaken on that land, including clearing land of fieldstone, brush, weeds, encroaching forests, or other symbols of wilderness. For some groups, these acts represent place protective behavior, and encroachments such as these are viewed as threats to farmland. Forfeiting the defense of these spaces amounts to a forfeit of one's sense of duty to past generations. Who are they to fail previous generations of their family by letting the wild return, unaddressed?

Consider the responsibility felt by longtime area residents Ray and Elizabeth for maintaining their family's land in agriculture. Nearing retirement age, this married couple owns 200 acres of tillable and forested land and invited us to their kitchen for our meeting. The sense of duty they felt to keep their land in agriculture, even when they themselves no longer farmed, came across in their response to our questions about their land management preferences. Ray detailed how he and his brother helped their father with dairy and then beef cattle, along with some cash crops, when they were young, but how later, after his father died and brother left for college, much of the tillable land was put into hay production. Ray eventually left farming full time after "getting a good job that pays pretty well . . . farming just didn't compete with it." But failing to keep his land—which is often wet but "makes pretty good hay"—in agricultural production was a "worry" for him.

To ease this worry and fulfill his commitment to earlier generations, he developed a relationship with his neighbor Josh, who kept Ray's land in production by growing switchgrass for Ernst Seeds, the local company venturing into bioenergy crops. As Ray explained, by having Josh take over management of his land, he no longer had "to worry about keeping the land open myself." When pushed to further clarify the "worry" he connected with his family's land, Ray explained that by having Josh maintain the land, "it's not growing up and that's all I care about . . . It's not growing up into weeds, red brush, briars, and trees, and it's just not growing up." When asked whether this would necessarily be a bad thing, Ray said, "I've put a lot of effort to get it where it is." At this point Elizabeth interjected: "His mom, who lives in the farmhouse over there [visible through their kitchen window], is eighty-six years old and worked her tail off all her life on this farm, and for her to see it grow up in weeds would rip her apart." Nodding in assent, Ray added: "My dad would look down on me with very sad eyes."

Other interviewees, including Richard, a retired veterinarian, echoed this theme. As a sixth-generation landowner who owned 200 acres originally settled and passed down by ancestors in the southern portion of the county, Richard shared the legacy struggle common to so many aging rural landowners: what will happen to my land if my children refuse to move home? While neither Richard nor his father personally farmed their family's land, a neighboring farm family who had leased their land—via several generations of handshake and oral agreements—had always kept the land in production. Now in his late seventies, Richard believed the grandson of the neighboring farmer his father dealt with was now in charge of the operation leasing his land. And the fact that Richard had not raised the price for leasing his land from what his father had negotiated many years prior reflects his sense of being beholden to these prior arrangements and the land use results they produced—all of which seemed to underpin his desire for continuity and his hesitancy toward change. As Richard explained, "Somebody told me that I should increase rent and so on. By the same token, I suppose I would be responsible for doing something else. The way it is, I don't do any of it; they do it all."

At the same time, some people we interviewed conveyed a *desire for autonomous control* over land use and land management decisions. While acknowledging the responsibility to continue farming their land, they also emphasized how important it was to them not to be beholden to the wishes, pressures, or plans of neighboring farmers, local community leaders, or, more abstractly, government regulation. In expressing this counterpoint to ones' sense of duty to past labors, these interviewees underscored that they valued being able to maintain flexibility in how they use their land and to plan for future land use. That is, the impulse toward control was cast in terms of land use practices—e.g., when, what, and where to plant—but also, and more generally, in terms of social preferences including when and where to work, who and what to work for, and how often.

The account given by Josh, neighbor to Ray and Elizabeth introduced above, of his life spent on his family's several-hundred-acre farm provides one striking example of how individual control confronts the moral obligations felt by many interviewees. Like Ray, Josh grew up on his father's farm and spent his youth helping his father and grandfather grow cash crops and raise beef cattle. After completing high school, Josh opted to work construction and took other odd jobs rather than farm up until his grandfather died.

He then began farming in earnest with his father and later with other partners until that experience, which he described as contentious, extinguished his remaining appetite for farming full time.

Now nearing retirement age, Josh lives on his farm with his wife, Elise, but has returned to a life of odd jobs that typically revolve around his owning hauling equipment. He also leases out much of his land, and his neighbor Ray's land, to neighboring farmers as well as to Ernst Seeds. While showing us around his property, Josh explained how his time spent taking irregular jobs related to transporting agricultural by-products, combined with leasing his land, provided the means to live life on his own terms. As he explained, "Working for people is real hard after you been self-employed all your life. I've done it!" His desire to be free of pressure from others to both live life and manage his land a certain way was most evident in our discussion of his relationships with his neighbors, Ernst Seeds, and others who lease his and his neighbor's land. As Josh put it, "It helps pay the bills. I still have control over it and I like control. I'm a control freak." He went on to illustrate this point with stories of how he could, and has, changed what was grown on his fields with short notice, but how by leasing out this land he does not need to spend all his time maintaining crops and can instead do as he wishes on most days.

However, Josh's desire for control ran up against his sense of duty to previous generations when he discussed his plans for the future of this land. As Richard's interview suggests, legacy issues and concerns were rife for Crawford landowners, and Josh's story was not unique: the future was uncertain in terms of what might happen to this land after he passed on. Unlike Richard, Josh has no children of his own. Josh's response to our inquiry into whether he had considered selling the land the land is especially telling: "I don't know that I've earned this piece of property, but I've been involved with it all my life." In reference to his father, he explained, "He played the game all his life; now he's going to hand us the ball and we're going to run with it." That is, Josh justified his sense of duty in terms of a feeling of inadequacy based on how he came to own this land, as compared with neighboring farmers: "Partly I don't want to change it because I inherited a lot [of the farm]. I didn't earn it in the way that a lot of farmers say, 'I bought all my ground.' Yeah you did and I didn't. But I worked my ass off when I was a kid and grown up as a teenager."

In sum, selling his property—a promise he had once made to Elise to provide for their retirement—was not something he felt he had the right to do.

And as he explained in discussing a recent sale of a small number of acres, this was okay to do because he had bought those acres himself, and the land was sold not to the highest bidder but to a party he was confident would continue using the land the way it ought to be used.

These brief accounts drawn from our interviews with Ray and Elizabeth, Josh, and Richard underscore a tension between, on the one hand, a sense of duty and obligation and, on the other, a desire for flexibility and control commonly stated across our interviews. As this section has illustrated, the cultural meanings and tendencies making up these counterpoints are rooted in the relationships people have with their farmland and other rural landscapes. Next, we turn attention to our second question: How do these cultural meanings guide the social psychological process of attributing meaning to energy crops in western Pennsylvania?

Cultural Counterpoints: Social Representations of Energy Crops in Western PA

Interviewees made sense of energy crops—constructed meanings for these new social objects—in terms of how growing them on their land (and all the new relationships this would entail) would "fit" with the cultural meanings they ascribed to that land. As shown above, ideas about the cultural meanings of land in this region are closely tied with the tension between moral duty and a desire for autonomy. In this section, we show how the social representations our interviewees attributed to energy crops were grounded in both ends of this tension in western Pennsylvania.

Landowners who emphasized their moral obligation to their ancestors saw energy crops as either a means for fulfilling one's duty to past labors or, alternatively, as an obstacle for doing so. Consider Karen, a landowner and part-time farmer in her thirties who, along with her husband, worked full time with one of the area's two bioenergy companies. Karen's job involved meeting with prospective clients to discuss opportunities for leasing their land with her company, and she discussed her and her husband's employment as a means to achieve a future in which her family would run its own farm operation full time.

As she described, both her personal commitment to farming, and the economic opportunities afforded to farmers like herself by the bioenergy company she worked for, provided means to protect and enhance the rural, farming character of her community. These considerations were important for Karen as she felt farmland and rural life in the area were threatened by myriad negative changes attributed to creeping urbanization in the area—most strikingly, an influx of new residents moving in to poorly planned housing developments built on former farmland:

> I look at land as being a precious resource. They are not making any more of it.
> Once it's bought and somebody puts a house on it, you're never going to get that back. Not in my lifetime.

Protecting working landscapes was essential to Karen, which was a challenge with the area's aging farm population, the necessary toil and labor involved with dairy and row crop production, and the high cost and risk of investing in the equipment necessary to compete as a farmer. However, Karen echoed other interviewees who contracted with area bioenergy companies in viewing energy crop production as a new opportunity for continuing on with a farming lifestyle even in the face of such threats:

> I look back on how hard my great-great grandparents worked for what they had. The people since them have worked so hard for what they had and all it took was one bad decision. Their equipment got repossessed and everything . . . if there would have been something like [our bioenergy company] where somebody was going to come in and willing to sign a long-term lease with you, that may have helped them in some way.

Karen's case shows how one version of the obligation many feel for continuing to farm their land—and for Karen this extends beyond her land into a duty to defend farming and farm life in her community—anchors the meaning she attributes to energy crops and the area's bioenergy industry.

For others, energy crops were part of the problem rather than the solution to continuing on with a familiar, but increasingly underappreciated and misunderstood way of life. This outlook was true for Patrick, a local landowner with a small beef cattle herd who worked in agricultural pharmaceutical sales regionally. During our breakfast meeting, Patrick cast energy crops, and the industry's success in the area, as threatening not only how he used his land but also how the area's farmland had once been used and ought to continue to be used. As Patrick emphasized, Ernst and Aloterra were successful farming-related enterprises, but at whose expense? Speaking from his personal experience working with clients across Pennsylvania, he interpreted the growing success of bioenergy crop production in the area as yet another threat to farmers producing traditional row crops, livestock, and dairy: "Even locally I hear my brother-in-law talk about it getting hard to find rented ground because of goddamn [local bioenergy crop firm]. 'They are renting up all the damn ground.' That's frustrating to me when you got dairy farmers that want ground and that's limiting them and they are saying we can't compete for ground anymore."

As this example suggests, salient symbols of obstacles to continuing long-standing agricultural and rural lifestyles—overly burdensome regulation, encroaching development pressure, and the like—also *objectify*, to use S. Moscovici's (1988) term, what it is bioenergy crops represent, even when those same crops are simultaneously seen as supporting area farming enterprises. Patrick's complaint was echoed by both some of the active and retired dairy farmers interviewed for this study, who made it clear that their enterprise was already under duress and that new schemes such as energy crops had no role to play in helping their dire situation. In short, a sense of duty to continue farming and support farm life in the community underpins both positive and negative social representations of bioenergy crops evident in western Pennsylvania.

In similar fashion, the desire for control that our interviewees expressed also connected with their representations of energy crops. A desire for control manifested repeatedly, with some variations, in which some stressed the freedom to make choices regarding their own land and others described control more abstractly as freedom from the constraints of existing farm production practices common for the area.

Our conversation with Jeffrey and his wife, Hanna, a couple in their late thirties who lived and farmed part time on sixty acres they had recently bought from her grandparents, provides a clear example of how desire for autonomy can converge with a positive interpretation of energy crops. While their stated purpose in maintaining their small mix of beef cattle, pigs, chickens, and horses was not to yield a profit, their operation did need to be self-sustaining. This need is where leasing land to Ernst for switchgrass fit into their enterprise. In further discussing his relationship with his land, Jeffrey also invoked the language of stewardship, suggesting the symbol through which he found meaning for energy crops: "Ernst got [our] property because there was a local farmer who farmed this for a while after grandpa died and he just raped it. He took out and never put in. Ernst came back and got it up to speed."

Keith, a retired Cooperative Extension employee who leased 200 of the acres he owned to an area bioenergy company, cast his decision to do so in terms of the opportunity that bioenergy development could provide to area farmers for escaping the yoke of production agriculture in an unproductive landscape—if only they would be willing to try something new. Here, alternative farming practices were possible, and bioenergy crops were one such possibility. For Keith, energy crops "solve a lot of problems within agriculture; it gives people alternative industries to go to. It's like we've been talking, once people got out of the dairy business, it's like what now? I'll row crop. No, you're not going to row crop; you're going to go broke doing it competing with Iowa."

Others who emphasized autonomy, however, objectified bioenergy by connecting perennial energy crops with earlier government biofuel programs, including corn ethanol, which symbolized government- rather than market-led agricultural development. This view was clear for Rex, latest in several generations of farmers to manage his family's 400 acres, which he had transitioned into commercial-scale egg production. While reticent to participate in the area's current bioenergy initiatives, which he believed involved undue government encroachment into agricultural markets, Rex viewed area bioenergy firms as making some strides toward economic self-sufficiency. He revealed that he had even weighed leasing some of his land for energy crop production:

> If [an area bioenergy firm] came to me and said, I'll rent your ground for 70 bucks an acre or 100 bucks an acre—which is two and three times what I'm getting for it now—then I would say I may have to spread manure on it once in a while. If they agree to that, then I'd let them have it.

In short, what kept him from investing his land in energy crop production was the concern that he might need that land for alternative purposes in the future, and, similar to the complex web of regulation he already navigated, bioenergy crops could limit his autonomy.

In sum, these various examples demonstrate how the meanings and symbols attributed to energy crop production in western Pennsylvania are grounded in the cultural counterpoints salient for landowners in this area.

Discussion and Conclusion

This chapter examined how the social-psychological processes suggested in Social Representation Theory can be examined in terms of a process whereby members of a group draw from across the continuum of cultural resources available to that group—what Erikson terms *axes of variation* (1976)—to invest meaning in novel or unfamiliar phenomena. In our case study of western Pennsylvania landowners, we found that the cultural resources salient for anchoring and objectifying bioenergy crops were *embedded in land meanings* community members drew on in order to make sense of new bioenergy crops and their implications for their land and their communities.

We conclude by offering three points about the implications of this research for extending Social Representation Theory with cultural theory and applying this synthetic approach to the question of public responses to renewable energy technologies such as bioenergy. The first point recapitulates our core argument: for Moscovici, people anchor and objectify new objects by drawing on extant meanings and symbols (1988). This finding implies that researchers should try to understand what meanings resonate for the population in question in terms of the relationships and practices that renewable energy developments require of landowners, farmers, or other relevant groups. Bioenergy crop production in rural places such as western Pennsylvania is envisioned by many economists and other social scientists as hinging on an individual landowner's willingness to grow or adopt specialty crop production systems on their personal land, where willingness or acceptance is studied in terms of a combination of economic motivation and structural factors including farm size (Eaton et al. 2018). This study contributes to this literature by demonstrating an additional influence on farmer interest in planting energy crops—whether and how potential growers connect cultural meanings for land with bioenergy crops. Accounting for the cultural meanings circulating in particular locales can be a daunting task; consider the dynamic, diverse, and seemingly contradictory land meanings we uncovered in western Pennsylvania. But avoiding such inquiry risks missing that the meanings people attribute to new energy technologies are rooted more broadly in what people in these places see as possible, ideal, or necessary for these landscapes.

Second, we have also shown how the store of meanings important for the social psychological process of selecting what new objects represent can be conceived in terms of Erikson's axes of variation (1976). Following Erikson, people draw from a store of cultural tendencies and meanings arrayed along a continuum bounded by two counterpoints. In this study, we found landowners experience a tension across, on the one hand, obligations to previous generations and, on the other, a desire for autonomous control of over land use and management decisions. And as we argued above, the tempo of change in a group's social setting may influence whether populations draw more heavily from either counterpoint. We found that in a slow change context, people draw somewhat more haphazardly across cultural continuums as they respond to a seemingly less certain future. When applying and further developing Social Representation Theory in the renewable energy context, future research should investigate locally salient meanings in terms of cultural counterpoints. Doing so could clarify which meanings are most active in anchoring and objectification processes, and shed light on how seemingly shared cultural resources within a specific group underpin competing views on what new social objects represent.

Our third point relates directly to our findings that showed how each counterpoint uncovered in western Pennsylvania underpinned both positive and negative representations of energy crops. Neither counterpoint aligned exclusively with either positive or negative representations of energy crops. That is, no matter which counterpoint was emphasized, landowners constructed energy crops as either *threatening* or *enhancing* their purpose, vision for, or obligation felt for their land and land in their community. This pattern resonates with Darrick Evensen and Rich Stedman's findings that diverse cultural meanings of the "good life" (2018) can be associated with both support for *and* opposition to fracking.

Others examine this puzzle in terms of the "symbolic fit" between place and technology (Devine-Wright 2011; McLachlan 2009). Our findings suggest further inquiry may be advisable, starting from the perspective of landowners, whose representations of new objects are important in terms of social responses, as well as energy technology and development promoters and experts, who envision social acceptance as key for their projects' success. As we have shown elsewhere (Eaton 2016; Rossi and Hinrichs 2011), renewable energy projects such as bioenergy crops are typically framed by promoters as providing a *new* means to solve what is envisioned as another rural problem, a frame captured in the phrase *rural revitalization*. By starting instead from the perspective of landowners, as we have intentionally done here, we uncovered little in the way of a desire for the *new*. Instead, landowners sought reason to believe that "the new thing" brought by promoters to their rural community could *enhance* already articulated goals and ongoing practices on their farms and the larger rural landscape. This distinction is more than semantic. Viewing a new technology as either threatening or enhancing a familiar way of life can underpin how people make sense of and respond to that new technology—either trying it, embracing it, or rejecting it altogether.

References

Batel, Susana, and Patrick Devine-Wright. 2015. "Towards a Better Understanding of People's Responses to Renewable Energy Technologies: Insights from Social Representations Theory." *Public Understanding of Science* 24 (3): 311–325.

Beamish, Thomas. 2002. *Silent Spill: The Organization of an Industrial Crisis*. Cambridge, MA: MIT Press.

Black, Brian. 2000. *Petrolia: The Landscape of America's First Oil Boom*. Baltimore: Johns Hopkins University Press.

Burnham, Morey, Weston Eaton, Theresa Selfa, Clare Hinrichs, and Andrea Feldpausch-Parker. 2017. "'The Politics of Imaginaries and Bioenergy Sub-niches in the Emerging Northeast U.S. Bioenergy Economy." *Geoforum* 82 (June): 66–76.

Buttel, Fredrick H., Olaf F. Larson, and Gilbert W. Gillespie Jr. 1990. *The Sociology of Agriculture*. New York: Greenwood Press.

Cochrane, Willard W. 1993. *The Development of American Agriculture: A Historical Analysis*. Minneapolis: University of Minnesota Press.

Crawford County Conservation District. n.d. Shenango River Watershed. Accessed October 10, 2017. http://www.crawfordconservation.com/watersheds/.

Creswell, John W. 1994. *Research Design: Qualitative and Quantitative Approaches*. Thousand Oaks: Sage.

Devine-Wright, Patrick. 2011. "Place Attachment and Public Acceptance of Renewable Energy: A Tidal Energy Case Study." *Journal of Environmental Psychology* 31 (4): 336–343.

Devine-Wright, Patrick, Susana Batel, Oystein Aas, Benjamin Sovacool, Michael Carnegie Labelle, and Audun Ruud. 2017. "A Conceptual Framework for Understanding the Social Acceptance of Energy Infrastructure: Insights from Energy Storage." *Energy Policy* 107 (August): 27–31.

Eaton, Weston M. 2016. "What's the Problem? How 'Industrial Culture' Shapes Community Responses to Proposed Bioenergy Development in Northern Michigan, USA." *Journal of Rural Studies* 45 (June): 76–87.

Eaton, Weston M., Morey Burnham, C. Clare Hinrichs, and Theresa Selfa. 2017. "Bioenergy Experts and Their Imagined 'Obligatory Publics' in the United States: Implications for Public Engagement and Participation." *Energy Research and Social Science* 25 (March): 65–75.

Eaton, Weston M., Morey Burnham, C. Clare Hinrichs, Theresa Selfa, and Sheng Yang. 2018. "How Do Sociocultural Factors Shape Rural Landowner Responses to the Prospect of Perennial Bioenergy Crops?" *Landscape and Urban Planning* 175 (July): 195–204.

Eaton, Weston M., and Wynne Wright. 2015. "Hurdles to Engaging Publics around Science and Technology." *Michigan Sociological Review* 29 (Fall): 48–74.

Erikson, Kai T. 1976. *Everything in Its Path: Destruction of Community in the Buffalo Creek Flood*. New York: Simon and Schuster.

Evensen, Darrick, and Rich Stedman. 2018. "'Fracking': Promoter and Destroyer of 'the Good Life.'" *Journal of Rural Studies* 59 (April):142–152.

Jovchelovitch, Sandra. 2007. *Knowledge in Context: Representations, Community and Culture*. New York: Routledge.

Lowe, Philip, Jeremy Phillipson, and Katy Wilkinson. 2013. "Why Social Scientists Should Engage with Natural Scientists." *Contemporary Social Science* 8 (3): 207–222.

McLachlan, Carly. 2009. "'You Don't Do a Chemistry Experiment in Your Best China': Symbolic Interpretations of Place and Technology in a Wave Energy Case." *Energy Policy* 37 (12): 5342–5350.

Moscovici, S. 1988. "Notes towards a Description of Social Representations." *European Journal of Social Psychology* 18 (3): 211–250.

Pellegrino, Margot, and Marjorie Musy. 2017. "Seven Questions around Interdisciplinarity in Energy Research." *Energy Research and Social Science* 32 (October): 1–12.

Rossi, Alissa M., and C. Clare Hinrichs. 2011. "Hope and Skepticism: Farmer and Local Community Views on the Socio-economic Benefits of Agricultural Bioenergy." *Biomass and Bioenergy* 35 (April): 1418–1428.

Stedman, Richard C. 2003. "Is It Really Just a Social Construction? The Contribution of the Physical Environment to Sense of Place." *Society and Natural Resources* 16 (8): 671–685.

Swidler, Ann. 1986. "Culture in Action: Symbols and Strategies." *American Sociological Review* 51 (April): 273–286.

USDA. 2012. "County Profile: Crawford County, Pennsylvania." *United States Department of Agriculture, National Agricultural Statistics Service*. Accessed December 8, 2017. www.agcensus.usda.gov.

Chapter 8 Summary

This chapter presents a theoretically informed qualitative case study of bioenergy development in western Pennsylvania's Crawford County. This region has been experiencing a slow transition away from traditional agricultural and energy extraction industries and toward new land uses including perennial bioenergy crop production on marginal or abandoned agricultural land. The study draws on interviews with both commercial and retirement/lifestyle farmers and with personnel at a local bioenergy crop enterprise, as well as on a focus group with local agriculture and environmental professionals. The chapter shows how the cultural meanings and norms people connect with the working landscapes they own or manage inform how they respond to the prospect of planting unfamiliar bioenergy crops on their land. Crawford County landowners experienced a tension across two cultural endpoints for managing working lands. On the one hand, landowners felt a moral obligation to continue managing their land in ways they believed prior generations of their family would expect them to and that continuing these traditions was the right thing to do. At the same time, landowners described a yearning for autonomy from those obligations in the form of flexibility and individual control over land use decisions in the present and into the future. The chapter shows how landowners in this region of Pennsylvania anchored their meanings for newly promoted bioenergy crops in accordance with the broader cultural endpoint they emphasized for their land. In short, meanings for bioenergy crops were forged in preexisting cultural understandings for the land on which these new crops would be grown. The chapter makes the case that we need to look for, rather than to bracket out, place-based meanings and cultures in energy technology transition projects.

KEY TAKEAWAYS

- Alongside the current focus on economic motivations and demographic and structural factors, future research on landowner willingness to grow bioenergy crops should study whether and how potential growers make sense of new crops in terms of extant cultural meanings for their land.
- Cultural counterpoint theory can help provide a more nuanced application of Social Representation Theory to questions of public responses to novel technologies. Identifying cultural tensions can help make explicit the store of meanings social actors draw from to anchor and objectify new things or ideas. Moreover, examining cultural tensions can shed light on how and why responses to novel technologies may vary across a group with a shared culture.

- Cultural endpoints did not shape responses to bioenergy crops in direct fashion. Instead, no matter which cultural counterpoint was emphasized (moral obligation to continue prior traditions vs. desire for flexibility and autonomy), landowners constructed bioenergy crops as either *threatening* or *enhancing* their vision or felt obligation for land management.
- Bioenergy crops are typically framed by promoters as providing change in terms of a new or novel solution to some long-standing rural issue. However, our findings suggest landowners that choose to grow bioenergy crops see these crops not as changing but as enhancing their livelihood, vision, or sense of responsibility for managing their land.

9

The Wider Array

A Qualitative Examination of the Social and Individual Impacts of Hydraulic Fracturing in the Marcellus Shale

CHRISTOPHER W. PODESCHI, LISA BAILEY-DAVIS, HEATHER FELDHAUS,
JOHN HINTZ, ETHAN R. MINIER, AND JACOB MOWERY

Energy "boomtowns" received considerable research attention starting in the 1970s. During this time, relatively isolated rural communities, most in the Rocky Mountain West, became host to personnel needed for nearby energy projects (see Cortese and Jones 1977). The increase of population and economic activity brought prosperity for some but also created challenges. Today, horizontal drilling coupled with hydraulic fracturing (HDHF) has become a prominent means of extracting oil and natural gas from shale formations in the United States. Like earlier boomtowns, this labor- and technology-intensive approach creates considerable impacts in communities that sit atop these formations.

Careful study of HDHF-impacted areas is necessary because HDHF has important differences from "classic" boomtowns. Foundational boomtown research examined communities in states such as Colorado, Utah, or North Dakota, where energy projects were miles away from communities on sparsely or unpopulated land. Australian research has similarly focused on energy projects in areas so remote that "purpose-built" towns were needed

DOI: 10.5876/9781646420278.c009

(see Lockie et al. 2009). Impacts in these "classic" boomtowns stem primarily from increased population and economic activity.

By contrast, since HDHF taps into expansive shale formations, it involves numerous well pads scattered throughout a region, often in close proximity to communities and homes. In fact, population densities in many HDHF zones (e.g., the Marcellus or Barnett Shale) are higher than anything seen near energy developments from foundational boomtown research (Jacquet and Kay 2014). This proximity means people in HDHF zones experience the "classic" population and economic impacts, but many also experience the industry and its business practices firsthand. As Colin Jerolmack and Nina Berman put it, "there is something distinct about how fracking reshapes the rhythms of community life and people's relationship to the land that has received scant attention . . . Companies routinely operate right in lessors' backyards. In this way, fracking is more *intimate*" (2016, 194; emphasis in original).

Our work contributes to a growing literature examining HDHF. During 2013 we conducted nine focus groups in three HDHF-impacted communities in Pennsylvania's Marcellus Shale zone. As qualitative research, this project shares impetus with a number of recent studies (see Anderson and Theodori 2009; Brasier et al. 2011; Fernando and Cooley 2015). The current study adds to a growing list of qualitative work on HDHF that has value for relying on data from the general public rather than from key informants (e.g., Ellis et al. 2016; Jerolmack and Berman 2016).

Our findings broadly support the model developed by Jacquet and David Kay; that is, this unconventional extraction technique creates an unconventional sort of boomtown (2014). Our participants talked about a wide array of impacts. Their experiences and perceptions resonate with "classic" boomtown research but also include challenges that clearly stem from the fact that the industry drills numerous wells over a broader area, bringing the energy industry into proximity with the places people call home. Of particular note are themes not given significant emphasis in the current literature, namely, participants' dim view of how the energy industry operates, concerns about environmental problems, and a sense of loss over transformed places. Given the fact that some landowners experience significant financial gains, economic inequality and related challenges appear to be another new aspect of these communities. These themes have been raised in the literature (e.g., survey researchers have measured concerns about inequality), but our data show clearly how these issues have mattered in people's lives.

After discussing methodology, we present our findings in two sections. First, we show connections between what our participants shared and "classic" boomtown impacts. Second, we examine what is distinct about HDHF's social impacts in higher-population density contexts such as the Marcellus Shale zone. Rather than reviewing literature ahead, in each section we show how our findings resonate with "classic" boomtown research and with recent work focused on HDHF.

Data and Methods

Three nonneighboring communities in Lycoming and Tioga County were chosen to host focus groups. Three were conducted in each community for a total of nine focus groups. Each community has a population of less than 5,000 and experienced a heavy wave of HDHF that was waning by the time of our focus groups. Potential participants were drawn from a random sample of publicly available data. Lists were drawn from the communities' primary and neighboring zip codes so that we sampled town-dwelling residents and residents who lived closer to HDHF activity. Participants were provided a $40 incentive gift card, sandwiches, and refreshments.

Ninety-six individuals agreed to participate, but 47 did not attend. Of the 49 individuals who attended, some brought spouses. In all, 68 people participated in the nine focus groups. A brief survey was administered during recruitment (spouses were not surveyed). Demographically, the surveyed participants were 54 percent male and were older and more educated than the population local to each community—the average age was fifty-seven and 78 percent had more than a high school education (US Census Bureau 2015a, 2015b). While these demographic differences from the local populations may have impacted responses, it is important to note that we also asked participants' views of HDHF. Results in table 9.1 clearly show that our participants arrived at these focus groups with a wide variety of opinions about HDHF.

Our ten-question focus group schedule asked participants to think about their community before asking them to consider HDHF. We asked for free responses to HDHF and then focused on personal and community-level changes participants believed were caused by HDHF. We also asked if there

TABLE 9.1. HDHF opinion summary

Feelings about HDHF (N = 47)				
Very Unhappy	Unhappy	Neutral	Happy	Very Happy
6%	17%	43%	23%	11%
Effect of HDHF on the environment (N = 48)				
	Extremely Risky	Somewhat Risky	Somewhat Safe	Extremely Safe
	13%	46%	35%	6%
Government should . . . (N = 47)				
	Ban HDHF	Heavily Regulate HDHF	Somewhat Regulate HDHF	Not Regulate HDHF
	9%	36%	47%	9%
Since HDHF began, are you . . . (N = 47)				
		Worse Off	About the Same	Better Off
		28%	51%	21%

was conflict about HDHF and what they felt the future held. Our semistructured approach allowed moderators to probe, avoiding "the tendency to follow a predetermined order of topics in rigid fashion" (Morgan 1997, 47). Effort was made to create an environment in which participants could speak freely, stressing that all perspectives were welcome and involving each participant as much as possible. New topics began with new participants, and three questions required participants to jot answers independently before anyone spoke.

Analysis was inductive, typical of a "grounded theory" approach. Effort was made to "follow the data" rather than code in a manner informed by past research. "Constant comparison" of dialogue generated categories and their properties (Glaser and Strauss 1967, 23). Open coding focused on broad topics. Data fitting initial codes were then subjected to further comparative analysis, revealing more specific properties. As a final step, our findings were compared to relevant literature (Glaser and Strauss 1967, 37).

Findings

Our focus groups showed that HDHF disrupted lives and communities and created worries and uncertainties about the future. Critical dialogue was prominent. In fact, the focus groups at times had the tenor of a group therapy session. There were participants supportive of HDHF, but only a very small number were uncritically so.

No Surprises

Much of what our participants raised would be expected in any boomtown that sees rapidly increasing population and economic activity. Rather than a condensed and abstract discussion of these findings, however, they are given full treatment here to preserve the authentic experience and perceptions of our participants and to communicate the significance of these impacts. This approach seems important in light of the argument that the regional nature of HDHF may diffuse and limit at least some of the expected impacts (Jacquet and Kay 2014).

Discussions of positive economic impacts served as an exception to the generally critical tenor of the sessions. Whether an energy boom truly improves a local economy is uncertain (Freudenburg and Gramling 1998; Tsvetkova and Partridge 2016). It is, however, unquestionable that booms bring increased economic activity and prosperity for some. Many participants were appreciative of this. More common than talk about windfalls for property owners, participants emphasized how HDHF brought jobs, many with good wages. Attention was also given to benefits for local businesses that provided goods or services to the natural gas industry or its workers. The local service industry was commonly identified. For example, Mrs. Jones,[1] from the Southwestern community (SWC), said the owners of the "mom and pop" store in town were "becoming like millionaires."

Participants also, however, raised economic caveats. Some found HDHF jobs problematic for their hours or working conditions or because the best jobs were going to outsiders, something reinforced by research (Christopherson 2015). Worries about the looming "bust" or lack of long-term benefit was raised in seven focus groups. Furthermore, the downturn in gas industry activity by the time of our focus groups was producing problems. Mrs. Brown of the Northern community (NC) said some local people obtained loans to make investments in vehicles or equipment but with demand for support services waning, "their trucks are sitting there or they had to sell them, or they went bankrupt." With limited capacity for absorbing population, small communities also see inflation during boom periods (Akbar, Rolfe, and Zobaidul

Kabir 2013; Fernando and Cooley 2015; Williamson and Kolb 2011). Some participants focused on prices for everyday goods and services, but claims about housing costs were most prominent, for example, mobile homes renting for $2,000 per month. Consequences of these increases, like increased homelessness, were raised in each community.

Increased population and economic activity can also harm quality of life by straining public and private services and infrastructure (Cortese and Jones 1977; England and Albrecht 1984; Ennis, Finlayson, and Speering 2013). Such challenges came up in seven focus groups, though it was not a topic that got sustained attention. Challenges for public schools were mentioned most, along with a handful of comments about things such as sewer systems, overcrowded restaurants, and fiscal challenges for municipalities that invested in infrastructure. There was no mention of strain on emergency services.

By contrast to these material issues, much boomtown literature focuses on the rural social fabric. Freudenburg emphasized how quality of life in these places rests on "density of acquaintanceship" or "the average proportion of the people in a community known by the community's inhabitants" (1986, 29–30). These ties are a foundation on which other aspects of community are built because people "know each other well enough to be willing or able to predict and depend on one another's behavior" (31). They also facilitate important social functions, including control of deviance and socialization of youth. As newcomers arrive, density of acquaintanceship declines and community can be harmed.

Researchers have found that boom periods harm quality of life or community social fabric, noting evidence of decline in support behaviors, trust, community attachment, community satisfaction, and the like (Brown, Geertsen, and Krannich 1989; England and Albrecht 1984; Petkova et al. 2009; Smith, Krannich, and Hunter 2001). Dialogue in our focus groups did raise these exact phenomena, but certain types of dialogue suggested as much. For example, participants often explained inflation not in terms of changing market conditions but in terms of the greed of local business owners and landlords. Mrs. Strong of the NC clearly showed reduced community satisfaction from the changes newcomers caused:

> I'm not sure that the people that are actually the natives in this part of the country really want this kind of an influx . . . maybe we don't want all this

> infrastructure, hotels and motels, and you know, all this stuff that's coming in. I like my town the way it was. I don't really want to see it change that much . . . And I don't know why every little town in the world has to suddenly try to become a big one. I don't want it. I really don't want it.

Similarly, a female participant from our Southeastern community (SEC) had considered moving away from her lifelong hometown—she did not want her kids to grow up in what was to her becoming a city. Others considered moving away too.

For some, the newcomers themselves were the problem. Mrs. Mitchell from the SEC emphasized that multigenerational families comprised her community and was concerned that newcomers were changing its makeup. According to Mrs. Moyer, from a different SEC focus group, newcomers were "people you're not familiar with, who don't care about the town, I guess, like the way you care about it." Others emphasized newcomers' reckless or irresponsible behavior. In the SWC and the NC there was talk about conflicts between newcomer and local youth in the schools. In each community participants spoke of rude behavior by gas industry workers, for example, partying, littering, behaving raucously in public spaces. Large white pickup trucks with out-of-state plates were symbols of the gas industry *and* aggressive driving. Rick Ruddell and Natalie Ortiz argue that while "antisocial behavior, physical disorder . . . and property damage, incivilities, and traffic offenses" (2015, 141) are not often considered in boomtown studies, they should because they harm quality of life. Our participants would appear to concur.

Research from "classic" boomtowns also finds evidence for increased rates of crime (Freudenburg and Jones 1991; Seydlitz et al. 1993). Not surprisingly, claims about increased crime were made in six of nine focus groups, representing each of the three communities. In addition to generic comments about higher crime rates, participants cited bar fights, knife fights, a shooting, stabbings, and increased drug use. Mrs. Walker of the SEC said there had been attempts at child abduction.

"Classic" boomtowns also see increased crime fear, even absent evidence of increased crime (Krannich, Berry, and Greider 1989; Smith, Krannich, and Hunter 2001). From the SWC, a natural gas industry official stoked Mrs. Jones's fear by telling her husband not to confront workers if they behaved poorly because the industry was "hiring them from all over" and did not know if they had criminal records. Mrs. Walker from the SEC was stressed because she had children and, in addition to her view that child abduction attempts had occurred, she claimed registered sex offenders were among the newcomers.

Resonating well with Freudenburg's density of acquaintanceship argument (1986), Mrs. Walker added, "It's not just a small-knit community anymore—now you have to worry about the person on Main Street whereas before it was just maybe somebody you graduated high school with." Mrs. Jordan of the NC said she no longer leaves her keys in her car or her house unlocked "because, like, there's a lot of people that we don't know. And they don't wave and they look unfriendly."

Unconventional Boomtowns

Our participants spoke of new impacts as well. Much of what we learned reinforces other recent work on HDHF, but some of our findings are unique or have not been given the emphasis our data suggest is needed.

Windfalls

Rather than one large energy project near the community, HDHF can involve hundreds of wells in and around a community on relatively densely populated rural land. This means the industry negotiates with numerous mineral rights owners, and many will benefit financially. As others have found (Fernando and Cooley 2015), participants spoke frequently about these windfalls. Echoing discussions of community development and "wealth retention" in the literature (Jacquet and Kay 2014), Mrs. Kelly from the SEC even saw it as "bust insurance": "We have had booms here before, but not like this. We had coal. We had timber. And generally speaking, the money from those resources went someplace else. I think more of this money is here because they had to sign leases. So, there was at least some initial money, and it has stayed here."

Likewise, there was talk of windfalls helping to save dying farms. Mr. Hart of the NC said some farmers attracted their children back home to work the farm because of improvements they had been able to make. Mr. Ford, from a different NC focus group, even predicted that by saving dying farms, HDHF might allow the rural setting to survive, albeit "in an odd way."

Water, Air, and Catastrophes

Environmental and related health concerns received much attention in our focus groups. Other kinds of large-scale energy projects certainly have environmental consequences, but HDHF's potential environmental impacts seem more salient because HDHF takes place in proximity to people's homes and water sources. Our participants are not NIMBYs; that is, they aren't fighting to keep HDHF *out of* their backyards. For many, HDHF is *in* their backyards. This location raises the stakes and makes environmental concerns a new source of difficulty.

By the time of our focus groups, HDHF had received considerable media attention primarily for the threat of groundwater contamination. It is thus no surprise that water was prominent in every focus group. The concerns, however, seemed authentic. Common to much dialogue about water was worry about current and future health problems. Mr. Thompson from SWC wondered if his recent cancer had been caused by contaminated water. Mr. Cruise from the NC worried about the secret proprietary chemicals being used,[2] and he said that even though the shale is deep underground, water contamination could eventually occur.

Participants also shared stories of known contamination events. Mr. Long of the SEC spoke of a truck that had a spill near multiple streams. The situation was troubling to him because the company responsible was slow to report the spill and he worried there was no way to test for the proprietary chemicals that had spilled. Mr. Kent, from the NC, told of an intentional wastewater spill that polluted a stream and killed a group of goats. Others spoke of household well contamination. A male participant from the SEC knew a family that came down with vomiting and soreness in the knees: "And none of this existed until like . . . they kinda put two and two together 'cause there was a well not too far away from them." From another SEC focus group, Mr. Becker spoke about a friend who had methane in his well water. According to Mr. Becker, his friend regretted leasing, saying, "I had my hand out. I wanted the money. Now I'm paying the price." The financial bind this development can put people in was also raised. Mrs. Stanley of the NC said the water supply had been destroyed on several farms in the area and added, "Nobody's going to buy that land . . . those people have lost their income *and* their assets are gone."

Worry about air pollution was fairly prominent as well. For some, diesel fumes from convoys of hauling and equipment trucks were the problem.

Mr. Jones of the SWC said topography mattered: "That's like them trucks going down our valley. You can't even breathe if you go out there in the day. The fumes from them trucks, man, it just hangs in the valley." Others discussed the intentional flaring of gas. A male participant from the SEC said, "don't tell me that, that just perfectly clean gas is being flared off." Mrs. Stone from the SWC expressed concern about possible cancer and said, "Who knows what health issues that's gonna create . . . they're burning it off, and it's going into the air that we breathe."

Participants feared catastrophic events, too. Mrs. Knight from the NC worried that her community would become like Centralia, Pennsylvania, a town abandoned because coal seams continuously burn beneath it. Mr. Horn, from a different NC focus group, worried about earthquakes and reported adding an "earthquake rider" to his homeowner's insurance. Participants from the other communities worried about explosions at well pads as well as collapsing mountains and sinkholes caused by the fracturing of the shale.

Mostly using survey results others have reported on environmental concern or worries about catastrophes in HDHF zones (Schafft, Borlu, and Glenna 2013; Stedman et al. 2012; Theodori 2009; Willits, Luloff, and Theodori 2013), though the findings of Ellis et al. (2016) and Jerolmack and Berman (2016) align more closely with ours. It is important to emphasize that even if a claim or concern is unfounded, the worry and reduced quality of life are real. Given how commonly environmental impacts were discussed, as well as how there is clearly real concern for near and long-term health impacts, our data suggest that this concern needs to be taken seriously as a potential impact on individual and community well-being.

Disturbed Country

Given the geographic spread of HDHF, day-to-day industry operations were bound to be important, something documented in numerous recent HDHF studies (Brasier et al. 2011; Fernando and Cooley 2015; Jacquet and Stedman 2013; Ladd 2013; Schafft, Borlu, and Glenna 2013; Theodori 2009; Willits, Luloff, and Theodori 2013). Not commonly raised in this literature is the workers themselves as a problem. Participants in the NC and the SEC reported discomfort when they noticed industry workers walking on their land unannounced. Presumably they were surveying or doing other work, but this practice created discomfort nonetheless. A more intrusive disturbance was light pollution from active well pads and flaring. Mr. Long of the SEC had recently moved to the area and said, "You can see the fire in the sky at night sometimes . . . This was not what I wanted." Complaints about noise were more common. From trucks to flaring to compressor stations to seismic testing, HDHF can be a major aural presence. When asked about health concerns, Mr. Taylor of the SWC spoke about noise:

> You know, noise has been proven to be a great stressor. And we're all being exposed to way more noise. Whether it's the helicopters, which I had trouble with, trucks up and down the road, drilling pads. I mean, it's just more intrusion by noise and I think probably for the short-term . . . health concerns, it's noise! It's the noise.

Industry vehicle traffic was the most identified source of noise. According to Mr. Thompson, from a different SWC focus group, "I'm always saying, 'Can you say that again when there's no truck?'"

Displeasure with all manner of natural gas industry vehicles was in fact one of the most common topics. In addition to noise and air pollution complaints, participants focused on three other aspects. First, many participants were scared and upset by what they saw as reckless driving. Comments focused on aggressiveness, high speed, and unexpected passing by the white pickup trucks commonly used by gas industry employees. Hauling and equipment trucks reportedly driving too fast on narrow roads created fear as well. In response to a question about health impacts of HDHF, Mr. Thompson of the SWC said health effects from wells did not worry him; he was concerned about getting hit by a truck. Second, there was talk of convoys of heavy vehicles that damaged rural roads. For some this was an annoyance, but we also heard of roads made impassable, creating major inconveniences for people. Finally, local roads could not handle the volume HDHF required, and so local roads would become clogged with convoys of hauling or equipment trucks. Mrs. Howard from the SEC spoke of a man with a lucrative lease but now also contaminated water and a sick spouse needing regular doctor visits. She remarked: "Waiting for the water trucks to go by while he's waiting to take her to an appointment and so on . . . I know that he is specifically seeing the other side of the coin."

In other boom scenarios, the welter of industrialization and extraction is distant from populated areas. In contrast, HDHF brings the industry's activities close to home. Jerolmack and Berman nicely sum up the way these challenges create disturbed country: "Like the traffic, the smells, noise, and light pollution from fracking do not stop at the property lines of those who hold or profit from gas leases. Everyone experiences a diminished capacity to access peace and quiet, unbroken vistas, dark skies, unhurried country roads, and the sounds of nature" (2016, 203).

Altered Land

Hydraulic fracking has a considerable footprint from the clearing of land for well pads, pipelines, and road building. Participants often spoke directly about changes to the character of their communities and familiar landscapes. Dialogue about these changes came up in every focus group, and several participants seemed truly anguished by changes to familiar views. Mr. Gilmore of the SWC said:

> I've looked at mountains for sixty some years, and now you look at the mountain and there's a great big scar going right up on it. A great big squath of trees all cut out and everything. They're not mountains anymore. You know what I mean? Now they got big squaths and pipelines going up over and everything like that. I just feel it's terrible that what's happening.

Also from the SWC, Mrs. Martin talked about a mountain road "that was the most beautiful to ride" but had been ruined by wastewater ponds and truck traffic.

Additionally, participants spoke of changes to special wild places where they liked to spend time, a theme prominent in Jerolmack and Berman's (2016) work as well. Mrs. Brown of the NC said: "It's been especially heartbreaking to be on my bike, my mountain bike, out in the middle of nowhere, where you never see anybody, and then there's a road big enough for a tractor trailer to traverse." Mr. West and Mrs. Stanley from the NC were upset about how the natural gas companies have access to state game lands and state forests. Mrs. Stanley spoke of loss of access to a place where she used to pick wild blueberries. Mr. West made his feelings about HDHF disrupting a special place very clear:

> Oh I mean, it was like, the Louvre in Paris. You know, nobody should touch that. It should be left like that. It's so perfect. And now they're blowing the

> road open. They're—I seriously thought about—what would they do, or what would I do if I knew that I had incurable cancer. I had six months to live. Would it really be worth it to get a gun and go up there and wait for some of those sonofabitches to come by and shoot them? You know, not hit the man, but to blow holes in the truck.

All of this suggests that place attachment can make HDHF's disruption more difficult. Familiar places contribute to our sense of self and evoke memories that imbue spaces with meaning (Bell 1997; Proshansky, Fabian, and Kaminoff 1983). This bonds people to places to an extent they may not realize until some change—HDHF, for instance—occurs. Furthermore, the relatively high population density in rural Pennsylvania where HDHF occurs has meant that farmland and forested mountains that many people know well get altered. Given this "footprint" and what our participants shared, along with evidence that attachment predicts concern for places (Vorkinn and Riese 2001), it is surprising that Jacquet and Richard Stedman found place attachment did not impact attitudes about HDHF (2013).

Business Practices

Natural gas industry business practices served as a new stressor as well. This sort of talk was abundant in our focus groups, something surprisingly not prominent in other recent HDHF research (exceptions include Ladd 2013; Wynveen 2011). Again, the nature of HDHF means companies will have business dealings with numerous people and numerous people will witness their on-the-ground operations. This means there is ample opportunity for opinions to form, and apart from a few exceptions, opinions were critical. Our findings may help clarify why Fern Willits, Albert Luloff, and Theodori (2013) found persistence of low levels of trust toward the natural gas industry (Stedman et al. 2012).

Some participants talked about irresponsibility in the process of drilling, and so the environmental fears and nuisances already discussed fit here. Many complaints about the industry, however, focused on communication or negotiations with landowners. Generally, dialogue portrayed the industry as savvy, secretive, and dishonest, often intentionally withholding information. In the NC, for example, the industry was faulted for failing to alert communities about the housing shock HDHF would bring. Some participants spoke of how the industry "knew the game" and easily took advantage of naive local people. From the SEC, Mrs. Mitchell's negative experiences led her to characterize the landowner–gas industry relationship as a "Samson and Goliath" situation. Mr. Horn of the NC said, "My experience with the gas companies is that—that they came in. They rushed in and they trampled over the local landowners. And everything they told me, just about, is a lie."

"Land men," who establish contracts with landowners, were common targets of criticism. Participants from the SWC reported some landowners signed for $2.50 per acre while others received $2,500 an acre. Mr. Ford of the NC described tactics: "The land man would actually say, 'Well, don't tell your neighbor, but I'm giving you a good deal.' [*laughs*] And people kind of went into that." In keeping with this tactic, Mr. Horn was very upset at the unequal terms he and his neighbor received. He said the land man insisted, "like he swore on a stack of bibles," that his company would not agree to more than $20 per acre: "So like a sucker, I signed it, and then they paid my neighbor $2,000."

"Fine print" was another sore spot. A participant from the SEC was frustrated when an industry employee "stuck a piece of paper" in his face that gave a company's permission to conduct seismic testing on his property on a Sunday afternoon. Participants from the SEC and the SWC also claimed that the industry advised people to have water tests, but they turned out to not be the sort that would hold up in court. Mr. West of the NC complained about royalties. He said leases "run for pages" with a small clause about postproduction costs for things such as pipeline construction or site remediation. According to Mr. West, "some of these people are now getting zero for royalties."

Tension and Conflict

Hydraulic fracking may threaten the social fabric because it engenders tension and conflict. Perhaps the wide array of stressors makes it polarizing. One the one hand, participants from each community spoke of verbal conflict over HDHF itself. Mrs. Carter from the SEC had to tell a coworker they needed to stop speaking about HDHF because the two individuals' opinions differed so much. Mr. and Mrs. Madison of the NC said the variety of opinions in their extended family led to fights, though Mr. Madison emphasized the wounds were short term. Mr. Harvey, from a different NC focus group, had a dimmer view: "It's not the peaceful, quiet community it used to be. It's a lot more stress. The neighbors against neighbors and family against family. That causes a lot of stress." Similarly, Jerolmack and Berman write about conflict stemming from "spillover effects," (2016), that is, situations in which activity on a property with HDHF harms the well-being of people on neighboring properties without HDHF.

Economic inequality created by the presence of lease money appears to strain relationships as well. While inequality is a minor theme in recent HDHF work (see Anderson and Theodori 2009; Brasier et al. 2011; Ladd 2013; Stedman et al. 2012), our findings suggest more attention is needed. As Jacquet and Kay put it, a newcomers–old-timers social divide was salient in "classic" boomtowns (2014), but in HDHF-impacted communities the key divide may be based on leasing revenue. The authors emphasize that this divide produces differences in attitudes about HDHF, but our findings point to the potential for more significant problems.

Some of our participants emphasized how landowners' windfalls had not strained community relations, but others revealed multiple ways in which inequality may generate social strain. In some cases, the problem was directly related to the money involved. Mr. Williams of the SWC mentioned "friction" in families. Mr. Hart of the NC knew of problems among adult siblings from farm families if one sibling stayed to work the farm and stood to benefit from a lease more than the others. Mrs. Richards, from the same NC focus group, worked at a local bank and was aware of conflict over how to divide money, of cases of elder financial abuse related to lease money, and of family members that were "greedy and trying to, you know, steal money from this person or that person." Jerolmack and Berman see another social strain stemming from these sorts of challenges: "The stakes of fracking—financial, environmental, or otherwise—can be so high that neighbors and kin now routinely filter everyday interactions through legal-rational frameworks" (2016, 204–205).

Inequality may reduce social cohesion via interpersonal tension too. Mentioned already, some landowners can be upset because others' leases are more lucrative. Others spoke of tension between those with and without leases. Mrs. Davis of the SEC knew of people with leases who had been the object of anger from coworkers at a local factory. Mrs. Strong from the

NC experienced this sort of thing personally. Because of her wells, she said certain people did not like her or her family anymore, and neighbors, friends, and even people she did not know were "trying to get into our business to find out what we're making." This clearly made her uncomfortable:

> I was at the eye doctor's office the other day, and one of the girls in the eye doctor's office said something to me. I mean I don't even know this girl. And she said, "Are you the person that has all the gas wells?" And I said, "Well, we have some, yes." And she said, "Well my husband works for the [pause] and he's been telling me all about your farm and how many you have and how much money you must be making."

She said that in other cases the tension was subtler, but "You can see what some people are thinking about you."

Some of this tension may stem simply from relative deprivation, but people without leases also get no financial benefit while having to deal with the challenges HDHF brings. In this vein, Ellis et al. (2016) found nonlandowners, especially low-income residents and seniors, have more negative experiences with HDHF than do others (see also Fernando and Cooley 2015). Inequality may be further problematic given how landholdings and thus lease and royalty income are concentrated in a relatively small number of hands (Kelsey, Metcalf, and Salcedo 2012).

Individual Well-Being

Studies find evidence that individual well-being can decline in a number of ways in "classic" boomtowns (Freudenburg 1984; Freudenburg, Bacigalupi, and Landoll-Young 1982; Osborne, Boyle, and Borg 1984; Shandro et al. 2011). It may be a foregone conclusion that HDHF does so as well given the various impacts already raised; that is, living with increased crime fear, noise, worry about well contamination, and other things can harm well-being.

Worth emphasizing again in this vein is the sense that sessions at times felt like group therapy. But we also heard explicit dialogue suggesting declines in individual well-being. Participants did not talk about serious personal problems, but some degree of mental health difficulty was acknowledged in each community. In response to a probe about stress and mental health, Mr. Jones of the SWC declared, "We're about stressed out, ain't we? We're all about stressed out of our limits." Then after a comment about truck traffic, he said something that may or may not have been an attempt at humor: "[My wife] wonders why I drink so much wine at night before I go to sleep. [*participants laugh*] I'm out working in the road with these monkeys." Mrs. Stanley of the NC was quite pointed: "When they first came, I felt nothing but rage. I mean—I mean I wasn't, like, running through the streets, but I really felt rage." Mrs. Jordan, from a different NC focus group, said her stress came from "when you have that feeling of helplessness." Both of these women mentioned improvement in their mental state—Mrs. Stanley from growing acceptance, Mrs. Jordan from declining HDHF activity—but they both experienced a real period of difficulty.

Conclusion

As much research on boomtowns is survey based, this and other qualitative accounts have value for giving voice to the conditions experienced. Focus groups allowed participants to tell their stories, explain what they thought and felt, and describe what it was like to live near HDHF. While we cannot confidently generalize from qualitative work, we have provided a window onto the variety of sentiment and experiences in HDHF communities.

Our findings have resonance with "classic" boomtowns characterized by distant energy projects in lower population density contexts. Expected challenges from the influx of population and increased economic activity include things such as inflation, fear of crime, and reduced community satisfaction. There are also impacts unique to HDHF that make the "classic" boomtown model inadequate (Jacquet and Kay 2014). Much of this has been given attention in recent literature (e.g., stress from noise and truck traffic), but our findings suggest some additional impacts that have arguably not been given their due. We have also provided a view not just of what mattered to people, but how those things mattered. The influx of wealth may be beneficial, but may also drive conflict, jealousy, and subtle-but-real social tensions. Many residents are burdened with worries about environmental threats because an industry perceived as risky operates nearby. All experience a transformation of their home environments and other familiar locales. Energy industry business practices themselves are a stressor and lead many to see the industry as a bad neighbor. The wide array of impacts is also significant because it could mean greater declines in community and individual well-being than what has been seen previously. Going forward, especially as energy prices rebound and lead to additional waves of HDHF activity, these communities deserve continued research attention.

Future work assessing the impacts of the energy industry should account for this breadth of impacts. Longitudinal work is needed since research suggests recovery is possible (see Brown, Dorins, and Krannich 2005). In fact, baseline community assessments should commence soon as we are in a lull of HDHF activity. Given that a primary cause of the new impacts appears to be the way HDHF spreads out over a landscape, touching places close to homes and places people care about, space and place should be a focus. Interdisciplinary work between social scientists and geographers will be of value and should include robust measures of proximity, density, visibility of energy industry activity, and so forth. Recent large-scale work with geocoded survey data by Boudet et al. (2018) should be complemented by geosocial studies that rely solely on data from people who live in HDHF zones.

The issue of increased economic inequality needs attention as well. Relative deprivation and other frustrations may reduce social cohesion, but inequality also predicts health problems, social problems, and increased crime and deviance (Bernburg, Thorlindsson, and Sigfusdottir 2009; Wilkinson and Pickett 2009). With more wells drilled in subsequent waves of activity, inequality may continue to increase and prove to be an influence on quality of life. Apart from guiding future research, the wide array of impacts suggested by this and other research may provide useful guidance for rural social workers and extension service professionals.

Notes

1. All names used are pseudonyms.
2. In addition to the "Halliburton Loophole" in the Energy Policy Act of 2005 that allows companies to not disclose proprietary chemicals used in HDHF, Pennsylvania's Act 13 allows companies to keep some chemicals used in HDHF secret if the list of chemicals is deemed by the company to be "confidential proprietary information" (Pennfuture 2012, 14).

References

Akbar, Delwar, John Rolfe, and S. M. Zobaidul Kabir. 2013. "Predicting Impacts of Major Projects on Housing Prices in Resource Based Towns with a Case Study Application to Gladstone, Australia." *Resources Policy* 38 (4): 481–489.

Anderson, Brooklynn J., and Gene L. Theodori. 2009. "Local Leaders' Perceptions of Energy Development in the Barnett Shale." *Southern Rural Sociology* 24 (1): 113–129.

Bell, Michael Mayerfeld. 1997. "The Ghosts of Place." *Theory and Society* 26 (6): 813–836.

Bernburg, Jón Gunnar, Thorolfur Thorlindsson, and Inga Dora Sigfusdottir. 2009. "Relative Deprivation and Adolescent Outcomes in Iceland: A Multilevel Test." *Social Forces* 87 (3): 1223–1250.

Boudet, Hilary S., Chad M. Zanocco, Peter D. Howe, and Christopher E. Clarke. 2018. "The Effect of Geographic Proximity to Unconventional Oil and Gas Development on Public Support for Hydraulic Fracturing." *Risk Analysis* 38 (9): 1871–1890. https://doi.org/10.1111/risa.12989.

Brasier, Kathryn J., Matthew R. Filteau, Diane K. McLaughlin, et al. 2011. "Residents' Perceptions of Community and Environmental Impacts from Development of Natural Gas in the Marcellus Shale: A Comparison of Pennsylvania and New York Cases." *Journal of Rural Social Sciences* 26 (1): 32–61.

Brown, Ralph B., Shawn F. Dorins, and Richard S. Krannich. 2005. "The Boom-Bust-Recovery Cycle: Dynamics of Change in Community Satisfaction and Social Integration in Delta, Utah." *Rural Sociology* 70 (1): 28–49.

Brown, Ralph B., H. Reed Geertsen, and Richard S. Krannich. 1989. "Community Satisfaction and Social Integration in a Boomtown: A Longitudinal Analysis." *Rural Sociology* 54 (4): 568–586.

Christopherson, Susan. 2015. "The False Promise of Fracking and Local Jobs." *Conversation*, January 27. http://theconversation.com/the-false-promise-of-fracking-and-local-jobs-36459.

Cortese, Charles F., and Bernie Jones. 1977. "The Sociological Analysis of Boom Towns." *Western Sociological Review* 8 (1): 75–90.

Ellis, Colter, Gene L. Theodori, Peggy Petrzelka, Douglas Jackson-Smith, and A. E. Luloff. 2016. "Unconventional Risks: The Experience of Acute Energy Development in the Eagle Ford Shale." *Energy Research and Social Science* 20 (October): 91–98.

England, J. Lynn, and Stan L. Albrecht. 1984. "Boomtowns and Social Disruption." *Rural Sociology* 49 (2): 230–246.

Ennis, Gretchen M., Mary P. Finlayson, and Glen Speering. 2013. "Expecting a Boomtown? Exploring Potential Housing-Related Impacts of Large Scale Resource Developments in Darwin." *Human Geographies* 7 (1): 33–42.

Fernando, Felix N., and Dennis R. Cooley. 2015. "An Oil Boom's Effect on Quality of Life (QoL): Lessons from Western North Dakota." *Applied Research in Quality of Life* 11 (4): 1083–1115.

Freudenburg, William R. 1984. "Boomtown's Youth: The Differential Impacts of Rapid Community Growth on Adolescents and Adults." *American Sociological Review* 49 (5): 697–705.

Freudenburg, William R. 1986. "The Density of Acquaintanceship: An Overlooked Variable in Community Research." *American Journal of Sociology* 92 (1): 27–63.

Freudenburg, William R., Linda M. Bacigalupi, and Cheryl Landoll-Young. 1982. "Mental Health Consequences of Rapid Community Growth: A Report from the Longitudinal Study of Boomtown Mental Health Impacts." *Journal of Health and Human Resources Administration* 4 (3): 334–352.

Freudenburg, William R., and Robert Gramling. 1998. "Linked to What? Economic Linkages in an Extractive Economy." *Society and Natural Resources* 11 (6): 569–586.

Freudenburg, William R., and Robert Emmett Jones. 1991. "Criminal Behavior and Rapid Community Growth: Examining the Evidence." *Rural Sociology* 56 (4): 619–645.

Glaser, Barney G., and Anselm L. Strauss. 1967. *The Discovery of Grounded Theory: Strategies for Qualitative Research*. Chicago: Aldine.

Jacquet, Jeffrey, and David L. Kay. 2014. "The Unconventional Boomtown: Updating the Impact Model to Fit New Spatial and Temporal Scales." *Journal of Rural and Community Development* 9 (1): 1–23.

Jacquet, Jeffrey B., and Richard C. Stedman. 2013. "Perceived Impacts from Wind Farm and Natural Gas Development in Northern Pennsylvania." *Rural Sociology* 78 (4): 450–472.

Jerolmack, Colin, and Nina Berman. 2016. "Fracking Communities." *Public Culture* 28 (2): 193–214.

Kelsey, Timothy W., Alex Metcalf, and Rodrigo Salcedo. 2012. "Marcellus Shale: Land Ownership, Local Voice, and the Distribution of Lease and Royalty Dollars." *Pennsylvania State University Center for Economic and Community Development Research Series*. State College: The Pennsylvania State University. Accessed April 28, 2018. http://aese.psu.edu/research/centers/cecd/publications.

Krannich, Richard S., E. Helen Berry, and Thomas Greider. 1989. "Fear of Crime in Rapidly Changing Rural Communities: A Longitudinal Analysis." *Rural Sociology* 54 (2): 195–212.

Ladd, Anthony E. 2013. "Stakeholder Perceptions of Socioenvironmental Impacts from Unconventional Natural Gas Development and Hydraulic Fracturing in the Haynesville Shale." *Journal of Rural Social Sciences* 28 (2): 56–89.

Lockie, Stewart, Maree Franettovich, Vanessa Petkova-Timmer, John Rolfe, and Galina Ivanova. 2009. "Coal Mining and the Resource Community Cycle: A Longitudinal Assessment of the Social Impacts of the Coppabella Coal Mine." *Environmental Impact Assessment Review* 29 (5): 330–339.

Morgan, David L. 1997. *Focus Groups as Qualitative Research*. Thousand Oaks: Sage.
Osborne, J. Grayson, William Boyle, and Walter R. Borg. 1984. "Rapid Community Growth and the Problems of Elementary and Secondary Students." *Rural Sociology* 49 (4): 553–567.
Pennfuture. 2012. "Pennsylvania's New Oil and Gas Law (Act 13): A Plain Language Guide and Analysis." Harrisburg, PA: Citizens for Pennsylvania's Future. Accessed December 16, 2014. http://pennfuture.org/UserFiles/File/MineDrill/Marcellus/CitizenGuide_Act13_2012.pdf.
Petkova, Vanessa, Stewart Lockie, John Rolfe, and Galina Ivanova. 2009. "Mining Developments and Social Impacts on Communities: Bowen Basin Case Studies." *Rural Society* 19 (3): 211–228.
Proshansky, Harold M., Abbe K. Fabian, and Robert Kaminoff. 1983. "Place-Identity: Physical World Socialization of the Self." *Journal of Environmental Psychology* 3 (1): 57–83.
Ruddell, Rick, and Natalie R. Ortiz. 2015. "Boomtown Blues: Long-Term Community Perceptions of Crime and Disorder." *American Journal of Criminal Justice* 40 (1): 129–146.
Schafft, Kai A., Yetkin Borlu, and Leland Glenna. 2013. "The Relationship between Marcellus Shale Gas Development in Pennsylvania and Local Perceptions of Risk and Opportunity." *Rural Sociology* 78 (2): 143–166.
Seydlitz, Ruth, Shirley Laska, Daphne Spain, Elizabeth W. Triche, and Karen L. Bishop. 1993. "Development and Social Problems: The Impact of the Offshore Oil Industry on Suicide and Homicide Rates." *Rural Sociology* 58 (1): 93–110.
Shandro, Janis A., Marcello M. Veiga, Jean Shoveller, Malcolm Scoble, and Mieke Koehoorn. 2011. "Perspectives on Community Health Issues and the Mining Boom-Bust Cycle." *Resources Policy* 36 (2): 178–186.
Smith, Michael D., Richard S. Krannich, and Lori M. Hunter. 2001. "Growth, Decline, Stability, and Disruption: A Longitudinal Analysis of Social Well-Being in Four Western Rural Communities." *Rural Sociology* 66 (3): 425–450.
Stedman, Richard C., Jeffrey B. Jacquet, Matthew R. Filteau, Fern K. Willits, Kathryn J. Brasier, and Diane K. McLaughlin. 2012. "Marcellus Shale Gas Development and New Boomtown Research: Views of New York and Pennsylvania Residents." *Environmental Practice* 14 (4): 382–393.
Theodori, Gene L. 2009. "Paradoxical Perceptions of Problems Associated with Unconventional Natural Gas Development." *Southern Rural Sociology* 24 (3): 97–117.
Tsvetkova, Alexandra, and Mark D. Partridge. 2016. "Economics of Modern Energy Boomtowns: Do Oil and Gas Shocks Differ from Shocks in the Rest of the Economy?" *Energy Economics* 59 (September): 81–95.
US Census Bureau. 2015a. *Selected Social Characteristics in the United States, American Community Survey 5-Year Estimates Data Profiles*. Accessed August 27, 2020. https://

data.census.gov/cedsci/table?g=0400000US42_0500000US42081&y=2015&d=ACS%205-Year%20Estimates%20Data%20Profiles&tid=ACSDP5Y2015.DP02.

US Census Bureau. 2015b. *ACS Demographic and Housing Estimates, American Community Survey 5-Year Estimates Data Profiles.* Accessed August 27, 2020. https://data.census.gov/cedsci/table?g=0400000US42_0500000US42081&d=ACS%205-Year%20Estimates%20Data%20Profiles&tid=ACSDP5Y2015.DP05.

Vorkinn, Marit, and Hanne Riese. 2001. "Environmental Concern in a Local Context: The Significance of Place Attachment." *Environment and Behavior* 33 (2): 249–263.

Wilkinson, Richard G., and Kate E. Pickett. 2009. "Income Inequality and Social Dysfunction." *Annual Review of Sociology* 35: 493–511.

Williamson, Jonathan, and Bonita Kolb. 2011. "Marcellus Natural Gas Development's Effect on Housing in Pennsylvania." *Center for the Study of Community and the Economy* 31. Williamsport, PA: Lycoming College.

Willits, Fern K., A. E. Luloff, and Gene L. Theodori. 2013. "Changes in Residents' Views of Natural Gas Drilling in the Pennsylvania Marcellus Shale, 2009–2012." *Journal of Rural Social Sciences* 28 (3): 60–75.

Wynveen, Brooklynn J. 2011. "A Thematic Analysis of Local Participants' Perceptions of Barnett Shale Energy Development." *Journal of Rural Social Sciences* 26 (1): 8–31.

Chapter 9 Summary

The lands above the Marcellus Shale have much higher population densities than those near energy projects from classic boomtown studies or in other current shale plays (e.g., the Bakken Shale in North Dakota). This proximity means people living in the Marcellus zone experienced a wide array of social and personal impacts during the massive wave of fracking that recently ended. Not only did the influx of industry personnel and economic activity change their communities in expected ways, but also the industry was frequently operating in relatively close proximity to peoples' homes, typical travel routes, special places, and so on. As Jerolmack and Berman (2016, 194) put it, "fracking is more *intimate.*" Conversations with focus group participants from three different communities made clear that this "intimacy" leads to a wider array of impacts and that because of them, the declines community and individual well-being during waves of fracking need to be taken very seriously. In fact, our focus groups often had the tenor of group therapy sessions even though survey data gathered prior to the sessions shows that our participants were a fairly balanced mix of people who supported, opposed, or had mixed feelings about fracking. Some of the impacts we learned about have been emphasized in recent literature, but much of what we learned needs greater emphasis and future research attention. Furthermore, our rich qualitative data help to bring into clearer relief how these issues mattered for people's lives.

KEY TAKEAWAYS

- People discussed impacts that would be expected in any boom scenario. Increased population and economic activity meant participants emphasized things such as positive economic changes, worry about a bust, strain on local services, declines in community, and fear of crime.
- Other impacts seem to be unique to fracking in relatively high-population rural areas. These are what really create the "wider array."
  - Participants worried a lot about the environment and health. Groundwater contamination was prominent, but so were worries about air pollution and catastrophes such as earthquakes or collapsing mountains.
  - Much of what our participants said amounts to what could be labeled concerns for disturbed countryside. Traffic, light pollution, and noise pollution were commonly discussed.
  - Changes to the character of communities and familiar landscapes came up in every focus group, and several participants seemed truly anguished by changes to familiar views and special places.
  - Because the natural gas industry operated in people's home environments and had to engage in business dealings with landholders, many people came to view the natural gas industry as a bad neighbor because of a variety of business practices they found problematic.
  - Tension and conflict within families and among community members appear to also threaten community well-being. Fights about the mere presence of the industry were cited. Other residents spoke of trouble within families caused by lease and royalty windfalls. Inequality among leaseholders and between those who do and do not have leases creates some tension as well.

10

Drilling Impacts

A Boom or Bust for Schools? A Mixed Methods Analysis of Public Education in Six Oil and Gas States

NATHAN RATLEDGE AND LAURA ZACHARY

Introduction

Since its initial expansion in the early 2000s, unconventional oil and gas have fundamentally shifted the international oil and gas marketplace.[1] The boom's benefits have not come without costs. At the local level, a community may see revenues rise from severance taxes or other state or local taxes on the sectors causing the boom. These revenues can benefit communities by being spent to increase the quantity and quality of public services or to reduce local tax rates. At the same time, the economic boom often increases demand for public services such as water, sewage, roads, and schools—which adds additional burdens to these often small and rural communities (Raimi and Newell 2014). Residential housing and rental prices can also increase with the boom (Price et al. 2014; Headwaters Economics 2008; Jacquet 2009) generating increased rents for the owners and pushing out existing, lower-income residents.

Looking specifically at educational impacts is fundamentally important in and of itself and as a proxy for the broader economic health of a community

DOI: 10.5876/9781646420278.c010

both in the short and long term. In the long term, resource booms can influence career-based decision making and educational attainment—two drivers of a diverse and robust local economy (Marchand and Weber 2015; Measham and Fleming 2014; Rickman, Wang, and Winters 2017). Therefore, it is in the public interest to ensure that the net effect of a resource boom on education is at worst neutral and ideally a positive factor for individuals and communities.

A limited number of studies have assessed the short-term effect of resource booms on public school metrics, including budgets and student population, or on student performance metrics, such as standardized test scores or graduation or dropout rates (Cascio and Narayan 2015; Marchand and Weber 2015; Rickman, Wang, and Winters 2017).

In the broadest sense, our research seeks to answer the following near-term question: Did public school districts in regions with high levels of oil and gas production during the recent unconventional energy booms fare better or worse in terms of financial and educational outcomes than comparable school districts in regions that did not experience a boom?

To answer this question, we used a mixed methods research approach, combining statistical analysis of a quasi-experiment combined with semi-structured interviews. The quasi-experiment uses a difference-in-difference (DID) design. To complement the statistical analysis and probe deeper into related issues, we conducted over seventy interviews with education stakeholders across the states included in the study. Notably, detailed quantitative data analysis at the district level and complementary interview responses reinforce and better explain the results, increasing confidence in our findings. Our mixed methods research extends the existing body of literature related to US commodity booms and public education. As far as we are aware, this is the first study to use extensive interviews in conjunction with a DID analysis to analyze the effects of unconventional oil and gas on public education across multiple plays.

Amid a series of compelling and complex conclusions, several important themes rise to the surface. The statistical results and interview responses illustrate a general bifurcation between the Bakken and the Marcellus/Utica in regard to student populations, teacher demands, and revenue and expenditures. Yet neither region reported concerns with increased dropout rates or effects on academic achievement. Colorado districts generally fell between the two, though they shared more common characteristics with the Bakken despite having robust oil (Denver-Julesburg) and development driven by natural gas (Piceance). This divergence of trends across regions gives us pause about overgeneralizing impacts of resource booms.

Generally, interviews did not contradict the statistical results. However, while this chapter finds little evidence that the recent boom affected student learning outcomes in the short run, we encourage future examination of long-term impacts on student performance. Issues such as classroom congestion and high rates of teacher turnover in western rural districts suggest a potential for lower educational achievement in the long run that is not captured in existing data.

Regarding student population, North Dakota had a statistically significant increase during the boom, whereas Pennsylvania and the larger Marcellus/Utica region experienced the opposite effect across most grade levels. Similarly, student-teacher ratios (STRs) rose in North Dakota boom districts and dropped in the Marcellus/Utica. Of importance, the trend in STR implies that despite budget challenges across Pennsylvania, on average teachers were retained in Marcellus/Utica boom-area public schools.

Interviews with education professionals revealed insights that were not apparent when looking only at the official data. Annual student enrollment totals underestimate the real increase in new Bakken students because of high levels of student mobility throughout the school year, which are not captured in the official population data recorded in October each year. For example, consider a school that reported twenty new students in one year, a significant amount for many rural districts. It would be reasonable to find that 40 new students arrived and 20 students left—resulting in the net reported increase of 20. Addressing the needs of 40 new students requires twice as many resources as 20 new students, and yet this discrepancy is lost in the official numbers. Teachers and staff throughout North Dakota and Montana boom districts reported this "revolving door" phenomenon as a key challenge for the schools. Education professionals also reported the revolving door effect in Colorado but to a lesser degree than in the Bakken.

Another insight revealed through interviews concerned teacher retention. On one hand, conversations indicated a negligible concern over teachers or staff leaving the educational sector for higher-paying industry jobs in any of the regions. On the other hand, most western districts reported challenges with teacher recruitment and retention because of being so remote. The rapid rise in student population during the boom years exacerbated this situation.

As with student population estimates, there were parallel, though somewhat less pronounced, east-west divergent trends in per pupil total revenue. North Dakota boom districts experienced a weakly statistically significant decrease compared with in-state nonboom districts. Marcellus / Utica districts experienced the opposite effect and saw per pupil revenue rise compared with nonboom districts. Statistical results in Colorado were insignificant. Yet, the experience of boom districts in the state are illustrative of how boom impacts can exacerbate questionable state policies.

On the spending side of the ledger, total per pupil expenditures show a complementary scenario to revenue; however, the analysis did uncover additional nuances. Capital spending in boom districts increased significantly in North Dakota compared to the Marcellus / Utica, which did not exhibit a statistical difference from nonboom districts, indicating potential further diverting resources from in-classroom use in the Bakken.

Interviews with educational professionals across all regions commonly reported stress related to uncertainty about the future of school funding. In particular, interviewees in western oil districts vocalized this concern because of the salience of the recent downturn in global oil prices. As a result, several administrators suggested it was safer to spend money on one-time investments such as new computers than on recurring costs such as teacher pay raises.

A final theme was the limited concern expressed toward student dropout rates or negative effects on educational attainment in boom districts. Without the retracted quantitative data on dropout rates, interviews provided valuable, if potentially biased, insight.[2] Across almost all districts, interviewees did not feel that the boom increased the rate of student dropouts, commonly suggesting that this boom had higher technological demands than previous oil and gas development periods.

Analysis of standardized test data from third and eighth graders, as well as SAT and ACT scores, in Marcellus / Utica states did not reveal any overarching conclusions, as results were largely mixed. (DID analysis in other states was precluded by the limited time series of available data.) Interviews reinforced the statistical outcomes, suggesting there was little impact on student performance in the Marcellus/Utica. However, a lack of measurable short-term impacts does not preclude impacts on long-term educational attainment.

This chapter proceeds as follows. We first review the study's sample choice, data sources, and methodology. We then present our primary econometric results in parallel with interview findings. We also discuss the importance of a mixed methods approach, shortcomings in current socioeconomic research of resource development, and important implications garnered from this study for policy makers.

Mixed Methods: Difference-in-Difference Estimation and In-person Interviews

This study uses a mixed methods research design, combining statistical analysis and in-person interviews, to explore the effects of resource booms on school finance and education metrics. The statistical analysis relies on a DID regression design that uses binary indicators of boom time periods and boom districts to designate participation in the treatment. The DID approach simulates a natural experiment by examining the differential effect on a treatment group compared with a control group. It does so by estimating an unobserved counterfactual for the treatment group—based on the control group, and subtracting that estimate from the observed outcome in the treatment group.

The quantitative analysis (based on available data) spans school years 2000–2013, but not all metrics cover the entirety of the time series. The 2000–2013 period captures the production boom of both oil and gas in each of the six sample states and the gas price collapse in the late 2000s, but it does not catch the 2014 global oil price collapse. Through interviews with educators and staff, however, we gathered insight on impacts of the oil price drop and ensuing production decline.

The analysis covers 1,496 nonmetropolitan school districts in six states that overlie several major oil and gas formations: the Marcellus/Utica (Pennsylvania, Ohio, and West Virginia), the Bakken (North Dakota and Montana), and the Denver-Julesburg, Piceance, and San Juan Basin (Colorado). Throughout the chapter when we write Marcellus/Utica we are referring to the joint Marcellus *and* Utica region. We chose this six-state sample because it includes important energy development regions that are each dissimilar from the others, allowing for intra- and interplay comparisons. The Bakken is an oil play, whereas the Marcellus/Utica region is primarily a natural gas play. Colorado produces meaningful quantities of both natural gas and oil. Therefore, this study can compare impacts within the Bakken against general effects in the Marcellus/Utica, for example.

The bulk of our standardized education data comes from the National Center for Education Statistics (NCES). Within the US Department of Education, NCES collects extensive data on students, teachers, and school finances. In this study we use NCES data on student enrollment by grade, student-teacher ratios, student demographics, and participation in additional programs such as free and reduced lunches, as well as school revenues and expenditures. Rather than analyze public school effects aggregated to the county level or commuting zone, we evaluate impacts on a smaller, more detailed district-specific basis. Not only can this concentration detect the sometimes large differences among districts within the same county or region, but using districts also increases the sample size. Using nonmetropolitan districts precludes districts located in cities (with populations over 250,000) from biasing estimates.[3] Since the majority of unconventional oil and gas production occurs in rural regions, nonmetropolitan districts are more representative of the core population of interest.

Our study defines a boom district based on a district meeting three production metrics: (1) top average producing district during the boom (measured at the top 10 percent and top 20 percent in MMBtus over a four-year average), (2) exceeding national average percentage change in production from preboom to boom, and (3) having a positive change in the number of wells averaged over peak boom years.

Average boom production is estimated over a four-year period: the peak production year in each state and the three preceding years. This average is used to identify the two treatment groups. A producing district in the top 10 percent is defined as being in the *core treatment* group. *Total treatment* districts include those that are in the top 20 percent of production during the boom. We chose to explore the effects in both the top 10 and 20 percent to infer whether impacts were hyperlocalized or more widespread across a region.

Exceeding national average percentage change (2000–2003 preboom and the four-year boom average) was chosen to guard against legacy production regions that had high production rates but did not necessarily grow or benefit from the unconventional oil and gas boom. Examples of this scenario are several counties in West Virginia that had high historic natural gas production from coal bed methane. Similar to the second criteria, increasing well numbers over the boom period is meant to winnow away districts that saw production increases during the boom but not from new well development.

To account for unobserved variation in local characteristics and across years, we employ a combination of regional and year fixed effects. In addition to adding fixed effects, we cluster standard errors by district. We use common control variables including population, population density, per capita income, total employment, and unemployment rate.[4]

We varied a number of parameters in various regression runs in order to increase confidence in the statistical results of our study. First, we ran DID regressions with and without neighboring districts to explore the potential for spillover effects. Second, we ran each regression with varying controls. As the general narrative does not change when we alter these parameters or controls for the DID regression scenarios, we report regression estimates for the core treatment, dropping neighboring districts.

Realizing that a quantitative approach alone can miss nuanced parts of any story, we conducted over 70 predominantly in-person interviews across the regions involved in the study between February and August 2016. The interviewees involved teachers, school and district administrators, district staff including chief financial officers; school board members; and state government, nonprofit, and community college staff. Interviewees were primarily identified by their leadership of districts in boom regions or via recommendations by district leadership. The broad goals of these semistructured meetings were to better understand nuances in the data and to incorporate the on-the-ground experience before the boom, during the boom, and during the recent oil price decline (post-2014).

Although Montana did not ultimately experience a large increase in oil production, we included it in the qualitative analysis because of its proximity to North Dakota and to explore spillover effects in neighboring districts. Montana was not included in the statistical analysis of state data because of irregularities with data aggregation across primary and secondary school districts. Counties in western Colorado's Piceance Basin were also dropped from econometric analysis because of increasing levels of natural gas production immediately prior to the introduction of unconventional gas. However, both regions are included in the qualitative analysis and contribute important insights into the effects of rapid oil and gas development on public schools.

Results and Discussion

Tables 10.1 and 10.2 summarize estimated effects of being in a boom district across a series of key metrics. In order to examine regional variation, we ran regressions on a variety of geographic groupings: pooling five states, the Marcellus/Utica as a region, and state specific outputs. Of significance, our findings caution against overgeneralizing results from one play to a broader collection of resource-rich regions. Looking at the differences across regions with high shale development also reveals important lessons for how to tailor public policies to address varying needs.

Several of the initial five-state pooled model results were unexpected. For example, the pooled results suggested that on average boom districts experienced a decline in student population compared with nonboom districts, which contradicted state-specific findings in western districts. Further comparing the pooled five-state results and the state-specific outputs revealed a number of other dissimilar impacts across regions. As such, we base our statistical conclusions primarily on results of state-specific and regional model runs. The five-state pooled run is shown for comparative purposes.

We hypothesize that different levels of remoteness likely explain much of the statistical heterogeneity across states. Simply, the Bakken—and to a slightly less extent, parts of western Colorado—are much more remote than rural regions in the Marcellus/Utica. For perspective, from Williston, North Dakota—informally referred to as Boomtown USA—it is a five-hour drive to Billings, Montana, the closest town with over 100,000 people. This remoteness likely accounts for much of the growth in total population and the magnitude of industrial expansion, including support services.

An additional explanation for diverging trends may be the relative value of oil versus gas and the potential premium for pay in an area such as the oil-rich Bakken. A third potential factor, the level of secondary development—for roads, facilities, and support activities—was simply that much larger in the Bakken. Unlike the Marcellus/Utica, the development of industrial parks, corporate headquarters, and supporting businesses was abundantly obvious while visiting the Williston region.

TABLE 10.1. Effect of being a top-producing district on public school populations

	# of students							
	Pre K & K	*1st & 2nd*	*3rd & 4th*	*5th & 6th*	*7th & 8th*	*9th & 10th*	*11th & 12th*	*Student-Teacher Ratio*
5-State Pooled	7.021	−7.961	−15.11[c]	−17.02[c]	−17.99[c]	−22.26[c]	−6.445	−0.0428
	(5.496)	(5.053)	(4.480)	(4.575)	(5.372)	(5.737)	(5.684)	(0.148)
Colorado (D-J)	−16.73	−1.514	−19.61	−18.88	−9.758	−12.41	−32.84	1.391
	(43.40)	(63.80)	(59.34)	(61.02)	(58.85)	(58.72)	(68.62)	(1.189)
North Dakota	1.946	8.392[b]	6.965[c]	5.451[a]	5.739[a]	8.164[c]	2.517	1.120[c]
	(2.233)	(3.469)	(2.586)	(2.848)	(3.207)	(3.120)	(2.717)	(0.365)
Marcellus Region	7.826	−7.994[a]	−15.63[c]	−16.77[c]	−17.82[c]	−20.76[c]	−9.415[a]	−0.325[b]
	(6.009)	(4.304)	(4.243)	(4.555)	(5.654)	(6.016)	(5.608)	(0.135)
Ohio	−4.179	−12.56[a]	−14.00	−14.33	−11.69	−11.74	−7.279	−0.0843
	(7.707)	(7.360)	(9.362)	(9.317)	(8.257)	(7.474)	(10.01)	(0.393)
Pennsylvania	−6.445[a]	−14.94[c]	−21.14[c]	−15.06[c]	−11.07	−10.04	−21.21[c]	−0.292[a]
	(3.717)	(4.904)	(4.947)	(5.335)	(7.099)	(6.948)	(7.044)	(0.151)
West Virginia	−28.98	3.876	−28.76[a]	−33.74[a]	−37.50	−43.82[b]	−24.37[a]	−0.612[c]
	(48.12)	(11.51)	(16.99)	(19.53)	(26.26)	(21.65)	(14.06)	(0.0990)

[a] $p < 0.1$
[b] $p < 0.05$
[c] $p < 0.01$

Student Population and Teachers

As shown in table 10.1, North Dakota experienced statistically significant increases in student population for grades 1–10. Conversely, and despite striking growth in natural gas production, Marcellus/Utica districts experienced a statistically significant decline in student population for grades 1–12 in boom districts.

On the front end of the boom, interviewees in each region expressed similar expectations of large influxes of workers and families. The population changes however, turned out very differently across regions. Bakken teachers and staff reported an initial heavy influx of young men, and, eventually, families with mostly elementary-age children followed. Several interviewees noted that either most men coming to work the fields were too young to have children in higher grades or those workers with high-school-age children left them back home to finish school. With the boom bringing mostly young

TABLE 10.2. Effect of being a top-producing district on public school finance

	$ per Pupil (in $2014)					
	Total Revenue	*Local Revenue*	*Property Tax Revenue*	*State Revenue*	*Education Spending*	*Capital Spending*
5-State Pooled	−127.6	310.5[a]	95.30	−367.3[c]	−201.5	218.9
	(223.9)	(181.1)	(144.3)	(119.6)	(151.5)	(190.6)
Colorado (D-J)	−1,759	−310.2	88.85	−1,341	−1,112	340.1
	(1,657)	(1,065)	(753.7)	(1,033)	(1,062)	(1,085)
North Dakota	−1,498[a]	−390.4	−1,195[c]	−1,212[c]	−1,451[c]	872.6[a]
	(807.2)	(470.2)	(376.1)	(250.7)	(548.4)	(466.4)
Marcellus Region	374.8[b]	138.2	62.56	205.7[a]	240.3[b]	127.5
	(176.1)	(117.7)	(109.9)	(123.0)	(112.0)	(208.6)
Ohio	525.5	428.4[c]	420.9[c]	75.42	199.9	424.3
	(469.6)	(149.8)	(158.3)	(403.5)	(178.6)	(605.6)
Pennsylvania	161.3	−2.188	−29.00	103.7	64.84	90.21
	(162.4)	(130.2)	(117.5)	(93.21)	(122.5)	(267.6)
West Virginia	559.3	963.6[b]	875.4[b]	−449.2[c]	532.9[b]	−182.5
	(474.7)	(461.9)	(410.7)	(123.9)	(211.9)	(198.4)

[a] $p < 0.1$
[b] $p < 0.05$
[c] $p < 0.01$

children, many schools reported a "bubble" of students from the boom moving through lower grade levels and into the upper grades over time. This bubble itself caused strain on schools—necessitating rapid increases in faculty and resources for certain grade levels. This situation made longer-term planning more difficult.

Although Marcellus/Utica staff and teachers also had expected a surge of workers and an associated inflow of students, the flood never came. In contrast to the Bakken, the gas boom in the Marcellus/Utica mostly brought temporary workers who maintained their families in home states such as Texas or Louisiana. In fact, our study finds that while most nonurban districts in the Marcellus/Utica continued on a steady long-term declining student population trajectory, top producing districts actually experienced a more rapid decline in enrollment. Finding this same decline in top producing districts in Pennsylvania, Kelsey et al. hypothesize that a rise of nonnative, temporary workers without families; an increasing demand for housing; and the subsequent increase in the cost of rent drove lower-income households with children out of these communities, contributing to the out-migration from Marcellus/Utica boom districts (2012).

In Colorado interviewees revealed an increase in student populations, but conversations suggested that not all new arrivals—especially in western Colorado—came as a result of employment in the oil and gas sector. Staff in the Piceance cited some new students coming due to growing regional tourism and ski industries. Staff in Denver-Julesburg also reported that while some new students had parents tied to the oil and gas sector, Colorado school choice laws allowing students to enroll in schools outside their zoned district complicates an analysis of the movement of students as a measured effect of the boom (C.R.S. 22-36-101 n.d.).

Student Mobility

In addition to a net increase in student populations, Bakken teacher and staff frequently stated that student mobility (or turnover) was much larger than the commonly reported population statistics made it appear. Several interviewees suggested that the actual number of new students could be twice as large as the net gain, meaning that a significant number of students were leaving the school on a regular basis. A handful of interviewees mentioned that students would come and go without notice; in some cases, a student reappeared after winter within a single school year when drilling resumed. This heightened level of student mobility reportedly disrupted classrooms, contributed to teacher fatigue, and was often cited as a greater concern than simply increasing enrollment numbers.

In contrast to the Bakken, elevated student mobility was not referenced as a problem in the Marcellus/Utica. Interviewees in western Colorado reported increased mobility concerns, but the phenomenon did not seem entirely related to gas development nor as acute as the Bakken.

Student-Teacher Ratios

Student-teacher ratio (STR) trends in North Dakota and the Marcellus/Utica also diverged. This contradictory trend, confirmed in both interviews and statistical results and displayed in table 10.2, is important to remember when considering both short- and long-term student achievement. Essentially, classrooms in North Dakota boom districts became relatively more congested compared with nonboom districts, whereas districts in the core boom of the Marcellus / Utica had more teachers per student compared with similar districts.

Colorado statistical results, mainly reflecting effects within the Denver-Julesburg, were insignificant. Educators in several western Colorado districts noted increasing classroom congestion. Once again, as with student population, Colorado teachers and staff felt that while gas activity and associated volatile local revenue were part of the problem, the boom did not account for all of the changes to classroom atmosphere. Interviewees heavily referenced declining state funding in the postboom period due in large part to the "negative factor"—a state-imposed reduction in school funding stemming from the national recession.

Teacher Recruitment and Retention

Bakken interviews revealed an additional problem of rapid influx of students—the critical challenge of hiring and retaining qualified candidates. In one extreme case, a North Dakota district had to hire twelve full-time teachers within two weeks of the school year commencing. Administrators reported difficulty finding qualified teachers given the small applicant pool in the rural Bakken. Conducting nationwide searches for new talent and finding candidates were just the first hurdle. Highly elevated housing prices due to the boom meant that many new instructors simply could not afford rent. To overcome this obstacle, superintendents reported offering housing bonuses and even purchasing district-owned housing, thus becoming landlords themselves. In one dire case, a superintendent reported striking a deal with local homebuilders to house teachers until the new homes were sold.

Once teachers were hired, trained, and living in the Bakken, schools then faced the challenge of retaining them. With the remote geography and lack of social outlets, it was reported that new teachers frequently left after only two to three years. This cycle of finding, hiring, training, and replacing added significant soft costs to Bakken districts. Furthermore, districts could often only recruit first-year teachers, which represents a potential threat to the quality of classroom education.

Due to the challenges in recruiting teachers, some Bakken schools reported increased use of teaching aides. Several conversations noted that wives of oil workers were a source of aides, however, they noted that the new aides were less qualified and prone to unexpected moves, like the students.

Pennsylvania districts reported the opposite experience as the Bakken. Teachers often came from the local area and stayed for their entire career. In fact, the only substantial threat to retention was teachers leaving for nearby districts that offered higher pay. This up-trading occurred across state lines in the Marcellus/Utica. West Virginia educators, in particular, stressed facing competition from districts in southeast Pennsylvania and Ohio—neighboring states that pay higher wages and are only a few miles away. Administrators in West Virginia noted that any increase in teacher pay over the last three years came exclusively from individual districts and not the state.

Colorado school staff in the Denver-Julesburg and Piceance Basin reported both a challenging hiring environment and the loss of teachers to wealthier districts. Western Colorado administrators stressed that the difference in facilities and funding from school foundations made it extremely difficult for Piceance districts to compete with the Aspen or Denver-Boulder districts. Housing prices were also referenced as a limiting factor, though in general, these problems appeared less severe than those experienced in the Bakken.

Interestingly, in contrast to studies in other shale regions, none of the states included in this study reported concern with teachers leaving for higher-paying oil and gas industry jobs. The only commonly reported threat to school workforce poaching across all regions was related to bus drivers, who could earn significantly more with their commercial driver's licenses.

Education Finance

As shown in table 10.2, our study also found diverging financial effects across boom regions.

Per Pupil Revenue

Districts in the Marcellus/Utica region and the Bakken experienced opposite revenue trends. Table 10.2 shows that on average Marcellus/Utica boom districts, with relatively shrinking student bodies, saw a statistically significant increase in total per pupil revenue. When broken down, state-specific revenue source effects where somewhat different. As an example, note that Ohio and West Virginia saw increases in local funding, where Pennsylvania did not. This difference is not surprising, since neither counties nor school districts in Pennsylvania were able to tax oil and gas production as local property.

Conversely, North Dakota boom districts had a statistically significant decline in per pupil revenue, which is not entirely unexpected given the remarkable growth in student enrollment. While some funding is based on student numbers—and therefore could theoretically track linearly with enrollment, this is not the case for all monetary and programmatic support. Thus, as student population increased, per pupil funding declined.

Colorado core districts experienced a weakly negative statistical effect. We hypothesize that the decline in core districts was not statistically significant, because districts around the state were also hit by widespread budget cutbacks. The revenue changes experienced in Colorado provide an important insight into how booms and busts can exacerbate already tenuous financial situations. In Colorado, the primarily complicating factor is the taxpayer bill of rights (TABOR), which essentially limits a public entity, such as a school district, from fully capturing local revenue and saving excess funds. In short, mill levies in boom districts were automatically ratcheted down as the value of oil and gas being produced increased. When oil and gas production declined, the mills were not required to go back up toward their original rate, thereby creating a financial gap for school budgets. Without the ability to smooth income over time, the western Colorado boom and bust translated into a desperate financial situation for many local schools. These districts were also impacted by the statewide rollback in funding supporting, known locally as the negative factor. As a result, several boom districts have cut programs and even taken the notable step of moving to a four-day school week.[5]

Per Pupil Expenditures

Statistical analysis of per pupil spending in the three regions tells a similar story between Bakken and Marcellus/Utica public schools. Total Marcellus/

Utica per pupil expenditures saw a weakly positive increase in boom areas. Educational spending, a component of total per pupil spending, was significantly positive. This relative increase in educational spending means that more money per student was being spent on teacher salaries or educational materials than in nonboom districts. North Dakota districts saw a distinctly negative effect on per pupil educational spending, at the 99 percent level. Conversely, capital spending was statistically positive. Together, the North Dakota results imply that while expenditures decreased per pupil, available funds were commonly allocated to capital improvements, including school expansions or constructing new buildings.

Near Williston, North Dakota, high levels of capital spending were commonly reported. Given the significant rise in student population, expansions were essential in many cases. As an example, a new state-of-the-art high school was built near Watford City and opened in 2016. Not all Bakken districts were as successful as Watford City in gaining voter approval for new bonds to support school construction. One district reported being unsuccessful on the ballot three times. While new classrooms are a positive impact in most cases, their financing through taxpayer-backed bonds also represents a risk to taxpayers over the long term if oil development slows down over long stretches.

When we visited the Bakken in spring 2016, oil prices had been near record lows for over a year, new well completions were vastly diminished, and industry layoffs had begun. Many teachers and staff wondered aloud what would happen if the drilling did not return. While the acuteness of the global oil glut was clearly felt in the Bakken oil fields, the common refrain of financial uncertainty was a point echoed across the Marcellus/Utica and Colorado.

As noted briefly above, Colorado schools had compounding reasons to be concerned with long-term school financing other than simply the oil and gas busts, notably, state funding reductions and issues related to TABOR (Ratledge and Zachary 2017). Pennsylvania districts faced similar uncertainty due to a bitter and prolonged state budget battle that left some schools within weeks of closing their doors. In both cases, as well as in eastern Ohio districts, declining production-based revenue exacerbated the common stress of financial security. In a broad sense, this uncertainty refrain calls into question the design of school funding mechanisms and to what extent they restrict administrators from optimizing financial strategies.

Academic Performance and Dropouts

A chief takeaway from interviews was the lack of concern about dropout rates and academic performance across the Bakken, Marcellus/Utica, and Colorado. Unfortunately for the DID analysis, NCES retracted its dropout and high school completion data during the research phase of this project. Efforts to collect dropout data from state agencies did not prove fruitful either, as the data were inconsistent over time in most states. Fortunately, interviews in each region provided valuable insight.[6] Across the Bakken, Denver-Julesburg in Colorado, and the Marcellus/Utica, the consensus of interviewees runs counter to previous studies (e.g., Cascio and Narayan 2015; Rickman, Wang, and Winters 2017) and suggests that the boom had not meaningfully increased dropout rates. Interviewees in different regions commented that the oil and gas development was technically advanced compared with prior boom periods and that it was not conducive to high school dropouts looking for employment, though the effect on college enrollment writ large is uncertain.

Interviews across the Bakken and the Marcellus/Utica also reported limited concern about impacts on academic performance. Because of poor data quality from state agencies, changing test structures over time, and a limited time series from the Stanford Education Data Archive (SEDA), we could only conduct a DID regression analysis on test outcomes for Marcellus/Utica states. Analyzing third grade through eighth grade standardized test scores showed mixed results for English and math scores, with a weakly negative trend across the five grade levels.[7] Additional data from supplementary DID analysis of SAT and ACT scores in Pennsylvania and Ohio did not return any statistically meaningful difference when compared with nonboom districts. In contrast to the findings in the Bakken and Marcellus/Utica, several interviewees in western Colorado expressed appreciable concern with student performance related to oil and gas development and the ensuing industry retraction. Interviews suggested that new students were sometimes less prepared academically and reported that funding declines were negatively affecting specific educational program. Unfortunately, reliable data were not ultimately available for comparative analysis.

To be clear, from the available data and extensive interviews, it is not obvious that the oil and gas boom in the Bakken or Marcellus/Utica has had a distinctly negative effect on educational outcomes in the short term. However, it is important to note that interview responses could have been influenced by several common cognitive biases, including the availability heuristic or optimism bias. Furthermore, a lack of evidence in the short term does not preclude long-term impacts, which are not currently measurable (Haggerty et al. 2014; Weber 2014). For example, classroom congestion, higher STRs, and teacher turnover in the Bakken are reasonable red flags for negative long-term effects on educational performance.

Of interest, when discussing academic attainment and potential educational effects stemming from the oil and gas boom, interviews in each region referenced the positive impact of area community colleges that were providing training to prepare local candidates for higher-paying work in the oil and gas sector. For example, a scholarship fund in North Dakota stemming from local oil revenue provides tuition-free community college to any high school graduate in the five-county region. While the increasing role of community colleges was a constant story, it is unclear whether the community college route is diverting students away from a four-year degree or increasing educational attainment among students who would not have otherwise gone to college.

Major Policy and Research Implications

This research is both timely and significant given the consistent headwinds facing many oil and gas boom regions, ongoing policy challenges within each state, and the new and in-depth observation of regionally specific impacts. In order for local and state policy makers to help communities capitalize on potential benefits of a resource boom and mitigate potential negative impacts on public education, we discuss several critical policy implications of our findings. In addition, to improve future research, we examine the benefits of using a mixed methods approach as well as the shortcomings in this and other existing studies.

Lessons for Policy Makers

Unconventional oil and gas booms can have diverse impacts on public schools in boom regions. As shown, student enrollment and mobility, student-teacher ratios, and financial trends can vary across development regions. At the same time, other trends can be common across diverse development areas.

Examples include a common low concern with dropout rates, infrequent reporting of teachers and school staff leaving for higher-paying industry jobs, and a heightened sense of unease regarding financial volatility.

The primary drivers of the variation across boom regions appear to be growth in population from industry-driven in-migration and the size of industrial and infrastructure expansion in a locality. These two hypotheses may account for the divergent trends but do not limit some common effects. For example, it is not surprising to see greater student enrollment in the Bakken, given the remoteness of the region and substantially larger build-out in support services. Conversely, a heightened sense of financial uncertainty could be common across energy regions experiencing volatility in resource-based income.

Taken together, the divergent trends captured in the statistical analysis and interviews between the Bakken and the Marcellus/Utica warn against overgeneralization with respect to the effects of natural resource booms. Researchers and policy makers should be wary of applying broad statements and conclusions across all unconventional oil and gas development areas within the United States. Consider, for example, the stark contrast between the extremely remote and sparsely populated Bakken region and the relatively populated development regions near Pittsburgh (Marcellus/Utica), Denver-Boulder (Denver-Julesburg), or at perhaps the farthest extreme, the Barnett Shale, located near the heavily developed and highly populated Fort Worth region in Texas. It would be entirely unsurprising to find that some impacts on schools would be markedly different among regions.

While the differing trends observed herein provide vital insight for school administrators and policy makers in boom states, the commonalities are equally, if not more, insightful when fully understood. The interview results reporting minimal effects of increased high school dropouts across all states contradict the historic resource economics literature, which found increased rates of student dropouts in mining boom regions (Black, McKinnish, and Sanders 2005; Emery, Ferrer, and Green 2012; Marchand and Weber 2015). Several interviewees explained that while they worried about decreases in graduation rates as energy development began to increase in their districts, those fears never materialized, leading interviewees to conclude that the recent oil and gas booms required a higher level of training than what most potential high school dropouts possess. We find this explanation plausible given the highly technical nature of unconventional oil and gas development, similar responses across regions, and conversations with community college administrators, who cited an increased enrollment in technical training programs related to the oil and gas industry.

However, neither the lack of concern about high school dropouts nor the lack of evidence of an impact on student achievement in the short run alleviates long-term concerns over student educational attainment in some states. In fact, observations of reduced per pupil educational spending, congested classrooms, rapid teacher turnover, and low levels of experience for many new teachers in the Bakken raises red flags regarding long-term student success areas. With most of the new boom-related households arriving in the Bakken in the late 2000s, and a majority of these young families having students enrolling in grade school, it may take several more years for potential impacts of the boom on student achievement to become measurable.

Interviews across all states also reported infrequent teacher attrition due directly to the oil and gas industry. Despite this conclusion, it does appear that increasing industry wages are indirectly affecting teacher quality via challenges with acquisition and retention of new teachers. In particular, high housing prices—driven by increasing demand and the high ability of industry workers to pay—made it challenging to hire and retain new teachers who were not already living in rural parts of Montana, North Dakota, and western Colorado. Notably, the increase in housing and rental costs in parts of Pennsylvania may have contributed to the more rapid student enrollment declines in boom regions by pushing out lower-income families.

A final commonality across regions may be the most important for policy makers—increased concern with financial volatility. In each region we visited, including those in the Marcellus/Utica that did not experience outsized student enrollment increases, many staff expressed concern about future funding and suggested that a decline in fossil-fuel-based revenue would be detrimental to school budgets. Notably, interviews stated that this uncertainty was influencing near-term decisions as well as long-term financial commitments. Several interviewees noted that the financial stress was limiting increases in teacher pay and causing reductions in nonessential education programs. Instead, some administrators referenced committing only to short-term expenditures.

Certainty, it could be the case that financial stress is affecting schools throughout the six-state sample area. However, the stated volatility and uncertainty from fossil fuel revenue would only intensify this trend. Colorado is a prime example, as many districts are dealing with state funding reductions but some boom districts are also facing local funding shortfalls and increased future uncertainty from fuel revenue. Overall, we find it to be a distinctly negative sign for public schools that financial stress from uncertainty has grown with increased fuel production.

From more than seventeen interviews, only a handful of interviewees suggested their districts were strictly better off because of the influence of unconventional oil and gas. Although oil and gas revenue has certainly created a variety of new upsides and opportunities for local economies in the short term, this common response from school staff is discouraging for the health of public education in many areas.

Therefore, we encourage local and state policy makers to consider revisions to education funding procedures that would allow for greater savings opportunities in times of excess revenue generation; allow for boom districts to capture larger *net* revenues from local resource extraction; and provide long-term, clearly stated commitments on state-based funding. Each of these updates would allow for consistency in revenue and expenditure smoothing over time, thereby providing districts a better opportunity to allocate resources based on their changing needs.

Future Research Needs

Our conclusions reinforce several of the prevailing long-term concerns associated with resource-based economic specialization. However, we also found confirmation against other common concerns, notably, a lack of evidence supporting increased dropout rates. The nuanced and divergent effects noted herein also add significant insight into a resource boom's comparative effect on public education and local economic health between disparate boom regions. However, substantial further research is needed to address several remaining questions.

We suggest three important lines of inquiry for future analysis. First, the varying designations or measurements for defining a boom create ambiguity in interpreting results across studies. Resource economists and other academics should work to standardize what constitutes the start of a resource boom and ways to measure a boom treatment effect.[8] Another approach would be to retroactively compare results of published papers if a different definition for boom were employed. As an example, a paper with a boom defined such as ours would benefit from being reevaluated based on a boom treatment derived from geographic information systems.

Despite the relative consensus stemming from our inquiries, we also believe further statistical analysis is needed regarding the effect of boom development on dropout rates. For starters, our study covered only six states. Moreover, the retraction of NCES data in August 2016, poor-quality state data, and potentially insufficient specificity from ACS all raise concerns regarding the validity of past analyses. Notwithstanding states with high-quality data, researchers may need to wait for the data quality and standardization of metrics on a national scale to improve before tackling the dropout question via statistical analysis across a number of producing states.

Recognizing the important role that recent papers have had in evaluating the effects of historic booms, we also recommend that effects on standardized test scores would be good to revisit at some point in the future. Furthermore, an analysis of the effects on career choices would be enlightening.

Importance of Mixed Methods Research

We encourage future researchers exploring similar questions to consider employing a mixed method analysis that includes an interview- or survey-based component. The interviewees' contributions to this analysis were essential to disentangling details in the data and unearthing hidden stories. Without the interview component, we likely would have misinterpreted or underestimated the magnitude or nuances of several important storylines such as the booms' impact on student mobility and the degree to which uncertainty of future income inhibits school planning.

Furthermore, the interviews helped build deeper insight into the practices being used among school districts to cope with the resource booms and why their efforts are or are not working in each situation. One illustration of this point is the challenge of acquiring and retaining new teachers. Administrators in many western communities were forced into nationwide searches, offered housing bonuses, and in some cases even became landlords to reduce the cost of housing. Despite these efforts, many young teachers in western districts often left after just a few years. This common scenario increases the soft costs of finding teachers, training them, and integrating them into the school community. Being able to illustrate these types of situations in detail helps confirm or contradict statistical results and aids in understanding the true magnitude of costs and benefits stemming from the unconventional oil and gas boom.

Conclusion

To conclude, we return to the central question guiding this paper: Did public school districts in regions with high levels of oil and gas production during the recent unconventional energy booms fare better or worse in terms of financial and educational outcomes than comparable school districts that did not experience a boom? First, we think it is important to restress the fact that study regions had distinctly different experiences with respect to specific metrics, such as student enrollment; thus, we caution against overgeneralizing any singular trend from one development area to the next. However, when summarized at the broadest level, we find that despite some distinct local benefits—such as new revenue for capital projects—the average impact in our sample has not been strictly positive. This conclusion is reinforced by conversations in each of the six study states. Notwithstanding dedicated and sustained efforts by district staff and communities to mitigate the challenges, it appears that the majority of districts struggled to manage elements of the myriad impacts of rapid growth and volatility in the short term. Furthermore, with rapidly changing student numbers, especially in very rural regions, and prevalent revenue ambiguity across most high-producing regions, students enrolled in boom school districts remain at risk of long-term negative impacts. One positive highlight from the interview component was the low concern with increasing high school dropouts. We hope future research will be able to further assess this conclusion as well as the boom's effect on college enrollment and career decision making. Perhaps most important, we encourage policy makers to update tax and school funding policies to provide school districts with greater longer-term financial certainty and resiliency against rapid financial swings.

Notes

1. We use *unconventional oil and gas* to refer to the marriage of hydraulic fracturing, horizontal drilling, and 3D seismic surveys to access previously uneconomic geologic formations to produce oil and gas. Often this resource is referred to as tight oil, shale oil, shale gas, or fracked gas.

2. The dropout data source identified for this study was retracted by the National Center for Education Statistics (NCES) during the course of our analysis.

3. As an example, this distinction eliminated Pittsburg's school district but not the surrounding suburban districts.

4. Many of these controls are reported on a county basis. Thus, for any district that straddles two or more counties, we attribute controls based on which county contains the largest percentage of the district by land mass. Local employment, personal income, and total population data come from the Regional Economic Accounts of the US Bureau of Economic Analysis (BEA). Total land area of each county comes from the US Census Bureau, and unemployment statistics come from the US Bureau of Labor Statistics. Oil and gas production were accessed through Drilling Info.

5. For more detailed analysis of public school finances issues in Colorado, see Nathan Ratledge and Laura Zachary (2017).

6. Several studies (Cascio and Narayan 2015; Rickman, Wang, and Winters 2017; Weber 2014) use the American Community Survey (ACS) to measure dropouts or completions. Because of large gaps in the time series and the use of multiyear averages for populations under 65,000, we question the robustness of using ACS data on dropouts.

7. The Stanford Education Data Archive aggregates and standardizes a nationwide database of district-level test scores by means and standard deviations to allow for comparisons on a subject and grade-level basis. We could not use SEDA data for North Dakota's or Colorado's DID analyses because both states lack preboom data within the dataset.

8. Major studies define boom treatment differently ranging from an interaction between geologic endowments and a time indicator; to changes in production outputs using Btus, number of wells, or change in monetary value of production; to changes based on the share of population deriving a majority of income from the resource (Bartik et al. 2016; Black, McKinnish, and Sanders 2005; Cascio and Narayan 2015; DeLeire, Eliason, and Timmins 2014; Fetzer 2014; Haggerty et al. 2014; Jacobsen and Parker 2016; Maniloff and Mastromonaco 2014; Michaels 2011; Paredes, Komarek, and Loveridge. 2015; Tsvetkova and Partridge 2016; Weber 2012, 2014; Weinstein 2014).

References

Bartik, Alexander, Janet Currie, Michael Greenstone, and Christopher R. Knittel. 2016. "The Local Economic and Welfare Consequences of Hydraulic Fracturing." SSRN, July 15. https://ssrn.com/abstract=2692197.

Black, Dan, Terra McKinnish, and Seth Sanders. 2005. "The Economic Impact of the Coal Boom and Bust." *Economic Journal* 115 (503): 449–476.

Cascio, Elizabeth U., and Ayushi Narayan. 2015. "Who Needs a Fracking Education? The Educational Response to Low-Skill Biased Technological Change." NBER Working Paper 21359. https://www.nber.org/papers/w21359.pdf.

C.R.S. 22-36-101. n.d. "Public Schools of Choice." Colorado Department of Education. Accessed July 23, 2020. https://www.cde.state.co.us/choice/openenrollment.

DeLeire, Thomas, Paul Eliason, and Christopher Timmins. 2014. "Measuring the Employment Impacts of Shale Gas Development." Working Paper. http://citeseerx.ist.psu.edu/viewdoc/download?doi=10.1.1.642.610&rep=rep1&type=pdf.

Emery, J. C. Herbert, Ana Ferrer, and David Green. 2012. "Long-Term Consequences of Natural Resource Booms for Human Capital Accumulation." *Industrial and Labor Relations Review* 65 (3): 708–734.

Fetzer, Thiemo. 2014. "Fracking Growth." *CEP Working Paper 1278*. London School of Economics and Political Science, UK.

Haggerty, Julia, Patricia H. Gude, Mark Delorey, and Ray Rasker. 2014. "Long-Term Effects of Income Specialization in Oil and Gas Extraction: The U.S. West, 1980–2011." *Energy Economics* 45 (September): 186–195. https://doi.org/10.1016/j.eneco.2014.06.020.

Headwaters Economics. 2008. "Impacts of Energy Development in Colorado: With a Case Study of Mesa and Garfield Counties." Bozeman, MT: Headwaters Economics.

Jacobsen, Grant D., and Dominic P. Parker. 2016. "The Economic Aftermath of Resource Booms: Evidence from Boomtowns in the American West." *Economic Journal* 126 (October 18): 1092–1128. https://doi.org/10.1111/ecoj.12173.

Jacquet, Jeffrey. 2009. "Energy Boomtowns and Natural Gas: Implications for Marcellus Shale Local Governments and Rural Communities." *Northeast Regional Center for Rural Development*. Paper No. 43. University Park: NERCRD, Pennsylvania State University.

Kelsey, Timothy W., William Hartman, Kai A. Schafft, Yetkin Borlu, and Charles Costanzo. 2012. "Marcellus Shale Gas Development and Pennsylvania School Districts: What Are the Implications for School Expenditures and Tax Revenues?" *Marcellus Education Fact Sheet*. Pennsylvania State University, University Park.

Maniloff, Peter, and Ralph Mastromonaco. 2014. "The Local Economic Impacts of Hydraulic Fracturing and Determinants of Dutch Disease." Working Paper No. 2014-08. Colorado School of Mines.

Marchand, Joseph, and Jeremy Weber. 2015. "The Labor Market and School Finance Effects of the Texas Shale Boom on Teacher Quality and Student Achievement." Department of Economics Working Paper, University of Alberta, Edmonton. Accessed April 29, 2016. https://sites.ualberta.ca/~econwps/2015/wp2015-15.pdf.

Measham, Thomas G., and David A. Fleming. 2014. "Impacts of Unconventional Gas Development on Rural Community Decline." *Journal of Rural Studies* 36: 376–385.

Michaels, Guy. 2011. "The Long Term Consequences of Resource-Based Specialisation." *The Economic Journal* 121 (March): 31–57.

Paredes, Dusan, Timothy Komarek, and Scott Loveridge. 2015. "Income and Employment Effects of Shale Extraction Windfalls: Evidence from the Marcellus Region." *Energy Economics* 47 (January): 112–120.

Price, Mark, Luis Basurto, Stephen Herzenberg, Diana Polson, Sharon Ward, and Ellis Wazeter. 2014. "The Shale Tipping Point: The Relationship of Drilling to Crime, Traffic Fatalities, STDs, and Rents in Pennsylvania, West Virginia, and Ohio." *Multi-State Shale Research Collaborative*. http://www.multistateshale.org/shale-tipping-point.

Raimi, Daniel, and Richard G. Newell. 2014. "Shale Public Finance: Local Government Revenues Associated with Oil and Gas Development." Durham, NC: Duke University Energy Initiative.

Ratledge, Nathan, and Laura Zachary. 2017. "Impacts of Unconventional Oil and Gas Booms on Public Education: A Mixed-Methods Analysis of Six Producing States." Resources for the Future Research Report. https://media.rff.org/documents/RFF20Rpt-Shale20Community20Impacts20Schools.pdf.

Rickman, Dan S., Hongbo Wang, and John V. Winters. 2017. "Is Shale Development Drilling Holes in Human Capital Pipeline?" *Energy Economics* 62 (February): 283–290.

Weber, Jeremy G. 2012. "The Effect of a Natural Gas Boom on Employment and Income in Colorado, Texas and Wyoming." *Energy Economics* 34 (5): 1580–1588.

Weber, Jeremy G. 2014. A Decade of Natural Gas Development: The Makings of a Resource Curse? *Resource and Energy Economics* 37 (August): 168–183. https://doi.org/10.1016/j.reseneeco.2013.11.013.

Weinstein, Amanda L. 2014. "Local Labor Market Restructuring in the Shale Boom." *Journal of Regional Analysis and Policy* 44 (1): 71–92.

Chapter 10 Summary

Our study pairs statistical analysis with extensive in-person interviews to evaluate the effects of the unconventional oil and gas boom on public education in six oil and gas producing states—Colorado, Montana, North Dakota, Pennsylvania, Ohio, and West Virginia. We find divergent trends in student enrollment, student-teacher ratios, and per pupil revenue and expenses between school districts in eastern and western regions, which are generally more remote. The primary drivers of the variation appear to be growth in population from industry-driven in-migration and the size of industrial and infrastructure expansion. Despite these disparate regional impacts, nearly all boom districts reported heightened stress from financial volatility and increased student mobility. Our study also uncovered several unexpected observations, including teacher fatigue. Although some distinct local benefits from the unconventional oil and gas boom have occurred within public education, the net impact has not been strictly positive. Despite dedicated efforts to mitigate the challenges, it appears that most regions in this study struggled to manage the impacts of uncertainty and volatility in the short term. Furthermore, with rapidly changing student numbers in the Bakken and prevalent revenue uncertainty across all high-producing regions, students in boom school districts remain at risk of long-term negative impacts.

KEY TAKEAWAYS

- *Student enrollment*, particularly in lower grades, was statistically higher in boom districts than in nonboom districts in North Dakota. Conversely, Marcellus boom districts experienced a statistically significant decline in student enrollment compared with nonboom districts, despite striking increases in natural gas production.
- *Financial effects* were also divergent between eastern and western regions. North Dakota boom districts experienced a statistically significant decline in per pupil funding, whereas Marcellus boom districts had a statistically positive increase in per pupil revenue.
- From an *expense perspective*, less money was spent on educational services within North Dakota, while statistically more money was spent on capital projects. Increased capital spending is not overly surprising, given the growth in student numbers in the Bakken; however, the decrease in per pupil educational spending raises red flags for long-term effects.
- Educators in all regions noted the challenge of high *student mobility*—a trend that was commonly referred to as a "revolving door." Not knowing when a student might arrive or leave created distinct challenges for budgeting and curriculum planning, with several teachers in western regions citing physical and emotional fatigue.
- In contrast to some historic literature and public anxiety, educators across all regions expressed minimal concern regarding increased *high school dropout rates*.

- Our analysis underscores the importance of pairing quantitative and qualitative research and cautions against overgeneralization of effects across disparate boom regions.
- Given the consistent headwinds facing many oil and gas boom regions, research on local public education impacts is *important to help address ongoing policy challenges* within each state, particularly in light of new and in-depth observation of region-specific impacts.

11

Effective Community Engagement in Shale-Impacted Communities in the United States

MYRA L. MOSS, NANCY BOWEN-ELLZEY, AND THOMAS MURPHY

Introduction

Community engagement in shale-impacted areas is critical to a community's long-term sustainability and an industry's ability to operate. In 2016, an interdisciplinary team of Extension professionals from the Ohio State and Penn State Universities began to explore their common experiences regarding community engagement in shale-impacted regions of Ohio and Pennsylvania and across the United States. The researchers wanted to identify best practices and lessons learned from community engagement experiences in order to share these insights with other shale energy-impacted communities, peers, and researchers and to seek answers to the following questions:

1. Is an engaged and informed community better positioned to address shale impacts and exploit shale benefits?
2. Are there examples of particularly beneficial / impactful engagement strategies or community benefit schemes?
3. What is social license to operate, and what role does the industry's need to obtain it play in community engagement?

DOI: 10.5876/9781646420278.c011

This chapter provides a theoretical piece based the authors' over three decades of observation and experience with shale-impacted community engagement. As Extension professionals, the authors have been working closely with communities in the Marcellus and Utica shale regions of Pennsylvania and Ohio as well as with Extension peers in other shale regions of the United States and throughout the world. We will share our reflections, identify our take on best practices, and share what we believe should undergo further research. This chapter will contribute to the larger field of energy research by providing unique experience-based insights into lessons learned and best practices for engaged shale-impacted communities, adding depth and direction to future social science research.

Shale Development Impacts Leading to Community Engagement

As shale oil and gas development continues to spread and gain significance throughout the United States, impacts of this relatively new industry sector are just beginning to be measured and understood. Benefits can range from employment to wealth building. Challenges can be environmental or social in nature and range from air and water quality to housing and crime. Impacts of shale development are measured in different ways, using various indicators. This section will review examples of impacts of shale development, including highlighted findings of a three-year study of a twenty-five-county shale region in eastern Ohio administered by the Ohio State University (OSU) Extension and sponsored by the Economic Development Administration (EDA), in addition to the review of other recent literature on shale development impacts. Authors of this chapter participated on a grant team for an EDA shale research project, a regional collaborative that elicited lessons from the field on community engagement and sustainability (Bowen, Civittolo, and Romich 2016).

Research-based analysis provides stakeholders with the insights they need to make informed decisions. In the case of OSU's EDA shale project, Extension educators collaborated with community stakeholders to help them understand the impact of shale development in their communities and identify implementation strategies in response to clear economic, social, and environmental changes. This project identified the need for Extension workers and others to play a role in engaging community stakeholders to build community capacity and plan for a sustainable future based on energy development.

Jobs and Economic Development

Anecdotes from residents and local businesses in Bradford County, Pennsylvania, clearly demonstrated that shale gas development had a major impact on local employment and income. Early estimates indicated Marcellus Shale created around 23,000 jobs in 2009, and the total employment impact in 2010 was approximately 44,000 jobs (Kelsey et al. 2012).

Ohio found similar results during the three-year EDA study of shale energy employment impacts within manufacturing sectors, with almost 81,200 jobs created or supported within five manufacturing sectors between 2010 and 2014. The five industrial sectors were energy, chemicals and chemical-based products, forest and wood products, metals manufacturing, and machinery manufacturing. A location quotient and shift share analysis of these manufacturing sectors revealed that 90 percent of the job creation in the five sectors occurred as a result of regional influences, most likely shale development, rather than national economic trends (Bowen, Civittolo, and Romich 2016).

Since shale development has the potential for creating many temporary jobs in a short period of time, the workforce "shock" can be a driving force in motivating communities to engage and plan. Rural communities that suddenly have the need for worker housing and infrastructure can start to feel the strain. Economic development spinoff effects also present opportunities that communities can benefit from, if they are prepared and responsive.

Research demonstrates that areas benefiting the most from economic development related to shale have existing unique competitive advantages associated with the energy sector. Thomas Tunstall indicates that the south Texas region is unique relative to other states in that it benefits from all phases of oil and gas production: upstream, midstream, and downstream (2015). As a result, communities in this region have greater opportunities to engage in the participation of economic activities along the supply chain.

Literature also points to economic challenges related to shale development. Dan Black, Terra McKinnish, and Seth Sanders examined how counties in West Virginia, Kentucky, Ohio, and Pennsylvania fared during the coal boom of the 1970s and subsequent bust in the 1980s (2005). In terms of employment effects, they found that the bust's negative effect was stronger than the boom's positive effect. According to Alexandra Tsvetkova and Mark Partridge, energy sector booms have negligible to negative effects in metropolitan areas, mostly due to metro areas being large and energy being a relatively small sector in terms of employment (2016). A literature review authored by Thomas Kinnaman described three studies that attempted to measure the economic impact of shale gas extraction and found that there is disputed evidence of significant economic impact related to shale development (2011).

Evidence of job creation and economic development impacts of the shale gas and oil industry is mixed. What appears to be clear however, is that communities that engage stakeholders to plan a future that is diversified and not strictly dependent on an oil and gas economy are able to leverage sustainable and long-term positive impacts.

Social Impacts

Shale development significantly affects predominantly rural communities with relatively low population density and little economic and social diversification (Lendel 2014). These communities cannot easily absorb change, and shale development has been associated with many challenges related to social and family services, emergency response abilities, and law enforcement. Increased communication capacity, enhanced equipment and training, and additional individuals with unique skills have been required to provide these services.

Daniel Raimi used a statistical analysis to determine whether there were significant relationships between drilling and crime rates (2012). In Colorado and Wyoming, increased shale development was correlated with high rates of violent crime, particularly aggravated assault. In North Dakota, Oklahoma, and Pennsylvania, no relationship was found between shale development and crime rates. In Texas, the data showed a significant relationship between increased shale development and lower nonviolent crime rates. The mixed results indicate that a broad generalization of the relationship between shale development and crime rates is difficult to make.

Another sizable social implication of shale energy exploration and production is regional workforce development. Research indicates that the greatest challenge facing the oil and gas industry over the next five to seven years is the "great crew change" (Lendel et al. 2015), highlighting the large gap in knowledge between baby-boomer-generation managers and the young workers who will replace them. Even Iryna Lendel et al.'s conservative scenario illustrates that 20,000 direct jobs in the build-out and maintenance of upstream and midstream infrastructure will require training of a wide variety of workers from environmental technicians, to heavy equipment operators and general laborers.

In her article "Unconventional Oil and Gas Development's Impact on State and Local Economies," Amanda Weinstein stated, "Unconventional oil and gas development has undoubtedly increased the employment and earnings in communities with shale resources" (2014, 6). Sustained, stable economic growth should be the goal of these communities. "Communities may be able to avoid or lessen the impact of a bust by using their newfound fortunes to prepare for the long run—by diversifying their economies, raising the skill level of their workforce, maintaining or improving their local services, and mitigating any other negative effects associated with drilling" (6).

Community Engagement in Shale-Impacted Communities

The anticipated impact of projected increases in shale energy production on our domestic energy future and on the communities where this growth is taking place will be substantial. Currently, sixteen states are directly impacted by shale development. The remaining thirty-four states are impacted to various degrees by growth in supplier networks and pipeline development. As a result of increasing shale oil and gas extraction, the United States is projected to be a net energy exporter by 2026 (U.S. Energy Information Agency 2017).

Predicted growth in the shale industry will continue to impact communities within the most active regions. The question then becomes: How can communities become fully engaged in discussions about their energy future; understand and anticipate social, economic and environmental results of energy extraction; and proactively plan to take advantage of the growing economic boom while mitigating adverse impacts?

What Is Community Engagement within the Context of Shale Development?

Throughout the literature on community engagement there are many definitions; however, certain common themes emerge that, combined with the authors' direct experiences, form a clearer understanding of engagement in shale communities. Community engagement is a process that does the following:

- Engages community stakeholders and shale companies in identifying and collaboratively addressing concerns that impact the well-being of each
- Results in trust that is built on a foundation of reciprocity, transparency, understanding, and respect
- Is intentional, necessitating ongoing communication and follow-through on commitments
- Occurs over time, requiring a consistent, long-term process tailored to stages in the shale life-cycle and specific needs of the host community
- Goes beyond one-way information sharing to two-way substantive communications that are timely and productive
- Furthers the community's vision and goals for a sustainable future, while permitting shale companies to continue to operate effectively

Community engagement in shale regions of the United States has largely been the domain of shale producers. Often, at the earliest stages of activity—exploration and securing private landowners' mineral rights—the wider community is largely uninformed. While engaging the host community is widely recognized as being an important step in shale development, this step, its timing and implementation, is usually dependent upon the interests and commitment of individual shale companies. A study of the Eagle Ford shale region in Texas disclosed that "serious and sustained communication and engagement was the exception rather than the rule . . . industry efforts to engage with community leaders to address local concerns vary widely. When present, these efforts appear to be especially effective" (Potterf et al. 2014, 3).

Engaging the host community is a critical component in a company's ability to obtain a *social license to operate*. The need for shale producers to secure social license has emerged as a topic of considerable discussion and concern among oil and gas producers, not only in the United States but also throughout the world. Many industry leaders recognize that improving the public's perception of their performance and building trust with host communities directly impacts their ability to operate successfully.

Social License to Operate and Community Engagement

Social license to operate is a *tool* used by the shale industry to "manage socio-political risk by conforming to a set of implicit rules imposed by their stakeholders. It is an ongoing social contract that allows a project to start and continue to operate within a community" (Smith and Richards 2015, 89). The public's perception of a particular shale company and how it operates, both within and outside of the community, can lead to either acceptance and approval or ongoing controversy and conflict. Corporations involved in energy development often do not fare well in the court of public opinion. In 2017 (as in previous years) there were no oil and gas companies listed in the top 100 trusted companies by the Reputation Institute. "Companies that are more open, more genuine and communicate more often have far stronger reputations" (Reputation Institute 2017, 17). The shale energy industry has not always been seen in a positive light, so working together and promoting a unified industry can put a positive spin on shale oil (Casey, Rasmussen, and Turner 2014).

Social license to operate is analogous to legal license to operate but within a different context. The absence of either can prevent a project from moving forward. Legal licenses are issued by regulatory agencies, while the social license to operate is "issued" by the host community and its stakeholders. The editor of the *Oil & Gas Journal* has written: "'License to operate' means not just legal permission to perform specific work but social sanctions for business activity . . . Judgments about it are rendered not in courts of law, but in the much less well-defined yet often more potent court of culture" (Smith and Richards 2015, 97). Anadarko Petroleum's manager of stakeholder relations said that social license is "tantamount to Relationship 101" and states that the risk of a shale resource remaining undeveloped is "not for a lack of legal license, but for lack of growing, earning, and maintaining a social license" (Smith and Richards 2015, 117).

Community engagement is the *method* by which companies seek to acquire and maintain their social license to operate within a community. The industry's role in engaging the community in a meaningful and sustained way is, by and large, voluntary, leading to considerable variation among producers. While some companies recognize engagement's importance and implement well-thought-out engagement methods, others are less responsive to community needs and issues. Ineffective methods, such as tasking employees who are expert in technical aspects of operations but do not possess the social/group process and communications skills necessary for outreach, can do more harm than good.

For those shale producers who do take seriously their role in community engagement, the label of socially responsible drillers applies. As noted by Ellis et al. (2015, 7), social responsibility has been described by industry leaders as "investing in communities and working to understand residents' concerns." One industry representative who participated in the study stated that social responsibility "is about 'being a human being' and knowing when to 'pick up the phone'" (7). Several company representatives "talked about 'being a good corporate citizen,' others talked about the 'social license to operate,' and still others wanted to be 'good neighbors'" (Ellis et al. 2015, 7). Involvement in community networks was viewed as important because it "allows [the companies] to better understand community needs, be transparent, set appropriate expectations, and ultimately work collaboratively to address concerns" (Ellis et al. 2015, 7).

Guidelines for Socially Responsible Shale Companies

Recognizing the importance of community engagement to the shale industry's social license to operate, the American Petroleum Institute (API), the industry's national trade association, issued *Bulletin 100-3, Community Engagement Guidelines* (API 2014). This bulletin provides engagement guidelines, operating principles, and specific suggestions for industry seeking to establish and maintain social license within host communities.

Basic industry principles identified by API include integrity, follow-through, safety, environmental responsibility, and effective communications. Specific outreach recommendations are offered for the five life-cycle stages of shale development (entry, exploration, development, operations/production, and exit) and suggest that adaptations might be needed to meet the unique situations and needs of each host community. Common themes include the need to do the following:

- Proactively engage community stakeholders during the earliest stages
- Anticipate and address stakeholder's needs and concerns
- Be transparent and intentional about changes in development plans

- Hire expert staff who meet ethical codes of conduct and professional standards
- Monitor and maintain regulatory and environmental compliance
- Design effective methods to share information and maintain open lines of communication throughout the life of the project: (1) tailoring communications to stakeholder's needs and expertise; (2) delivering communications through someone familiar with the community's norms, beliefs, culture, and values; and (3) establishing "just in time" communication platforms to immediately address emerging concerns

The American Petroleum Institute recognizes that the third phase of shale operations—development—has the greatest potential for community pushback due to the highly visible and impactful nature of activities. Problems frequently mentioned include truck traffic and degradation of roads, public safety, trash, light, noise, and dust during construction and drilling, as well as increased cost of living and perceived impacts to water and air quality (Alter et al. 2010; Potterf et al. 2014). The institute suggests active relationship building with the community to address issues of concern and create an early warning system to identify and correct problems before they turn into crises. Also recommended by the API are the identification of opportunities to build relationships, including targeted community investments, involvement on local boards, local contracting and hiring initiatives, and collaboration with educational institutions to identify career opportunities and promote job-training programs.

Community's Role: Authority and Influence over Shale Development

Local governments have some limited powers over shale development, though in the United States, state law usually takes precedence over local law, even in "home rule" states. Some states—including Pennsylvania, Ohio, and Colorado—have enacted legislation that prohibits local jurisdictions from banning oil and gas development but does leave the door open for requirements that are consistent with local land use plans (Minor 2014).

In Ohio, state laws preserve the regulatory power of local government to control most infrastructure development but prohibit local governments from exercising these same powers over oil and gas activity. Pennsylvania's

Act 13 implemented statewide regulation of shale development that restricts local control except for the design and enforcement of land use regulations and zoning plans that guide the siting of shale gas infrastructure. In Colorado, local governments have traditionally maintained broad control over local land use issues, but the state has ultimate control over oil and gas development and retains certain powers to preempt local ordinances. Still, communities in some states have found niches through which to exercise their land use authority while remaining compliant with state law. Examples include developing overlay districts, requiring special use permits, and designing setbacks and minimum standards relating to undesirable impacts of shale development (noise, light pollution, and watershed protection, for instance) (Squillace 2016). Some have promoted more substantive community involvement by state and federal regulators presenting a "framework for thinking about oil and gas regulation as if communities mattered" and to develop regulation that "promotes good planning that is orchestrated and directed proactively by regulatory agencies, with special attention to protecting the air, water and community resources that too often suffer when oil and gas resources are developed" (Squillace 2016, 559).

Local jurisdictions have the ability to plan well in advance of shale development, taking a proactive approach to addressing issues that may occur down the road. The Michigan State University Extension recommends that communities who seek to implement oil and gas regulatory initiatives should identify, through comprehensive planning, where shale development is likely to occur and carefully craft plans that address the potential impacts of these land uses. To be most effective, the community's comprehensive land use plan, supported by zoning, should be in place prior to the oil and gas industry entering the community, and must not be in conflict with state regulations (Solomon 2015). Nancy Bowen, David Civittolo, and Eric Romich (2016, 1) recommend that "Restricting development patterns can help constrain oil and gas mining operations to address infrastructure solutions." Land use controls are often controversial but may be necessary to avoid becoming "unwilling sacrificial zones" (Kelsey, Partridge, and White 2016, 206).

The challenge in Ohio, as in many other states, is that shale development often occurs in rural areas that are less likely to have land use tools in place such as comprehensive plans and zoning controls. The culture and values of many rural communities emphasize individual property rights over communitywide planning and zoning for the public good. In the authors' experience, convincing stakeholders to proactively plan well in advance of the shale oil and gas industry can be a difficult sell.

Comprehensive planning can provide an effective tool to engage the entire community. The planning process itself can be as important as the results, fostering resident engagement from initiation through implementation. Literature supports the need for proactive, sustainable planning to prepare for the potential "bust" that can occur in natural resource-based economies. Some scholars look beyond the economic impact of natural resource-based energy development, to examine how communities and regions are coping and planning for the future. They argue for the need for increased research to measure change and identify best practices that can be replicated in other regions affected by natural resources. Two planning strategies identified by Kelsey, Partridge, and White (2016) include (1) planning for diversification, and (2) planning for development of permanent funds. To effectively implement these two forms of planning, communities can capture revenues generated by ongoing energy development through severance, property, sales, or corporate income taxes, and/or impact fees to fund targeted activities.

Too often there is little advanced planning for large, landscape-scale change that occurs in communities throughout the world. Rural communities and small towns often view themselves through a historical lens of "what they have always been." These communities commonly exhibit bias toward an older population that is even more reluctant to plan for, and accept, abrupt change impacting lifestyle and/or community dynamics. The question is: Does the community sense the change while it is happening, and does it react in an organized and thoughtful manner? Or does it just wake up one day, realize something does not look quite the same, and ask what can be done to resolve the situation?

In conversations with one of the authors, Todd Bryan from RESOLVE, a Washington DC–based nongovernmental organization, addressed this point through the organization's community consultation process. Bryan contends that communities need to actively and continuously plan for change in order to make the successful transitions that are inevitable over time by considering the following questions:

- How does a community see itself historically, presently, and over the future?
- What does it hold dear and what is it not willing to compromise, such as an historical downtown district, or an active agricultural heritage?
- How does it plan for change to support the demands of younger residents?

The research undertaken by RESOLVE has shown that planning that is thoughtful, transparent, and inclusive has the greatest chance of success. It strengthens community relationships and increases community resilience by working toward common goals. As something of significant scale happens in this type of community, the dialogue has already taken place that allows residents to either welcome the opportunity or have answers for the question "If not this, then what?"

Extension's Role in Community Outreach and Engagement

Educating and engaging the broader community are a significant challenge and should be addressed by industry and various credible voices, early on, and repeatedly. Larger questions are who are the residents and stakeholders and when and how to them and on what topics, as the impacts can vary considerably over time and regions.

Community engagement and topics addressed should be planned to coincide with each life-cycle stage of shale development. A targeted approach to engagement is often the most impactful with audiences and seeks to not get ahead of the public interest or dialogue. Outreach activities should evolve as the public conversation seeks additional detail.

Public workshops, educational sessions, and multimedia outreach organized and moderated by an independent third party can provide early, successful public education and engagement at the community level. University professionals, such as those available through the Cooperative Extension Service, are experienced in group process and facilitation. They are perceived by the community and industry as both trustworthy and a good source of information (Alter et al. 2010). Having a range of industry and government officials also present during some of those educational sessions can be particularly impactful.

Targeted engagement with unique groups of community leaders adds value to successful outreach and education by creating advocates for the science of energy in all stages from local development to downstream utilization. Also, developing a relationship with media representatives who have an interest in the topic and provide balanced reporting of the issues can broaden Extension's educational reach.

Lessons from the Field

The authors' insights gained from direct involvement with shale-impacted communities prompt us to propose best practices in community engagement. The Dimock, Guernsey County, eastern Canada, and Marcellus/Utica case studies, based on our experiences as Extension educators, provide "real-world" examples that emphasize the engagement practices we believe work best. The Eagle Ford case study based on research conducted by Jeb Potterf et al. (2014) further supports our conclusions. We offer these insights to help community residents and stakeholders implement effective engagement, education, and outreach and to suggest to shale companies what they might do to gain and maintain social license to operate.

Promoting Industry-Community Communication and Engagement: Lessons from the Eagle Ford Region, Texas

There is wide variability among shale companies regarding their approach to communicating and engaging with communities. *Who* does the communicating on behalf of the company, *when* it happens, and *how* they conduct this outreach are key variables that impact success (Potterf et al. 2014). Research investigating community perceptions of shale industry operations in the Eagle Ford region was conducted by the Institute for Social Science Research on Natural Resources at Utah State University. They identified the—*who* and *how*—which isolated the most effective manner in which a company could engage with communities where they had commercial interests. This research observed three basic industry approaches: (1) the company hired a community liaison or economic development representative who was embedded within the region, (2) the company hired a community liaison who visited the region frequently but was based outside in a nearby city, and (3) the company relied on employees with other primary job duties to handle communications with the local community (Potterf et al. 2014).

The first option, establishing an embedded liaison, was the most effective in building open lines of communication. This employee understood the local culture and was able to develop personal, close relationships with key stakeholders. The liaison attended community events and participated on boards, becoming well known locally. The person was a conduit for community issues and concerns, and her or his local presence enabled issues mitigation and problem solving (Potterf et al. 2014). This last point is crucial. Solving problems in the community before they escalated helped to build trust. "Trust is hard to earn, easy to lose, and very difficult to recover once lost" (Smith and Richards 2015, 93).

Industry representatives who lived outside of the region and visited when needed were found to be less effective in engaging stakeholders, due to a lack of complete local knowledge leading to an erosion of credibility over time. The third option, using an employee with other primary responsibilities, was often shown to be harmful to the creation of relationships. As stated by Ellis et al. (2015, 12), "Not having someone who fully understands the basic issues faced by communities may lead to counterproductive community engagement and philanthropic efforts."

Study results in the Eagle Ford region "point to the need for industry to view communication efforts as part of a comprehensive 'best practices toolkit'. While the deployment of technical solutions . . . are appropriate and necessary, it is also important to see ongoing communications with communities as a complementary way to maximize benefits, minimize negative impacts, and improve public acceptability of energy development" (Potterf et al. 2014, 34).

Marcellus Shale Region: Industry Community Engagement Gone Wrong

Dimock, Pennsylvania, made famous in the 2010 antidrilling movie *Gasland* (Fox 2010), provides an example of industry-initiated community engagement gone wrong. When a large shale energy company featured in the film initiated drilling wells in this rural community, their public liaison was an attorney tasked with avoiding corporate liability. When the energy company was accused of a well-publicized wellbore integrity issue, the liaison's immediate response was denial, ignoring the intense concerns of impacted landowners and their elected officials. The outcome was rapidly elevated mistrust within the community and an untenable operating environment for the company. Responding to this threat to social license, the company reorganized its staff and hired a new manager skilled in outreach. Under his leadership the company held public outreach sessions during the county fair, rented the fairgrounds for a series of public education events, held sessions for students in local schools, and explained the shale development process to targeted stakeholders. These educational and outreach efforts by the company led to widespread community support.

Extensive field observations lead the authors to conclude that a proactive approach to stakeholder engagement initiated by either community leaders or appropriate industry personnel is most effective in reducing conflict and fostering some level of desired social license. These observations concur with the Eagle Ford research that trustworthiness is significantly increased when the person(s) charged with engagement have the correctly matched skill sets to be effective and credible communicators.

Engaging Residents through Sustainable Planning: Guernsey County, Ohio

Guernsey County is located in the Utica/Point Pleasant shale region of eastern Ohio. In 2010 community leaders witnessed the impacts shale development was having on nearby communities in Pennsylvania and feared they would soon face similar challenges. The Chamber director initiated monthly shale community education forums for key stakeholders and the Guernsey Regional Planning Commission, Community Improvement Corporation, and Commissioners commenced an update of their comprehensive land use plan (Romich et al. 2015). The previous plan was completed in 2010, largely prior to the wave of shale development moving west from Pennsylvania and had few references to shale impacts and implications. Leaders recognized that visionary planning was needed to address potential impacts of shale development and leverage energy benefits.

In parallel, a team of Ohio State University Extension educators started a project to test the applicability of sustainable comprehensive planning methods to shale-impacted communities in eastern Ohio. Funded through a small grant from the North Central Regional Center for Rural Development, the Extension team was invited by Guernsey County to update the county's plan. Early in 2014 the planning project was launched, and the final plan was approved by the Planning Commission and County Commissioners in 2015.

Cornerstones of sustainability suggest broad-based community engagement in the development and management of planning. To oversee plan development, Guernsey County established a steering committee comprising representatives from the various sectors of the community. During the first six months, broadly representative workgroups including social services, economic development, education/workforce, built and natural environment, housing, agriculture, health care, tourism, technology, and safety services were created and tasked with engaging their sector of the community, gathering input on the challenges and benefits of shale development. Common themes emerging from this inclusive engagement process were used to develop focal themes and goals. Extension coordinated regularly held public educational sessions, inviting shale regulators, companies, landowners, and researchers as speakers/discussants on social, economic, and environmental topics.

Throughout the development of the plan, local residents and leaders were actively involved in determining the future of their county. The plan's capstone chapter linked together social, environmental, and economic goals to insure sustainable, multidimensional goals that were widely supported by community leaders and residents.

Plan outcomes have included new jobs and increased private investment, new or expanded shale-related businesses, creation of a regional business/education council to address workforce issues, infrastructure improvements (water, sewer, roads), and expanded community engagement and outreach efforts by local organizations. Guernsey County leaders and residents remain mostly positive about shale development in their community. Benefits, such as increased tax revenues and philanthropic giving by both companies and residents, are being used to support comprehensive plan goals.

Community Discussion of "If Not This, Then What?" in Eastern Canada

Even before seismic testing began in 2012, a preemptive workshop was held in the province of New Brunswick to focus on proposed shale energy development. Gas resources were owned by the Canadian Crown, licenses were issued to a company for exploration and development, and access to the minerals was through private property owned by surface residents. Sponsored by two provincial universities, the workshops' participants composed over 200 residents representing diverse groups active in the public debate. Key attendees included elected officials who permitted shale development as well as government regulators charged with enforcing that decision.

A turning point in the discussions was the realization that the province's traditional industries—forestry and fishing—were declining, the age of the population was increasing as younger families were leaving for work elsewhere, and the percentage of residents receiving most of their income from government social programs was growing. Faced with a potentially emerging shale gas industry, the community attempted to answer the "if not this, then what" question.

The workshop sessions enabled participants to evaluate their community over time and better plan for their future, allowing for broad-based engagement in decision making and ownership of outcomes and decisions based on data, not perceptions. Politicians were held accountable by their constituents for their decision to move forward, regulators explained regulations instead of defending the industry, and the industry had the opportunity to make its case regarding production technologies and regional economic benefits.

Extension and the Evolution of Stakeholder Education and Engagement: The Marcellus/Utica Region

An independent third party, such as Cooperative Extension, can best moderate successful and early public education and engagement at the community level. Extension professionals are experienced in group process and facilitation and are trusted by both the industry and the community. They also have expertise in the applicability of different forms of media to educate, engage, and inform the public.

In the Marcellus and Utica shale region of the eastern United States, related educational outreach coincided with the energy industry's earliest moves as it approached landowners for access rights to drill on private property (Murphy 2016). Although the Appalachian basin was the site of the first commercial oil and gas wells in the world, the advent of high-volume shale energy production was without precedent—a nonlegacy activity (Joel et al.

2016). Community leaders had little knowledge about horizontal drilling and hydraulic fracturing, the workforce was not skilled in shale technologies and institutional capacity for training did not exist (Brundage et al. 2011), gas pipeline transport capacity was insufficient, and there was little appreciation for regulatory changes needed to address pending impacts to air, water, land, and people. Additional constraints included a lack of needed housing (Kelsey and Murphy 2011), deficient road infrastructure, and little local supply of essential goods and services required 24/7 by the energy industry (Lendel et al. 2015).

In 2005 the Cooperative Extension Service, housed in the region's land grant universities and serving local communities, recognized this rapidly emerging public issue and proactively developed community engagement initiatives to coincide with the emergence of the shale industry. Robust programming began in Pennsylvania, Ohio, and West Virginia, the focal point of development. New York and Maryland initiated some Extension programs but then moved to shale drilling bans or moratoriums (Kriss 2013).

Extension's mission has historically been to apply the research conducted at the university to the critical needs of communities. Commonly invited across political boundaries, Extension professionals shared shale-related experiences and research outcomes with those making collective public policy decisions, formulating new legislation and mitigating impacts to local infrastructure. Programming for landowners and families attempting to navigate more personal impacts of shale development was also provided.

Extension's early programming occurred through well-attended town hall meetings targeting the broader public, who were attempting to understand the impacts of shale energy on a number of different levels—collectively as a state and nation and individually as members of an affected community. As dialogue matured, internet capacity improved, and the interests and abilities of the public evolved, Extension's methods to convey current information on the science of shale energy and the regulatory, economic, environmental, and health implications have also changed from "high-touch" venues to use of digital platforms such as websites, Twitter, Facebook, podcasts, webcasts, and so on. Both methods have their advantages and drawbacks (see table 11.1).

Extension experiences in Pennsylvania and Ohio demonstrate that no single method for educational outreach to the public was best. Numerous examples illustrate the positive use of tailored social media to connect educational messaging to community leaders and residents in those same communities

TABLE 11.1. Advantages and disadvantages of education/outreach methods

Face-to-Face Forums/Workshops		*Social Media/Web-Based Education*	
Advantages	*Disadvantages*	*Advantages*	*Disadvantages*
Provides community access to verified experts	Can provide forum for antishale disruptors	Time efficiencies: reaches broader audiences quickly	Outside influences' ability to disrupt/hijack local community discussions
Opportunities for two-way dialogue	Takes time and effort to organize	Faster deployment of information	Faster deployment of misinformation and rumors
Can create ongoing relationships	Information may not keep up with rapid changes	Flexibility in accessing and conveying information—mobile devices	Anyone's potential to be an "expert"—no credential oversight
Can target participants	Can attract "outsiders" who do not reside in community	Supports transparency and frequent engagement	No oversight over content
Allows access for non-tech-savvy residents	Effective facilitation skills needed to manage Q&A	Supports formation of "Communities of Interest"	Restricts usage to tech-savvy individuals
Can attract a large group of community residents	Not everyone's concerns can be aired or addressed	Global scope and reach	Challenge to direct information to intended audiences

(Levick 2013). Extension discovered that using town hall meetings in concert with social media created synergies that enhanced both methods.

Final Observations

While community engagement is important and consensus maintains it should occur, it is not part of the US permitting process and is often not on the radar of individual shale producers or community leaders. At the earliest stages of land aggregation, local officials are often unaware of shale activities, so community engagement can occur "after the fact," increasing the likelihood of community pushback and challenges to the company's continued operations.

Companies have become increasingly aware that social license to operate is crucial for their activities, and community engagement is the method, albeit voluntary, by which they secure and maintain this social license. In comparison to the rest of the world, most landowners in the United States own their surface rights along with the minerals underneath. Thus, a patchwork of competing energy companies strive to lease mineral rights from private landowners in order to stitch together a commercially viable, contiguous land position. By default, this patchwork commonly means many more companies operating in any one region with many versions of engagement with community stakeholders.

Based on the authors' experiences, certain characteristics of community engagement are more effective in creating relationships that allow communities to realize the benefits of shale development and companies to achieve and maintain social license to operate. We offer the following observations to assist impacted communities and shale companies in creating more effective models of engagement and communication. Without effective stakeholder engagement, shale industry's social license to operate may be jeopardized and host communities may be prevented from pursuing the future they envision.

Early, Inclusive Community Engagement

Extensive field experience reinforces that it is imperative communities address the issue of potential shale oil and gas development well before development occurs. By actively and inclusively engaging the community in "if not this, then what" conversations that are reflected through comprehensive plans and land use standards, issues that may arise can be addressed early on by an informed populace. Workforce development, business growth, economic diversification, and increased local revenues are potential opportunities of shale development. By engaging the community through inclusionary planning, strategies to exploit those benefits can be formulated and priorities can be identified.

Match Right Answers to Actual Questions

A key disconnect in the public discourse around shale energy is matching the right answers to the actual questions being asked in many of the community-based dialogues (Handy 2013). Quite commonly, energy company staff, local officials, and regulators answer citizen questions by explaining "how" the drilling process works, "how" the regulations apply and would be enforced, or "how" risk mitigation at all levels would be implemented. For those who are entering the discussions for the first time, more personal questions arise around "whether or not" shale energy should be developed near their homes due to perceived safety, environmental, or economic (effect on real estate values) concerns. Some take a different, more ideological approach, and question "whether or not" shale energy should be developed at all, especially when viewed through a lens of climate change. Individuals or communities asking this type of "whether" question need an avenue to have them addressed before "awarding" a social license to operate. Stakeholder engagement that addresses longer-term energy policies as well as shorter term localized impacts needs to take place early on through a broad and consistent outreach process. Tailoring outreach and education to the life-cycle stages of shale development can help to provide responsive answers that do not get ahead of the public dialogue.

Importance of Company Soft Skills

Equally challenging is *who* provides answers. In most regions of the United States minerals are privately owned and the right to develop them has primacy, so a contract is negotiated between the landowner—private or public—and the company. Once the company satisfies the landowner(s) and obtains the needed regulatory permits, development can begin. There is usually no legal requirement to address concerns of adjoining neighbors or the community at large, though it is good public relations to do so. Often the company—run by people with a technical, legal, or financial focus—has a strong economic motivation to move projects forward and is not interested in spending time on the community's "whether or not" questions. Company representatives who understand the local culture, are involved in the community, are adept at group process, and are effective, responsive communicators can build trust in company-community relationships.

Neutral Educator/Facilitator Needed

There needs to be a voice separate from government and industry in order to seek resolution to communication disconnects. Academics have a natural fit in that space. University specialists, such as those available through the Extension system, have the trust, credibility and influence needed to bring together diverse interests. Without having vested interests on either side of the dialogue and commonly enjoying credibility with large segments of society, they can speak to the larger issues of energy policy while also addressing the more personal and local implications of shale development. They can help to match the answers to the questions.

One Size Does Not Fit All

In effective community engagement one size does not fit all, and there is no "right" way. Engagement must be tailored to locality's culture, norms, issues, and circumstances. Critical partners in effective engagement include industry, community stakeholders, regulators, nongovernmental organizations and citizens. Each has an important role to play (Houston Advanced Research Center 2016).

Communities Issue Social License

Communities should recognize their "power" to issue social license to operate for shale companies within their jurisdiction and make their expectations known regarding company-community communications, outreach, and engagement.

There are a number of gaps in the current research regarding shale energy development and the communities in which they operate. Research that would help to inform the development of policies, procedures, and best practices that communities could use to encourage and participate in effective community engagement with shale companies would be valuable for stakeholders and residents. While the American Petroleum Institute has produced Community Engagement Guidelines for shale companies, there does not exist a similar, best practice research-based "handbook" for stakeholders to use as they attempt to navigate the intense changes brought about in their communities by shale development. With the anticipated growth in shale oil and gas exploration, the need for engaged community planning and effective community-industry relationships will not diminish anytime soon.

References

Alter, Ted, Kathy Brasier, Diane McLaughlin, Fern K. Willits, Teri Ooms, and Sherry Tracewski. 2010. "Baseline Socioeconomic Analysis for the Marcellus Shale Dvelopment in Pennsylvania." Institute for Public Policy & Economic

Development at Wilkes University. https://www.institutepa.org/perch/resources/marcellusshalestudy08312010.pdf.
American Petroleum Institute. 2014. "Community Engagement Guidelines." *ANSI/API Bulletin 103-3*. http://www.api.org/oil-and-natural-gas/wells-to-consumer/exploration-and-production/hydraulic-fracturing/community-engagement-guidelines.
Bowen, Nancy, David Civittolo, and Eric Romich. 2016. "Community Planning Strategies for Energy Boomtowns." Ohioline, Ohio State University Extension. CDFS-SED-6. https://ohioline.osu.edu/factsheet/cdfs-sed-6.
Black, Dan, Terra McKinnish, and Seth Sanders. 2005. "The Economic Impact of the Coal Boom and Bust." *Economic Journal* 115 (503): 449–476. https://onlinelibrary.wiley.com/doi/abs/10.1111/j.1468-0297.2005.00996.x.
Bowen, Nancy, Eric Romich, David Civittolo, et al. 2016. "Final Report: Building Sustainable Communities in Ohio's Shale Region." Ohio State University Extension. https://energizeohio.osu.edu/sites/energizeohio/files/imce/TB%20-%20EDAGrantProjectFinalReport.pdf.
Brundage, Tracy L., Timothy W. Kelsey, Janice Lobdell, et al. 2011. "Pennsylvania Statewide Marcellus Shale Workforce Needs Assessment. Marcellus Shale Education and Training Center." Pennsylvania College of Technology and Penn State Extension. http://pasbdc.org/uploads/media_items/pennsylvania-statewide-marcellus-shale-workforce-needs-assesment-june-2011.original.pdf.
Casey, Kevin, Kevin Rasmussen, and Sarah Turner. 2014. "An Engagement Guide for Shale Energy Supply Chain Companies, Energy Equipment and Infrastructure Alliance." https://www.wm.edu/as/publicpolicy/documents/prs/shale.pdf.
Ellis, Colter, Gene L. Theodori, Peggy Petrzelka, and Douglas Jackson-Smith. 2015. "Socially Responsible Drilling: Perspectives of Industry Representatives in the Eagle Ford Shale." Huntsville, TX: Center for Rural Studies, Sam Houston State University. https://www.shsu.edu/dotAsset/170a5b8f-6554-46a1-b41b-fcab3f5be2f3.pdf.
Fox, Josh. 2010. *Gasland*. Brooklyn: International WOW Company.
Handy, Ryan Maye. 2013. "Most Colorado Voters Support Fracking, A New Voter Poll Shows." *Coloradoan*, December 2, 2013. https://www.coloradoan.com/story/news/2013/12/02/most-colorado-voters-support-fracking-new-voter-poll-shows/3828089/.
Houston Advanced Research Center. 2016. "Appendix G: Community Issues/Public Perception (Task 5.3.5), G.1 EFD-TIP Societal Team Final Report." http://efdsystems.org/pdf/G.1_EFD-TIP_Societal_Team_Final_Report.pdf.
Joel, Burcat R., Stephen W. Saunders, Amy Barrette, et al. 2016. *The Law of Oil and Gas in Pennsylvania*. 2nd ed. Mechanicsburg, PA: PBI Press.
Kelsey, Timothy, and Thomas Murphy. 2011. "Economic Implications of Natural Gas Drilling in the Marcellus Shale Region." *Cascade* 77 (Spring/Summer). https://www.philadelphiafed.org/community-development/publications/cascade/77/01_economic-implications-of-gas-drilling-in-marcellus-shale#:~:

text=The%20economic%20implications%20from%20the,companies%20in%20exchange%20for%20access.

Kelsey, Timothy W., Mark D. Partridge, and Nancy E. White. 2016. "Unconventional Gas and Oil Development in the United States: Economic Experience and Policy Issues." *Applied Economic Perspectives and Policy* 38 (2): 191–214.

Kelsey, Timothy W., Martin Shields, James R. Ladlee, et al. 2012. "Economic Impacts of Marcellus Shale in Bradford County: Employment and Income in 2010." University Park, PA: Marcellus Shale Education and Training Center. https://aese.psu.edu/research/centers/cecd/archives/marcellus/economic-impacts-of-marcellus-shale-in-bradford-county-employment-and-income-in-2010.

Kinnaman, Thomas C. 2011. "The Economic Impact of Shale Gas Extraction: A Review of Existing Studies." *Ecological Economics* 70 (7): 1243–1249. https://doi.org/10.1016/j.ecolecon.2011.02.005.

Kriss, Erik. 2013. "New Yorkers Opposed to Fracking: Poll." *New York Post*, March 20, 2013. http://www.nypost.com/2013/03/20/new-yorkers-opposed-to-fracking-poll/.

Lendel, Iryna. 2014. "Social Impacts of Shale Development on Municipalities." *Bridge* 44 (2): 47–51.

Lendel, Iryna, Andrew R. Thomas, Bryan Townley, Thomas Murphy, and Ken Kalynchuk. 2015. "Economics of Utica Shale in Ohio: Workforce Analysis." *Urban Publications*. 0 1 2 3 1330. http://engagedscholarship.csuohio.edu/urban_facpub/1330.

Levick, Richard, 2013. "Colorado Rejects Fracking: The Money's Not Talking; Social Media Is." *Forbes*, November 7, 2013. https://www.forbes.com/sites/richardlevick/2013/11/07/colorado-rejects-fracking-the-moneys-not-talking-social-media-is/#7b435f98487a.

Minor, Joel. 2014. "Local Government Fracking Regulations." *Stanford Environmental Law Journal* 33 (1): 61–122. https://papers.ssrn.com/sol3/papers.cfm?abstract_id=2485889.

Murphy, Thomas. 2016. "U.S. Shale Trends—Economic and Global Implications." *Journal of Physics: Conference Series* 745 (2): 022004. University Park, PA: IOP Publishing. https://iopscience.iop.org/article/10.1088/1742-6596/745/2/022004/pdf.

Potterf, Jeb E., Peg Petrzelka, Douglas Jackson-Smith, Colter Ellis, Gene L. Theodori, and Cameron Alex Carmichael. 2014. "Community Perceptions of the Oil and Gas Industry in the Eagle Ford Shale Play." Logan, UT: Institute for Social Science Research on Natural Resources. http://efdsystems.org/pdf/G.3_Perceptions_in_the_Eagle_Ford_Executive_Summary.pdf.

Raimi, Daniel. 2012. *The Potential Social Impacts of Shale Gas Development in North Carolina*. Master's thesis, Sanford School of Public Policy, Duke University, Durham, NC.

Reputation Institute. 2017. "2017 Global Reptrak 100 Report." RepTrak. February 28. https://www.rankingthebrands.com/PDF/Global%20RepTrak%20100%20Report%202017,%20Reputation%20Institute.pdf.

Romich, Eric, Nancy Bowen-Ellzey, Myra Moss, Cindy Bond, and David Civittolo. 2015. "Building Sustainability in Gas- and Oil-Producing Communities." *Journal of Extension* 53 (3). http://www.joe.org/joe/2015june/iw1.php.
Smith, Don C., and Jessica M. Richards. 2015. "Social License to Operate: Hydraulic Fracturing-Related Challenges Facing the Oil and Gas Industry." *Oil and Gas: Natural Resources, and Energy Journal* 1 (2): 81–163. http://digitalcommons.law.ou.edu/cgi/viewcontent.cgi?article=1006&context=onej.
Solomon, Dean. 2015. "Local Government Roles in Oil and Gas Regulation: Part 2." *Michigan State University Extension*, March 15, 2015. http://msue.anr.msu.edu/news/local_government_roles_in_oil_and_gas_regulation_part_2.
Squillace, Mark. 2016. "Managing Unconventional Oil and Gas Development as If Communities Mattered." *Vermont Law Review* 40: 525–560. http://scholar.law.colorado.edu/articles/22.
Tsvetkova, Alexandra, and Mark D. Partridge. 2016. "Economics of Modern Energy Boomtowns: Do Oil and Gas Shocks Differ from Shocks in the Rest of the Economy?" *Energy Economics* 59 (September): 81–95.
Tunstall, Thomas. 2015. "Recent Economic and Community Impact of Unconventional Oil and Gas Exploration and Production on South Texas Counties in the Eagle Ford Shale Area." *Journal of Regional Analysis and Policy* 45 (1): 82–92.
U.S. Energy Information Agency. 2017. "Energy Outlook for 2017." *U.S. Energy Information Administration, (EIA)*, January 5. http://www.eia.gov/outlooks/aeo/pdf/0383(2017).pdf.
Weinstein, Amanda L. 2014. "Unconventional Oil and Gas Development's Impact on State and Local Economies." *Choices: The Magazine of Food, Farm and Resource Issues, Farm, and Resource Issues* 29 (4): 1–7. http://ageconsearch.umn.edu/bitstream/189797/2/cmsarticle_398.pdf.

Chapter 11 Summary

Based on the authors' three-plus decades of direct experience in shale-impacted regions, this chapter explores the importance of an engaged and informed community for residents and stakeholders as well as for the shale industry's ability to operate. As Extension professionals, the authors have been working closely with communities in the Marcellus and Utica shale regions of Pennsylvania and Ohio as well as with Extension peers in other shale regions of the United States and colleagues throughout the world. This chapter shares our reflections, identifies our take on best practices, and provides unique experience-based insights into lessons learned for engaged shale-impacted communities. It also suggests approaches and techniques that will help industry foster effective engagement with stakeholders and enable companies to continue to operate effectively within the community context. Observations based on case study examples to assist impacted communities and guide shale companies in creating more effective models of engagement and communication are offered.

KEY TAKEAWAYS

- Early, inclusive community engagement in planning is needed to effectively address challenges and capture benefits of shale development.
- Stakeholder engagement that addresses longer-term energy policies as well as shorter-term localized impacts needs to take place early on through a broad and consistent outreach process.
- For those tasked with representing the company to the community, "soft" human resource facilitation and effective communication skills are as important as technical skills.
- Company representatives who understand the local culture, are directly involved with the community, are adept at group process, and are effective, responsive communicators can build more trusting relationships.
- A neutral party—one separate from government or industry—can be effective as a trusted educator and facilitator of community-industry dialogue. University specialists, such as those available through the Extension system, have the trust, credibility, and influence needed to bring together diverse interests.
- In effective community engagement, one size does not fit all and there is no "right" way.
- Engagement must be tailored to locality's culture, norms, issues, and circumstances.
- Communities "issue" social license to operate that companies need to initiate and continue operations. Social license can be as important to the company in meeting its goals as legal licenses to operate (permits). Communities aware of this can open the door to communications that meet the needs and goals of both parties.
- There are a number of gaps in our current understanding regarding shale energy development and the communities in which it operates. Research that would help to inform the development of policies, procedures, and best practices that communities could use to encourage and participate in effective community engagement with shale companies would be valuable for stakeholders and residents.

12

A Framework for Sustainable Siting of Wind Energy Facilities

Economic, Social, and Environmental Factors

RONALD MEYERS, PATRICK MILLER, TODD SCHENK,
RICHARD F. HIRSH, ACHLA MARATHE, ANJU SETH, MARC J. STERN,
JISOO SIM, AND SEVDA OZTURK SARI

Introduction

Our interdisciplinary team has developed a framework for conceptualizing the sustainable siting of wind energy facilities (WEFs) in the United States. This chapter places public response to WEF siting within the larger framework of sustainability, addressing economic, social, and environmental aspects of sustainability. This framework guides our efforts to draw upon multiple disciplines to respond to the complex issues involved in siting of WEFs.

This work draws upon the social sciences, including public affairs and business economics; the arts, landscape architecture, and history; the natural sciences and conservation, and a variety of collaborative academic enterprises. Approaches incorporate use of geographical information systems (GIS) for site screening models, visual and aural simulation, the decision sciences, and agent-based modeling. Our experience is that working across disciplines has been challenging, and we continue to develop our framework and research questions to be truly interdisciplinary—that is, to use the unique insights of

DOI: 10.5876/9781646420278.c012

multiple disciplines to create new knowledge and synergies between those disciplines. We may contribute to field of energy research social science in multiple ways. For example, our work to integrate the social sciences with GIS/visualization technologies is generating ways to further test the "distance hypothesis" (Jacquet 2012) to assess the acceptability of proposed WEFs by communities and affected stakeholders. In the longer term, we anticipate efforts to expand the list of factors that predict the acceptability of proposed WEF sites to include a wider suite of economic, social, and environmental factors. Specifically, these efforts may be of use to wind energy developers and informing meaningful public engagement for identifying appropriate WEF sites. Moreover, if successful, these approaches may serve as useful models for others engaged in energy research and practice.

The chapter, then, addresses questions and issues about the magnitude of the WEF siting issue, introduces the concept of a sustainable WEF siting system, and describes how we have adapted the three spheres of sustainability (i.e., economic, social, and environmental) analytical approach as a heuristic for organizing siting questions and identifying positive goals for siting WEFs. Next, we address the economic dimensions of a more sustainable approach to WEF siting, then the social, and finally the environmental dimensions. Our aim herein is to develop a framework that seeks to advance WEF siting processes that focus significantly on meeting public concerns so that engaged parties can make decisions on best available data rather than developing a framework that "overcomes" public opposition. Research questions to improve the usefulness of this framework are raised throughout.

Magnitude of Wind Energy Facility Siting Issues

As of 2017, 114 gigawatts (GW) of solar and wind power are online in the United States. By 2020, about 21 GW of new wind and solar photovoltaic (PV) capacity are projected under a "business as usual" future scenario (US Energy Information Administration 2017). Meeting these goals will require significant additional construction of utility-scale wind and solar facilities in the United States. To better understand the magnitude of the WEF siting issue, it may be valuable for research to be conducted to characterize the potential number of turbines/projects that need to be constructed to meet this goal. This construction must be accomplished in ways that minimize adverse environmental and social impacts, reduce industry costs from delays and cancellations due to social opposition to perceived and/or actual adverse environmental, and reduce social impacts of renewable energy facilities (WEFs). We recognize there are many locations where opposition to WEFs is warranted due to some combination of poor public engagement in the siting process, significant environmental impacts, inequities in the distribution of costs and benefits, and visual and auditory impacts. We seek to understand how these concerns might be adequately addressed in some locations through improved siting (including processes that are perceived as procedurally and substantively just) and operations practices. This point is further addressed in the social sustainability section. A primary concern is that opposition to wind energy development as presently expressed may be prevent the United States from building adequate WEFs to meet renewable energy goals. Indeed, Rolf Wüstenhagen, Maarten Wolsink, and Mary Jean Bürer (2007) argued that public opposition to WEFs may be among the most limiting factors to successfully site WEFs. Steve Pociask and Joseph Fuhr Jr. found that 45 percent of proposed WEFs in the United States were halted at the local community level (2011).

Sustainable Development and Wind Energy Facility Development

The 1992 Rio Declaration focuses on creating development that provides sustainable economic growth, environmental protection, and social equity. Sustainable renewable energy development is defined as development that seeks to provide success along each of these dimensions. Our interest in advancing the development of WEFs led us to draw upon the theoretical framework of sustainable development. Similar to Sovacool and Ratan's effort to identify factors influencing the support for WEFs in Europe, researchers in the United States need to identify the myriad social, environmental, and economic factors that are most important to creating a sustainable system for siting and operating WEFs (2012).

Sustainable Wind Energy Siting System

It is our assessment that the number of WEFs projected to be built necessitates the adoption of a *system*, or approach, for constructing and operating them that is socially, environmentally, and economically viable for our nation and communities hosting WEFs. A significant challenge to creation of a national system is that the United States has wide variation in siting approaches at the state and local levels. Thus, any national system needs to be a voluntary framework of best management practices, standards, and guides. How to develop a voluntary system would appear to be a valuable research project. The approach in this chapter focuses on how the aggregate of siting processes for individual WEFs affects the national effort to site sufficient WEFs in order to meet global greenhouse gas (GHG) reduction goals. Unfortunately, we have found little additional research beyond Steve Pociask and Joseph Fuhr Jr. (2011) to identify how public acceptance has affected the percentage of proposed renewable energy facilities that are built. Important research questions to address include the need to better characterize the degree of support/opposition to the siting of WEFs, trends in opposition, how approval varies by sociodemographic variables, including salient regional variation, and other variables identified below.

Economic Sustainability

What is an economically sustainable system for the siting of WEFs? The costs per kWh of deploying WEFs are decreasing, the result of increased research and development efforts. Our work is intended to help reduce the costs of deploying renewable energy by increasing the likelihood of successful and rapid deployment. Increased siting success would reduce per project costs and is envisioned to help reduce costs for overall through industrywide risk reduction in the construction of WEFs. Further research on how to characterize the "success rate" for proposed WEFs, and how project delays due to public opposition increase project and industry costs, is necessary. The research may be quite challenging, however, due to the difficulty in obtaining what is often trade or proprietary data from energy firms.

It appears likely that project delays and cancellations increase not only the costs of energy produced from each WEF delayed or cancelled, as well as the overall larger effort to site WEFs. Increasing opposition inevitably leads to increased financing costs, as lenders perceive increased monetary risks associated with WEFs. Adverse environmental impacts to wildlife from the construction and operation of WEFs remain a highly problematic issue. Furthermore, in general, public opposition continues to be widespread throughout many regions and affected communities, a position that may be due to perceptions that WEFs are sited "unfairly," with firms sometimes using corporate power to site in unwilling or dispossessed communities that then must bear the costs of being near WEFs without adequate compensation. Further, research should model these interactions in ways that help us better understand them, identify the most salient factors—that is, the factors most amenable to improvement—and share the information to support decision making.

Social Sustainability

The social aspects of siting WEFs are enormously complex, and they have been investigated from many disciplines. We posit that when public participation affects public perceptions and behaviors toward the WEF site, it should be viewed as part of social sustainability. In this section, we briefly review selected literature on social aspects of energy facility siting and causes of opposition to WEFs, including work on unarticulated reasons for opposition and support for WEF's and the inequitable distribution of costs and benefits. Then we engage public participation through the lens of public engagement processes, including consensus building, trust ecology, and *joint fact-finding*. Next, we integrate landscape architecture's approach to visual impacts as part of social sustainability, since they affect public attitudes and behavior toward proposed WEFs. Finally, we explore potential for the use of agent-based modeling as a research technique to help integrate these many aspects of social sustainability to potentially identify which factors are most salient—and for siting success.

Those researchers examining siting impacts often explore opposition to proposed REFs through markets and sociopolitical dimensions (Wüstenhagen, Wolsink, and Bürer 2007). An important reorientation for the field began when Benjamin Sovacool and Pushkala Lakshmi Ratan (2012) and Patrick Devine-Wright et al. (2017) observed that social science research often focused too much on opposition to siting. Rather, these authors suggested developing means to identifying factors affecting acceptance of WEFs. As noted earlier, we seek to identify how to improve acceptance of WEFs not by overcoming opposition but by encouraging the development of an approach to siting that creates a siting system that is socially acceptable and sustainable. Elinor Ostrom's polycentric approach to investigating siting social dynamics identified different scales of actors: macro- (or national) actors, meso- (middle actors who mediate between national and local actors), and microlevel (individual) actors (2010). Each level comprises a group whose dynamics need investigation, and the groups interact in both bottom-up and top-down fashions. In 2012, Sovacool and Ratan added a crosscutting fourth dimension—the political.

UNARTICULATED OPPOSITION AND SUPPORT FOR WEFS

In understanding public attitudes toward wind turbines the social science approach of looking at unarticulated opposition can serve as the basis for one element of this ongoing research. In their examination of public responses to wind turbine plans, Richard Hirsh and Benjamin Sovacool expand upon the traditional reasons (environmental destruction, noise, etc.) for explaining resistance to the siting of wind turbines (2013). They suggest communities and stakeholders may object to renewable energy technology for subtle reasons difficult to empirically express, such as animosity to big-city businesspeople who appear eager to exploit rural residents. Most important, researchers hypothesize that turbines lose support from some people because the technology makes obvious a previously "invisible" electrical infrastructure that people would prefer to avoid confronting.

From a policy perspective, this insight has practical value. Perhaps most obviously, it confirms that stakeholders do not always make rational decisions based on economic or other quantifiable factors. While visually obtrusive to some, wind turbines exist within a broader energy system that includes large (but often segregated) generation stations and high-voltage transmission lines that have large environmental impacts as well as other large anthropogenic structures and landscape alteration. The knowledge that support or opposition to wind turbines are not always rational leads us to question whether educational efforts can help provide better contexts within which people can make decisions about renewable energy. Hirsh and Sovacool (2013) ask if people better understand that no energy production system has zero environmental impacts—even "earth-friendly" solar and wind technologies—will they find ways to consume less energy and therefore avoid the need for new facilities?

Science research has also helped policy makers better understand how unarticulated feelings may provide support for wind turbines. Brinkman and Hirsh have employed textual and discourse analyses to help understand that stakeholders—even those who do not benefit economically—often view the new technologies as symbols of modernity and technical sophistication (2017). For policy makers, understanding this unarticulated notion of how renewable energy technology fits into stakeholders' sense of identity (along with an explanation for the strong rural-urban conflict inherent in land use in seminatural, rural landscapes) could help as we work toward developing common ground on WEFs development.

SOCIAL SYSTEMS AND PUBLIC ENGAGEMENT

A critical step in improving acceptance for proposed WEFs is engaging the public and understanding its interests, needs, and priorities. The traditional model of proposing projects and then vigorously defending them against opposition with minimal stakeholder involvement tends to lead to unproductive and often adversarial public meetings. This approach is clearly ineffective in many cases. Fortunately, a range of best practices around more productive stakeholder engagement has emerged. The following sections introduce several novel approaches. Further research examining their efficacy, and the efficacy of other approaches for public engagement in different contexts is necessary.

Meaningful Public Engagement. Meaningful public participation is integral to the sustainable siting of renewable energy facilities. Aside from showing nor mative commitments to democratic principles, projects can only succeed in being socially just and renewable over the long term if externalities, including social and environmental costs, have a full accounting. Economic tools can help understand the potential magnitude of such costs and the degree to which these externalities have been internalized in the past. Engaging the public can foster support for projects, avoiding costly legal battles and project delays. Finally, stakeholders bring ideas to the discussion, creating opportunities for shared learning, mutual gains, and project optimization.

Unfortunately, public participation in siting of renewables can be weak and tokenistic in practice (Hendriks, Bolitho, and Foulkes 2013; Roberts 2004). At their worst, processes fail to truly engage stakeholders, leading to lost opportunities, wasted resources, and strained relationships among involved parties. Due to real or perceived fears, many people feel so disenfranchised that they avoid public engagement opportunities. Others attend public meetings and leave frustrated because they believe that decisions have already been made without their input. Too often, public hearings devolve into unproductive controversy with little substantive resolution and consensus. In many cases, opponents turn to litigation that, while often unsuccessful, adds to project delays and costs. Fortunately, public participation methods that reduce conflict exist. One model applied to a wide range of public policy and planning disputes, including around renewables, is the consensus building approach (Raab 2017; Susskind and Field 1996; Susskind, McKearnen, and Thomas-Larmer 1999).

Consensus Building and Joint Fact-Finding. Consensus building efforts gather stakeholder group representatives together to deliberate on matters; the emphasis is on uncovering interests and seeking creative solutions that maximize mutual gains. When applied, these efforts can work to keep constituencies not at the table informed and engaged, regularly conveying information and seeking feedback.

Consensus building efforts are not appropriate in all situations. It is integral that the key decision makers—whether government agencies or project proponents—take the process seriously and commit fully to considering outcomes. *Situation* or *conflict assessments* often are conducted by professional neutrals, before efforts are initiated, to identify key stakeholders and their interests, assess the viability of any effort, and propose an initial plan for how such an effort might proceed (Schenk 2007; Susskind, McKearnen, and Thomas-Larmer 1999).

Scientific and technical information often is central to disputes emanating from facility siting. For example, data on wildlife mortality or noise levels often are seen as authoritative and thus used as a factual basis for conflict between parties that support or oppose a project. However, disagreements and competing sets of data can complicate decision making. Joint fact-finding (JFF) processes seek to develop shared sets of information that stakeholders accept as "salient, credible and legitimate" for the purposes of their decision making (Matsuura and Schenk 2017). Rather than each party bringing their own data to the table, the stakeholders collectively define the research questions, select the methods and experts best able to address questions, monitor the research process, interpret the results, and disseminate factual findings on which decisions will be based. Joint fact-finding alone cannot provide the definitive solution, as differences in interests, priorities, and risk tolerance will remain, but it can provide a shared understanding and reduce disagreements on factual data. Joint fact-finding has been employed in renewables siting processes, including the controversial Cape & Islands Off-Shore Wind Stakeholder Process in Massachusetts (Raab 2017). Joint fact-finding efforts can be integrated with other methods, including geodesign, to increase stakeholders' acceptance of technical information.

Although there is strong, case-based literature for applying the consensus building approach to resolve disputes around renewables siting and comparable issues, robust and systematic evaluation has been limited. The evaluations conducted look very promising, and frameworks for more rigorous evolution have been proposed, but much more is necessary to assess the worth of public participation in siting processes and improve best practices (Chess 2000; O'Leary and Bingham 2003). Important questions include: Who should be in the discussion? What are the nature and appropriate mechanisms of decision making? What should the relationship to formal decision-making be? How should information be processed and utilized? And, last, what metrics can assess the likely success of the approaches?

Trust Ecology and Acceptance of WEFs. Stakeholders involved in renewable energy projects are numerous, including energy providers, government agencies, politicians and political groups, environmental groups, and local citizens. With diverse interests at play, trust is likely to play a key role in the success or failure of siting and development.

Trust can be defined as the willingness of one party to accept vulnerability to other entities in the face of uncertainty (Stern and Coleman 2015). This willingness enables what might have otherwise been considered risky behaviors, such as openly sharing one's interests, discussing weaknesses, or making concessions to another party. It also enables active listening and efforts to create mutually beneficial solutions based on the development of shared criteria for evaluating outcomes, all essential elements of successful negotiation (Fisher, Ury, and Patton 1991). Trust is critical to creating safe spaces for collaboration, for reducing stakeholder risks, and for enabling the reasoned consideration of different proposals.

The concept of trust ecology provides a useful tool for considering stakeholder dynamics over time, defining four key types of trust and elucidating their sources: dispositional trust, rational trust, affinitive trust, and systems-based trust (Stern and Baird 2015; Stern and Coleman 2015).

Dispositional trust describes individuals' preexisting inclinations toward trust. For example, one might consider him- or herself to be a trusting person, regularly accepting the assertions of others. Alternatively, one might be consistently skeptical of others and basing evaluations on the premise that people need to prove trustworthiness. For WEF projects, some engaged might occupy the skeptical dimension of this spectrum based on the political nature of the work. Prior research suggests that people with high degrees of trust in public officials or project purveyors rarely participate in public meetings (Smith et al. 2013). Thus, it is typically safer to assume low levels of dispositional trust in these situations. Dispositional trust is a preexisting condition, rather than a type of trust that can be actively manipulated over a short time frame. The other three forms of trust are actionable.

Rational trust emerges from evaluations of the predicted outcomes of the expected behaviors of others. In other words, if one expects another to perform in a way for shared benefit, rational trust is more likely. These expectations are typically grounded in perceptions of competence based on prior performance, consistency and follow-through, or reputation. Building rational trust typically requires the opportunity to demonstrate expertise, competence, or consistency in actions that align with a person's interests.

Affinitive trust is based on the development of affinity for a potential trustee. This type of trust can emerge from numerous sources, including charisma of the trustee; the sharing of positive social experiences; assumptions of similarity in values, backgrounds, or membership in shared social groups or categories; or demonstrations of active listening, caring, or responsiveness. The presence of affinitive trust signifies positive emotional connections between stakeholders and can be highly predictive of subsequent collaborative behavior.

Systems-based trust (also known as *procedural trust*) differs from the other forms of trust in that it entails trust for a system or set of rules that govern interactions, rather than trust in any specific individual. This trust in the system lowers the risk of participation, as long as participants believe that everyone will abide by the rules and that the rules are fair.

In the renewable energy context, these forms of trust may develop in various ways. For example, rational trust may develop as different parties demonstrate their areas of knowledge and expertise. Affinitive trust may develop as stakeholders express genuine concern for others or appreciation of their interests, even if they do not agree with their positions (Fisher and Shapiro 2005). Systems-based trust may develop as process leaders demonstrate fair treatment of all parties and demonstrate incorporation of all stakeholders' concerns. Similarly, trust can be eroded quite quickly as facts are misconstrued, misunderstood, or misrepresented or conflict with each other (rational distrust); ad hominem attacks are made that question the motives of other parties (affinitive distrust); or decisions are made behind closed doors or certain groups feel left out or unheard (systems-based distrust).

The theory of trust ecology argues that the development of sufficient stocks of all three forms of actionable trust—rational, affinitive, and systems-based—are necessary for long-term solutions to be satisfactory in multistakeholder environments, particularly when implementation involves long-term collaboration across stakeholder groups. Considering the siting and implementation of renewable energy projects from this perspective generates multiple pathways of inquiry with strong implications for enhancing these processes. For example, how do the different forms of trust function within this arena? Are certain forms more important than others at different stages? How does the distribution of different forms of trust and distrust impact the process and its outcomes? How does trust develop across political lines where moral values differ and social identity adherence create strong divisions? The science of trust and its different sources and impacts in public processes is still in its infancy. Examining these issues within the context of renewable energy projects could yield insights of value to multiple similarly contested public issues in addition to finding ways to make these processes more palatable and efficient in the near term.

VISUAL ACCEPTABILITY

The visual characteristics of the landscape have long been of interest to people. For example, a survey of North Carolina residents by Dennis Grady found aesthetics or visual pollution to be the most often cited barrier (54% of respondents) to the construction of wind turbines (2002). However, it was not until the 1960s and 1970s—with growing concern for the visual impacts of resource extraction on public lands—that visual characteristics of the landscape began to be considered in many landscapes (Sheppard 2001).

Those developing visual resource management systems believed that if these issues were to be taken seriously, then their systems would have the same rigor as those used to manage other resources. These *expert systems* are tools applied by experts in planning, designing, and managing the visual characteristics of landscapes. Experts used these tools to gain public acceptance for proposed landscape alterations. Expert systems were also useful in developing materials—such as maps, visual assessments, and visual simulations—to obtain public feedback on the visual acceptability of proposed WEFs.

Another group, led primarily by environmental psychologists, advocated for assessing the scenic value of the landscape by asking people how scenic they perceived the landscape to be (Kaplan 1979). This was called the "perceptual approach." Both the visual management tools (expert systems) and the perceptual approach (perceived scenic value) have applicability to renewable energy facility siting, but additional research is necessary if they are to be applied successfully.

From landscape architecture, *distance zones* are the distance from which a landscape alteration can be seen. There are three distance zones: Foreground is defined as between 0 and .4 to .8 km from the observer and is the distance at which the individual details of the landscape and landscape alterations can be seen (Litton 1968). Middleground is between .5 and .8 km and 5 to 8 km from the observer and is the distance where vegetation and topography patterns of the landscape are apparent. At this distance one is able to discern if the facility "fits" visually in the landscape. Background is from between 5 and 8 km to infinity from the observer and is the distance where the atmosphere begins to mute the visual characteristics of the landscape and, thus, reduce the visual acuity of any alterations. Jacquet found that five years after construction of an WEF, residents at 0–5.33 km were more strongly opposed to a WEF (37%) than residents at 5.34–10.66km (23%), and 10.67–16 km (22%), "demonstrating that landowner attitudes become slightly more positive the further away the property is located" (2012, 31). While this finding supports distance zones as

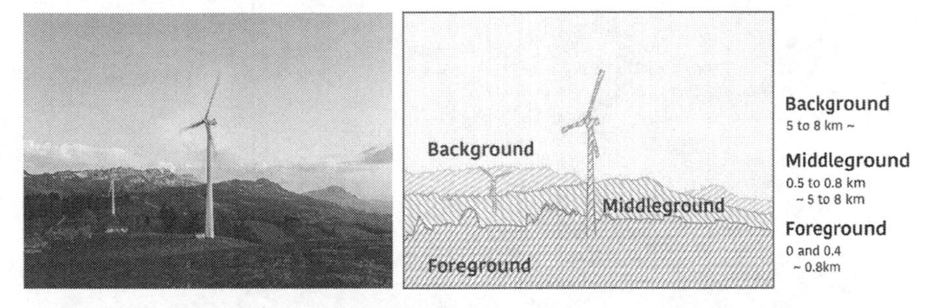

FIGURE 12.1. Picture of landscape depicting distanced zones. Photo credit: Jisoo Sim.

FIGURE 12.2. Photographs illustrating visual complexity. Photo credit (left): Tom Def. Photo credit (right): Jisoo Sim.

a tool in WEF, we suggest further research between the fields of renewable energy research and visual assessment tools used by landscape architects to draw upon the strengths of each; the results of this research will allow for better characterization of multiple factors affecting visual acceptability of turbines and facilities of different sizes viewed at different distances.

Scenic Quality. The scenic quality of the landscape is its perceived scenic value by the public (US Department of Agriculture 1974). The potential impact of an alteration is usually greater in a more scenic landscape. Visual complexity is often used as a measure of scenic quality. Visual complexity is the variation in the lines, forms, colors, and textures of a landscape (figures 12.1 and 12.2).

Traditionally, visual quality has been assessed in relatively undeveloped or natural landscapes of the US west. Because WEFs occur in many types of

FIGURE 12.3. Photographic example of high and low contrast. Photo credit (left): Carlos Koblischek. Photo credit (right): Aaron Cushi.

landscapes—including more populated regions, coastal areas, and mountainous regions with large viewsheds—there is a need to research factors affecting public perceptions of the scenic experience of proposed WEFs.

Visual Sensitivity. This is linked to the number of viewers in the viewshed and what activities they are engaged in at the time of viewing (US Department of the Interior 1984). A landscape would be more sensitive if able to be seen by many people or by people engaged in outdoor recreation, where aesthetics comprise an important component of the recreational experience. The greater the visual sensitivity of a landscape, the more likely people are to be concerned about a proposed WEF. Research on what factors influence landscape sensitivity will be important to advancing successful WEF projects.

Visual Acceptability. This measures how much a proposed alteration contrasts with the line, form, color, and texture of the landscape (US Department of the Interior 1986). The assumption is that greater contrast that the WEF would introduce means alterations are less acceptable to people (figure 12.3).

Visual Contrast. The size of an alteration is also an important factor in visual contrast. However, there is a fundamental difference between wind energy facilities and most other types of landscape alterations, and that is the significant motion of large blades. Wind turbines with spinning blades are not static elements that can be easily assessed, and they thereby attract attention. The motion of multiple turbines in close proximity to each other may exacerbate this effect. Although substantial research exists on attitudes toward WEFs, the effect of turbine blade motion on attitudes and perceptions has not been well researched and is needed to fill this gap in the literature.

Agent-Based Modeling

Renewable siting decisions are made as part of social systems. To better understand the dynamics leading up to siting decisions, it is important to understand the complexity of these social systems during WEF siting processes. Agent-based modeling identifies a set of agents and the relationships between them that define the underlying system. This process can be useful in analyzing the renewable siting dynamics and its relationship with the society. Nicholas Cain et al. used an agent-based model to simulate political, economic, and psychosocial factors that drive support or opposition to transmission siting projects (2011), and Mark Abdollahian, Zining Yang, and Hal Nelson (2013) used an agent-based model as well as game theory and predictive analytics to demonstrate how competing interests shape energy siting outcomes. In particular, the authors show that formation of community-based organizations and NGO involvement have a positive impact on citizen messaging and participation.

Agent-based modeling has been successful in helping understand other complex sociotechnical systems, such as transportation dynamics (Nagel, Beckman, and Barrett 1999), evacuation and disaster planning (Chandan et al. 2013), and reliability and vulnerability of infrastructures (Barrett et al. 2012; Beckman et al. 2013). In recent years, highly resolved agent-based models have been developed in which heterogeneous agents are endowed with a rich set of behavioral, familial, social, and demographic attributes (Parikh et al. 2013; Subbiah et al. 2017). This approach can be brought to bear on the sustainable WEF siting and operation challenge as well, to better understand complex interactions among the environment, individuals, communities, media, and government. Spatially explicit agent-based modeling can tie agents and their behaviors to the physical environment, land use, and built infrastructure, which is particularly useful in understanding the WEF decision-making process.

Because agent-based models are typically data hungry, in order to apply them to siting, data should be incorporated from past siting processes (i.e., economic incentives, built environment, load served with and without the renewable facilities, demographics of the population, the population's concerns and desires, etc.). This information will help calibrate and validate the models, critical steps to making them dependable.

Inequitable Distribution of Social Costs

Wind energy is reliant on energy sources that are public goods; that is, all parties have access to the good without reducing the quantity available to others. When WEFs are constructed and operated, adverse impacts (e.g., property value, peaceful use of property, visual and/or auditory impacts) on those experiencing the WEF are an external cost imposed by those who construct/operate the facility. The traditional cost-benefit calculus of economics and business decision-making focuses on the private costs of business entities, not the costs that may be borne by others. However, as highlighted by ecological economists, to properly inform these decisions, it is important to consider production and social costs, along with those who bear them.

Social costs are "all direct and indirect losses sustained by third persons or the general public as a result of unrestrained economic activities" (Kapp 1978, 13). This definition is consistent with the distinction between economic, environmental, and social factors outlined earlier in this chapter but also highlights the differences in incentives between different stakeholders. The producer (i.e., the business constructing the renewable energy site and generating profits from its operation) bears production costs. Previously, we indicated the need to research attitudes toward the siting and operation of WEFs. Our transdisciplinary approach builds upon insights from economics and business and seeks also to identify the external/social costs of WEFs and how compensation schemes and public policy can influence businesses to internalize these costs. Research is needed to explore how the magnitude of social costs, as well as who bears these costs, varies significantly between different types of renewable energy. These issues can have an important influence on the resistance by different groups of stakeholders to WEFs.

For example, wind energy facilities can create adverse impacts on homeowners, constituting social costs borne by local area residents. This may lead to socially inefficient choices by businesses because another party bears the social costs, thereby yielding a negative externality. Research needs to characterize the compensation arrangements by those who operate WEFs in the United States and elsewhere, how those compensation contracts affect the acceptability of the WEFs, and how each of these factors varies with salient socioeconomic variables. By doing so, we can better document what are considered "fair" terms for external costs for different groups, allowing for better bargaining and more equitable siting of WEFs.

A brief comparative case illustrates how compensation schemes can address these issues. For example, in 2009, Denmark enacted the Promoting Renewable Energy Act to establish a compensation scheme to address the loss of value to neighboring property due to the erection of wind turbines and to thereby promote local support (Jensen 2014). The institutional environment in Denmark requires businesses to bear the "social costs" of production of wind turbines that are effectively converted by law into production costs. The issues of the relative magnitude of social costs, as well as which stakeholder bears those costs, and how these are converted to private costs to businesses, are important questions to address with the successful siting for different kinds of renewable energy in the United States.

Environmental Sustainability: Wildlife

Impacts on wildlife from energy development have also challenged conservationists, producers, and regulators for decades. This concern is especially present where rare, threatened, and endangered, or high-profile species are involved. For example, coal, gas, and oil extraction have transformed millions of acres of highly productive and diverse forest communities into lands with poor or negative environmental quality across the United States. However, on both private and public lands in Appalachia, much of the region still contains large blocks of interior forest (> 20,000 ha) and watersheds that are globally significant for neotropical migratory songbirds, amphibians, fish, and freshwater mollusks.

Wind energy facilities require substantial loss of forest cover and contribute to landscape forest fragmentation from turbine placement, access roads, and power grid connections. High-elevation ridges where wind energy facilities are sited often contain regionally significant forest communities that support endemic plants and animals found nowhere else (Menzel et al. 2006). These forests are sensitive to permanent openings that provide entry points for exotic and invasive species into otherwise lightly impacted systems (Drohan et al. 2012). These areas are increasingly important as potential climate refuges for sensitive species.

Early concern about wind energy development focused on potential bird-strike mortality among neotropical migrants and raptor species. The National Environmental Protection Act requires planners to assess potential "take" under the Migratory Bird Treaty Act and the Bald and Golden Eagle Protection Act. Long-term avifauna monitoring programs have been established to assess mortality following construction (Kerns and Kerlinger 2004). Whereas bird strikes have been documented, biologists had failed to predict a far greater impact on bats here and with other facilities (Arnett et al. 2008). Bat mortality from strike and barotrauma (Baerwald et al. 2008), particularly of migratory bats such as the eastern red bat (*Lasiurus borealis*) and hoary bat (*Lasiurus cinereus*), has varied throughout North America but has been high throughout the East. These deaths been particularly common in the northern and central Appalachians, with substantial peaks in the fall and spring (Arnett et al. 2008; Frick et al. 2017).

Once wind energy impacts were recognized, preconstruction and siting requirements included bat surveys (Reynolds 2006). However, mismatched acoustic survey timing, or inability to correlate wind turbine presence as a structural attractant for foraging or breeding bats, meant that surveys failed to predict realized bat mortality (Jameson and Willis 2014). Calculation of effective population sizes and estimates of wind energy mortality have resulted in predicted population declines greater than 90 percent (Frick et al. 2017), which could be sufficient to warrant species listing under the Endangered Species Act, which, in turn, would trigger regulatory action. This type of impact is clearly an unsustainable environmental impact within our sustainability framework that needs additional examination.

Although mortality of nonmigratory bats has been several orders of magnitude lower than that of migratory bats (Arnett et al. 2008), wind energy sites in the Appalachians overlap with the summer or fall swarm/spring emergence distribution of the endangered Indiana bat (*Myotis sodalis*), endangered Virginia big-eared bat (*Corynorhinus townsendii virginianus*), and the threatened northern long-eared bat (*Myotis septentrionalis*) (Trani, Ford, and Brian 2007) and are considered additive stressors to the ongoing White-nose Syndrome disease event (Ford et al. 2011). The pre-siting presence of any or all of these species has resulted in preplanned mitigation measures such as summer curtailment of individual turbines and/or higher-than-industry-standard cut-in speeds. This effectively means that mitigation measures to reduce bat mortality, including stopping turbine activity during summer migration and only allowing blades to rotate when higher wind speeds are reached, are made part of the agreed-upon operating plan for the turbine (Arnett et al. 2011). Research is now attempting to fully document the timing and duration of important bat migration events with latitude, elevation and weather (Cryan, Stricker, and Wunder 2014; Muthersbaugh et al. 2016). Other work is exploring the potential for scaling up acoustic deterrents from the experimental scale to the operational scale (Arnett et al. 2013; Johnson et al. 2012). Although elimination of all wildlife impacts from wind energy development is likely not feasible, continued development of best management practices for mitigating negative wildlife impacts, especially for bats and birds, is ongoing (American Wind Energy Association 2015; Peste et al. 2015). Using best management practices from the forestry sector as a guide, conservationists and regulators are placing an emphasis on regional wind energy development.

Conclusion

From an interdisciplinary frame, we have identified some of the knowledge needed from the three spheres of sustainability that will provide the public with a better understanding of the environmental, social, and economic factors associated with WEF projects. We believe that to achieve a truly sustainable energy supply, public support must reflect adequate consideration of both positive and negative environmental, social, and economic impacts of WEFs. We anticipate that undertaking the research identified in this chapter will improve the processes used in siting WEFs and increase the likely success of proposed projects. Furthermore, we expect answers to these research questions are achievable and will engender greater public support in sufficient locations to increase WEF capacity, making a contribution toward meeting national goals for renewable energy power generation.

References

Abdollahian, Mark, Zining Yang, and Hal Nelson. 2013. "Techno-social Energy Infrastructure Siting: Sustainable Energy Modeling Programming (SEMPro)." *Journal of Artificial Societies and Social Simulation* 16 (3): 1–6. https://doi.org/10.18564/jasss.2199.

American Wind Energy Association. 2015. "Wind Energy Industry Announces New Voluntary Practices to Reduce Overall Impacts on Bats by 30 Percent." Accessed August 15, 2017. https://www.awea.org/resources/news/2017/wind-energy-industry-announces-new-voluntary-pract.

Arnett, Edward B., W. Kent Brown, Wallace P. Erickson, et al. 2008. "Patterns of Bat Fatalities at Wind Energy Facilities in North America." *Journal of Wildlife Management* 72 (1): 61–78.

Arnett, Edward B., Manuela M.P. Huso, Michael R. Schirmacher, and John P. Hayes. 2011. "Altering Turbine Speed Reduces Bat Mortality at Wind-Energy Facilities." *Frontiers in Ecology and the Environment* 9 (4): 209–214. https://doi.org/10.1890/100103.

Arnett, Edward B., Cris D. Hein, Michael R. Schirmacher, Manuela M. Huso, and Joseph M. Szewczak. 2013. "Evaluating the Effectiveness of an Ultrasonic Acoustic Deterrent for Reducing Bat Fatalities at Wind Turbines." *PLoS ONE* 8 (6): 1–11. https://doi.org/10.1371/journal.pone.0065794.

Baerwald, Erin F., Genevieve H. D'Amours, Brandon J. Klug, and Robert M. Barclay. 2008. "Barotrauma Is a Significant Cause of Bat Fatalities at Wind Turbines." *Current Biology* 18 (16). https://doi.org/10.1016/j.cub.2008.06.029.

Barrett, Chris, Karthik Channakeshava, Fei Huang, et al. 2012. "Human Initiated Cascading Failures in Societal Infrastructures." *PLoS ONE* 7 (10). https://doi.org/10.1371/journal.pone.0045406.

Beckman, Richard, Karthik Channakeshava, Fei Huang, et al. 2013. "Integrated Multi-network Modeling Environment for Spectrum Management." *IEEE Journal on Selected Areas in Communications* 31 (6): 1158–1168.

Brinkman, Joshua T., and Richard F. Hirsh. 2017. "Welcoming the PIMBY ('Please in My Backyard') Phenomenon: The Acceptance of Wind Turbines and the Culture of the Machine in the Rural Midwest." *Technology and Culture* 58 (2): 335–367. https://doi.org/10.1353/tech.2017.0039.

Cain, Nicholas, Hal Nelson, Mark Abdollahian, Brett Close, and Jake Hoffman. 2011. "Not on Planet Earth (Nope): An Agent Based Model Simulating Energy Infrastructure Siting Dynamics." Presented at APSA Annual Meeting. https://ssrn.com/abstract=1902414.

Chandan, Shridhar, Sudip Saha, Chris Barrett, et al. 2013. "Modeling the Interaction between Emergency Communications and Behavior in the Aftermath of a Disaster." In *Social Computing, Behavioral-Cultural Modeling and Prediction*, edited by Ariel M. Greenberg, William G. Kennedy, and Nathan D. Bos, 476–485. New York: Springer. https://doi.org/10.1007/978-3-642-37210-0_52.

Chess, Caron. 2000. "Evaluating Environmental Public Participation: Methodological Questions." *Journal of Environmental Planning and Management* 43 (6): 769–784. https://doi.org/10.1080/09640560020001674.

Cryan, Paul M., Craig A. Stricker, and Michael B. Wunder. 2014. "Continental-Scale, Seasonal Movements of a Heterothermic Migratory Tree Bat." *Ecological Applications* 24 (4): 602–616. https:/doi.org/10.1890/13-0752.1.

Devine-Wright, Patrick, Susana Batel, Oystein Aas, Benjamin Sovacool, Michael Carnegie Labelle, and Audun Ruud. 2017. "A Conceptual Framework for Understanding the Social Acceptance of Energy Infrastructure: Insights from Energy Storage." *Energy Policy* 107 (August): 27–31. https://doi.org/10.1016/j.enpol.2017.04.020.

Drohan, Patrick J., Margaret Brittingham, J. Bishop, and K. Yoder. 2012. "Early Trends in Landcover Change and Forest Fragmentation Due to Shale-Gas Development in Pennsylvania: A Potential Outcome for the Northcentral Appalachians." *Environmental Management* 49 (5): 1061–1075. https://doi.org/10.1007/s00267-012-9841-6.

Fisher, Roger, and Daniel Shapiro. 2005. *Beyond Reason: Using Emotions as You Negotiate*. New York: Penguin.

Fisher, Roger, William L. Ury, and Bruce Patton. 1991. *Getting to Yes: Negotiating Agreement without Giving In*. New York: Penguin.

Ford, W. Mark, Eric R. Britzke, Christopher A. Dobony, Jane L. Rodrigue, and Joshua B. Johnson. 2011. "Patterns of Acoustical Activity of Bats Prior to and Following White-Nose Syndrome Occurrence." *Journal of Fish and Wildlife Management* 2 (2): 125–134. https://doi.org/10.3996/042011-jfwm-027.

Frick, W. F., E. F. Baerwald, J. F. Pollock, et al. 2017. "Fatalities at Wind Turbines May Threaten Population Viability of a Migratory Bat." *Biological Conservation* 209 (May): 172–177. https://doi.org/10.1016/j.biocon.2017.02.023.

Grady, Dennis O. 2002. "Public Attitudes toward Wind Energy in Western North Carolina: A Systematic Survey." Paper presented at the NC Wind Summit, December 9, 2002, Boone.

Hendriks, Carolyn M., Annie Bolitho, and Chad Foulkes. 2013. "Localism and the Paradox of Devolution: Delegated Citizen Committees in Victoria, Australia." *Policy Studies* 34 (5–6): 575–591. https://doi.org/10.1080/01442872.2013.862450.

Hirsh, Richard F., and Benjamin K. Sovacool. 2013. "Wind Turbines and Invisible Technology: Unarticulated Reasons for Local Opposition to Wind Energy." *Technology and Culture* 54 (4): 705–734. https://doi.org/10.1353/tech.2013.0131.

Jacquet, Jeffrey B. 2012. *Landowner Attitudes and Perceptions of Impact from Wind and Natural Gas Development in Northern Pennsylvania: Implications for Energy Landscapes in Rural America*. PhD diss., Cornell University, Ithaca, NY.

Jameson, Joel W., and Craig K. Willis. 2014. "Activity of Tree Bats at Anthropogenic Tall Structures: Implications for Mortality of Bats at Wind Turbines." *Animal Behavior* 97 (November): 145–152. https://doi.org/10.1016/j.anbehav.2014.09.003.

Jensen, Søren Stenderup. 2014. "High Court Rules on Compensation for Noise from Wind Turbines." *International Law Office*, September 1, 2014. http://www.windaction.org/posts/41138-high-court-rules-on-compensation-for-noise-from-wind-turbines#.WirhVCOZPUJ.

Johnson, Joshua B., W. Mark Ford, Jane L. Rodrigue, and John W. Edwards. 2012. "Effects of Acoustical Deterrent on Foraging Bats (Research Note NRS-129)." Newtown Square, PA: US Department of Agriculture, Forest Service, Northern Research Station. https://www.nrs.fs.fed.us/pubs/rn/rn_nrs129.pdf.

Kaplan, Rachel. 1979. "Visual Resources and the Public: An Empirical Approach." In *Proceedings of Our National Landscape Conference: Gen. Tech. Rep. PSW-GTR-35*, technical coordination by Gary H. Elsner and Richard C. Smardon, 209–216. Berkeley, CA: Pacific Southwest Forest and Range Exp. Stn., Forest Service, US Department of Agriculture.

Kapp, K. William. 1978. *The Social Costs of Business Enterprise*. Nottingham, UK: Spokesman.

Kerns, Jessica, and Paul Kerlinger. 2004. *A Study of Bird and Bat Collision Fatalities at the Mountaineer Wind Energy Center, Tucker County, West Virginia: Annual Report for 2003*. McLean, VA: Curry and Kerlinger.

Litton, R. Burton. 1968. *Forest Landscape Description and Inventories—A Basis for Land Planning and Design (Research Paper PSW-RP-049)*. Albany, CA: Pacific Southwest Research Station, Forest Service, US Department of Agriculture.

Matsuura, Masahiro, and Todd Schenk, eds. 2017. *Joint Fact-Finding in Urban Planning and Environmental Disputes*. New York: Routledge.

Menzel, Jennifer M., W. Mark Ford, John W. Edwards, and Leah J. Ceperley. 2006. *A Habitat Model for the Virginia Northern Flying Squirrel (*Glaucomys sabrinus fuscus*) in the Central Appalachian Mountains (Research Paper NE-729)*. Newtown Square, PA: US Department of Agriculture, Forest Service, Northern Research Station. https://doi.org/10.2737/ne-rp-729.

Muthersbaugh, Michael A., Sara E. Sweeten, Alexander Silvis, W. Mark Ford, and Lauren V. Austin. 2016. "Fall Activity Patterns of Migratory and Cave Dwelling Bats Species in the Central Appalachians." *Abstracts of the 23rd Annual Wildlife Society Conference*, October 2016. Raleigh, NC.

Nagel, Kai, Richard J. Beckman, and Christopher L. Barrett. 1999. "TRANSIMS for Urban Planning." Presented at the Sixth International Conference on Computers in Urban Planning and Urban Management, September 1999. Venice, Italy.

O'Leary, Rosemary, and Lisa B. Bingham, eds. 2003. *The Promise and Performance of Environmental Conflict Resolution*. Washington, DC: Resources for the Future.

Ostrom, Elinor. 2010. "Polycentric Systems for Coping with Collective Action and Global Environmental Change." *Global Environmental Change* 20 (4): 550–557.

Parikh, Nidhi, Samarth Swarup, Paula E. Stretz, et al. 2013. "Modeling Human Behavior in the Aftermath of a Hypothetical Improvised Nuclear Detonation." In *Proceedings of the 12th International Conference on Autonomous Agents and Multiagent Systems*, edited by T. Ito, C. Jonker, M. Gini, and O. Shehory, 949–956. Richland, SC: International Foundation for Autonomous Agents and Multiagent Systems.

Peste, Filipa, Anabela Paula, Luís P. da Silva, et al. 2015. "How to Mitigate Impacts of Wind Farms on Bats? A Review of Potential Conservation Measures in the European Context." *Environmental Impact Assessment Review* 51 (February): 10–22. https://doi.org/10.1016/j.eiar.2014.11.001.

Pociask, Steve, and Joseph P. Furh Jr. 2011. "Progress Denied: A Study on the Potential Economic Impact of Permitting Challenges Facing Proposed Energy Projects." US Chamber of Commerce. Washington, DC. https://www.uschamber.com/sites/default/files/pnp_economicstudy.pdf.

Raab, Jonathan. 2017. "Energy and Climate Stakeholder Processes in the United States." In *Joint Fact-Finding in Urban Planning and Environmental Disputes*, edited by Masahiro Matsuura and Todd Schenk, 139–164. New York: Routledge.

Reynolds, Scott D. 2006. "Monitoring the Potential Impact of a Wind Development Site on Bats in the Northeast." *Journal of Wildlife Management* 70 (5): 1219–1227. https://doi.org/10.2193/0022-541x(2006)70[1219:mtpioa]2.0.co;2.

Roberts, Nancy. 2004. "Public Deliberation in an Age of Direct Citizen Participation." *American Review of Public Administration* 34 (4): 315–353. https://doi.org/10.1177/0275074004269288.

Schenk, Todd. 2007. *Conflict Assessment: A Review of the State of Practice*. Cambridge, MA: Consensus Building Institute.

Sheppard, Stephen. 2001. "Beyond Visual Resource Management: Emerging Theories of an Ecological Aesthetic and Visible Stewardship." In *Forests and Landscapes: Linking Ecology, Sustainability and Aesthetics*, edited by Stephen Richard John Sheppard and Howard W. Harshaw, 149–172. New York: CABI Publishing.

Smith, Jordan W., Jessica E. Leahy, Dorothy H. Anderson, and Mae A. Davenport. 2013. "Community/Agency Trust and Public Involvement in Resource Planning." *Society and Natural Resources* 26 (4): 452–471. https://doi.org/10.1080/08941920.2012.678465.

Sovacool, Benjamin K., and Pushkala Lakshmi Ratan. 2012. "Conceptualizing the Acceptance of Wind and Solar Electricity." *Renewable and Sustainable Energy Reviews* 16 (7) 5268–5279. https://doi.org/10.1016/j.rser.2012.04.048.

Stern, Marc J., and Timothy D. Baird. 2015. "Trust Ecology and the Resilience of Natural Resource Management Institutions." *Ecology and Society* 20 (2): 14. http://dx.doi.org/10.5751/ES-07248-200214.

Stern, Marc J., and Kimberly J. Coleman. 2015. "The Multi-dimensionality of Trust: Applications in Collaborative Natural Resource Management." *Society and Natural Resources* 28 (2): 117–132. https://doi.org/10.1080/08941920.2014.945062.

Subbiah, Rajesh, Anamitra Pal, Eric K. Nordberg, Achla Marathe, and Madhav V. Marathe. 2017. "Energy Demand Model for Residential Sector: A First Principles Approach." *IEEE Transactions on Sustainable Energy* 8 (3): 1215–1224. https://doi.org/10.1109/TSTE.2017.2669990.

Susskind, Lawrence, and Patrick Field. 1996. *Dealing with an Angry Public: The Mutual Gains Approach to Resolving Disputes*. New York: Free Press.

Susskind, Lawrence E., Sarah McKearnen, and Jennifer Thomas-Larmer. 1999. *The Consensus Building Handbook: A Comprehensive Guide to Reaching Agreement*. Thousand Oaks: Sage.

Trani, Margaret K., W. Mark Ford, and R. Brian. 2007. *The Land Manager's Guide to Mammals of the South*. Durham, NC: Nature Conservancy.

US Department of Agriculture. 1974. *The Visual Management System: National Forest Landscape Management* 2 (1). Agricultural Handbook No. 462. Washington, DC: US Government Printing Office.

US Department of the Interior, Bureau of Land Management. 1984. *Visual Resource Management* (Manual 8400). Washington, DC: US Government Printing Office.

US Department of the Interior, Bureau of Land Management. 1986. *Visual Resource Contrast Rating* (Manual 8431). Washington, DC: US Government Printing Office.

US Energy Information Administration. 2017. "Annual Energy Outlook 2017." *U.S. Energy Information Administration, (EIA)*. https://www.eia.gov/outlooks/aeo/pdf/0383(2017).pdf.

Wüstenhagen, Rolf, Maarten Wolsink, and Mary Jean Bürer. 2007. "Social Acceptance of Renewable Energy Innovation: An Introduction to the Concept." *Energy Policy* 35 (5): 2683–2691. https://doi.org/10.1016/j.enpol.2006.12.001.

Chapter 12 Summary

In this chapter, we introduce an approach for siting wind energy projects, rooted in the concept of sustainable development, that is environmentally beneficial, economically efficient (i.e., projects are not delayed or abandoned due to unnecessary conflict), and socially equitable. We emphasize the need for early and meaningful public engagement and joint fact-finding to foster collaboration instead of conflict. We also describe an interdisciplinary research agenda to help all parties understand siting needs and conflicts and to test the effectiveness of various strategies.

KEY TAKEAWAYS

- A significant number of new wind farms will be needed in the United States to meet climate change goals.
- More sustainable approaches to siting will improve outcomes from the social, environmental, and economic perspectives.
- All stakeholders involved in wind siting are intended to benefit from this approach.
- Concepts from sustainable development can be applied to assist in public engagement, research, and decision making. These include social sustainability.
  - Social sustainability:
    - We propose a research agenda to better understand drivers of public opposition and support.
    - We demonstrate the importance of building trust, and how to do it, based on the concept of "Trust ecology."
    - We suggest addressing concerns regarding visual impacts by applying concepts from landscape architecture, including the perceptual approach, visual resource management, and scenic quality.
    - We survey the inequitable distribution of social costs affecting siting success, along with strategies for addressing these.
  - Environmental sustainability:
    - We present challenges to siting from adverse environmental impacts on wildlife and environmental integrity.
    - We recommend strategies to improve monitoring.
    - We describe the importance of considering forest fragmentation.
    - The need to protect avifauna, and progress made in addressing concerns, is reviewed.
  - Economic sustainability:
    - We suggest that this approach can reduce siting costs for individual projects, the industry as a whole, and consumers.
    - We recommend identifying suitable sites and reducing siting costs by forming collaborative partnerships between the public, industry, and government.

About the Authors

ALI ADIL, PHD, is an assistant professor at the Kautilya School of Public Policy at GITAM University in Hyderabad, India, where his work extends and builds on his doctoral research at the University of Texas at Arlington. Ali's research interests broadly include equity implications of ongoing energy transitions for vulnerable populations, with particular emphasis on strategies to ensure an equitable, sustainable, and resilient energy future for all.

LISA BAILEY-DAVIS, D.ED., RD, is the associate director of the Geisinger Obesity Institute and an assistant professor in the Geisinger Department of Epidemiology and Health Services Research. Dr. Bailey-Davis's research interests include intergenerational prevention of obesity and related chronic diseases, multilevel interventions that integrate delivery care systems across settings, and translating the best available evidence into practice to advance population health goals.

NANCY BOWEN-ELLZEY is an associate professor and field specialist, community economics, for the Ohio State University Extension. Her areas of program specialization include Business Retention & Expansion (BR&E), economic impact analysis, community and regional planning, entrepreneurship, and renewable energy development. Nancy obtained a master of urban and regional planning from the University of New Orleans in 1990. In 1993 she became a certified economic developer (CEcD) through the International Economic Development Council. She has over thirty years' experience as practitioner and educator in the field of community economic development, including thirteen years as an economic development director and Extension educator for Van Wert County, Ohio.

MOREY BURNHAM, PHD, is a research assistant professor in the Department of Sociology at Idaho State University. He is a broadly trained environmental social scientist who works at the interface of climate change adaptation and vulnerability, agriculture, and water in arid and semiarid regions. Morey received his PhD in human dimensions of ecosystem science and management from Utah State in 2014.

WESTON M. EATON, PHD, is an assistant research professor in the College of Agricultural Sciences at the Pennsylvania State University. He studies transitions in renewable energy systems and conservation practices in rural spaces. Wes received his PhD in sociology from Michigan State University in 2015.

HEATHER FELDHAUS, PHD, is a professor of sociology in the Department of Sociology, Social Work, and Criminal Justice at Bloomsburg University and the director of Bloomsburg's Center for Community Research and Consulting. Her research focuses on community approaches to social problems.

FELIX N. FERNANDO is an assistant professor of sustainability in the Hanley Sustainability Institute at the University of Dayton. Dr. Fernando received his PhD in natural resource management with a focus on sociology and economics from the North Dakota State University. His research focuses on human dimensions and planning issues pertinent to energy and food systems.

EMILY GRUBERT, PHD, studies how we can make better decisions about large infrastructure systems, with a particular focus on energy and water systems in the United States. Grubert is an assistant professor of civil and environmental engineering and, by courtesy, of public policy at the Georgia Institute of Technology. She holds a PhD in environment and resources from Stanford University.

JULIA H. HAGGERTY, PHD, is associate professor of geography in the Department of Earth Sciences at Montana State University and holds a joint appointment at the Montana Institute on Ecosystems. At MSU, she supervises the research activities of the Resources & Communities Research Group focused on understanding the ways rural communities respond to shifting economic and policy trajectories, especially as they involve land and natural resource use.

C. CLARE HINRICHS, PHD, is a professor of rural sociology at the Pennsylvania State University. Her research, teaching, and public engagement center broadly on the social and political dimensions of sustainability transitions in food, agricultural, and energy systems.

JOHN HINTZ, PHD, is a professor of environmental, geographical, and geological Sciences at Bloomsburg University. His teaching and research focus on environmental geography, sustainable agriculture, and US federal lands management.

RICHARD F. HIRSH, PHD, is a professor of history in the Departments of History and Science & Technology in Society at Virginia Tech University. Professor Hirsh performs research on the deregulation and restructuring of the American electric utility system.

SEASON HOARD, PHD, is an associate professor jointly appointed in the School of Politics, Philosophy and Public Affairs and the Division of Governmental Studies and Services (DGSS) at Washington State University. She is the project manager at DGSS, which provides applied research, technical assistance, and training for governmental agencies and nonprofit organizations throughout the Northwest. Season has a PhD in political science from Washington State University, and her areas of expertise include applied research methods, survey research, program evaluation, secondary data analysis, and public policy.

JEFFREY B. JACQUET, PHD, is an assistant professor in the School of Environment and Natural Resources at The Ohio State University. One of the first sociologists to study the process of hydraulic fracturing, Jacquet has gone on to examine a range of renewable- and non-renewable-energy-related social and environmental systems at institutions including the University of Wyoming, Cornell University, and South Dakota State University. At Ohio State, Jacquet leads students through coursework and mentorship to examine the areas of energy, environment, and rural societies.

DR. TAMARA J. LANINGA, AICP, is an associate professor in the Environmental Studies Department at Western Washington University (WWU). She has a PhD in design and planning from the University of Colorado. Her research is situated at the knowledge-action boundary and is driven by a desire to address complex and time-sensitive issues. Her scholarship has contributed to the areas of renewable energy, natural resource and recreation management, and sustainable development.

ERIC C. LARSON, PHD, is an assistant professor in the Department of Rural Studies at Abraham Baldwin Agricultural College. He is also a coowner of the applied social science research consulting company PEER Social Science. Dr. Larson received his PhD in Rural Sociology and the Human Dimensions of Natural Resources in 2016 from the Pennsylvania State University. His research focuses on understanding and addressing natural resource, environmental, and development issues in rural communities.

ACHLA MARATHE, PHD, is a professor at Biocomplexity Institute of Virginia Tech and the Department of Agricultural and Applied Economics. At the Network Dynamics and Simulation Science Laboratory, she works with a transdisciplinary group of researchers who specialize in building individual-based models and advanced simulation methods to study social processes on large social networks.

NATALIE MARTINKUS, PHD, is a former assistant research professor at Washington State University and Heritage University. Her research focuses on sustainability of engineered and natural systems, the use of GIS for decision-making in agricultural and industrial systems, supply chain analysis, and machine learning/deep learning of aerial and optical imagery in agricultural systems for the development of farming decision support tools. She is now a business owner providing GIS and aerial imagery mapping and analysis services for agricultural clients.

SEVEN MATTES, PHD, is an assistant professor in the Center for Integrative Studies at Michigan State University. Seven's interdisciplinary research brings nonhuman animals to the forefront in disaster contexts, locating the multispecies nature of our vulnerabilites. Her work is aimed at building resiliencies for nonhuman animals and their associated humans via policy change, direct action with nonprofit organizations, and education regarding the significance of human-animal bonds.

RONALD MEYERS, PHD, is an assistant research professor and lecturer in the Department of Fish and Wildlife Conservation at Virginia Tech University's School of Public and International Affairs. His research interests include fish and wildlife conservation, hunter recruitment and retention, ethics, and environmental education.

PATRICK MILLER, PHD, FASLA, FCELA is a professor of landscape architecture in the College of Architecture and Urban Studies at Virginia Tech University. His research interests lie in the area of human attitudes and perceptions toward the environment and how the profession of landscape architecture contributes to human well-being through better design and planning. His research examines the ways in which humans can use the landscape to support their needs without harming critical social, cultural, and environmental systems. More recently he has been working on the visual implication of renewable energy facility siting and scenic viewshed management.

ETHAN R. MINIER, MS, has a bachelor's degree in sociology from Bloomsburg University and a master's degree in epidemiology from the University at Albany. He currently runs an outdoor and medical education business and is a field guide at True North Wilderness Program in Vermont.

MYRA L. MOSS is a professor and educator, community development, for the Ohio State University Extension. She also serves as a co-leader of the OSU Extension Energy Outreach Program, a statewide partnership providing education and outreach on topics related to energy development. Within the topic of energy, her teaching, research, and publications have focused on community engagement methods and outcomes in various shale plays throughout the United States. Her emphasis has been to help communities prepare for shale development through education and outreach. Moss earned her BA at Long Island University and an MA in sociology and an MBA at Ohio University.

JACOB MOWERY is a research project manager with the Geisinger Obesity Institute located at Geisinger Medical Center in Danville, Pennsylvania. Mr. Mowery works with multi-disciplinary teams, community and institutional partners to coordinate the implementation of research projects in an integrated healthcare system.

THOMAS MURPHY is the director of Penn State's Marcellus Center for Outreach and Research (MCOR), with over three decades of experience working with public officials, researchers, industry, government agencies, and landowners during his tenure with the Extension branch of the university. His work has centered on educational consultation in natural resource development, with more emphasis specifically in natural gas exploration and related topics. He lectures globally on natural gas development from shale, the economics driving the process, and its broad impacts including landowner and surface issues, environmental aspects, evolving drilling technologies, critical infrastructure, workforce assessment and training, local business expansion, resource utilization, financial considerations, and liquified natural gas (LNG) export trends.

JOHN R. PARKINS, PHD, is a professor in the Department of Resource Economics and Environmental Sociology at the University of Alberta. His current research addresses the sociology of energy transition in Canada, sustainable agriculture in Alberta, and food security in the global south. John's current work on energy transition addresses the social and economic challenges of wind power development in Alberta.

CHRISTOPHER W. PODESCHI, PHD, is an associate professor of sociology in the Department of Sociology, Social Work, and Criminal Justice at Bloomsburg University of Pennsylvania. He teaches courses on sociological theory, race and ethnicity, qualitative methods, and environmental sociology. Apart from his interest in the social impacts of the energy industry, Dr. Podeschi's research focuses on place attachment, environmental concerns, and depictions of nature and environmental issues in popular culture.

NATHAN RATLEDGE is a co-principal investigator of Resources for the Future's (RFF's) Shale Schools Project. Before teaming with RFF, Nathan was the executive director of the Community Office for Resource Efficiency. He holds an MPA from Princeton's Woodrow Wilson School of Public and International Affairs and is currently pursuing a PhD in Stanford's E-IPER program.

SANNE A.M. RIJKHOFF, PHD, is an assistant professor of political science in the Department of Social Sciences at Texas A&M University–Corpus Christi. She holds a PhD in political science from Washington State University and master's degrees in both social and organizational psychology and political science from Leiden University, the Netherlands. Her research focuses on political behavior, public opinion, and political communication. Specifically, her work draws from political science and social psychology in order to better understand the causes of political attitudes and their effects on political behavior.

KELLI ROEMER is a PhD student in earth sciences at Montana State University. She earned her MS in natural resources from the University of Idaho and served two AmeriCorps member terms in Helena, Montana, and Lakeview, Oregon. Her dissertation examines the implications of policy and planning on community resilience in coal-reliant communities in the US West.

SEVDA OZTURK SARI is a landscape architect and PhD student in the Architecture and Design Research program at Virginia Tech University. She received her BLA from Ankara University and her MLA from Istanbul Technical University. Sevda's current research focuses on the impacts of green roofs on employee health and well-being in workplaces.

TODD SCHENK, PHD, is an assistant professor in the Urban Affairs and Planning Program of the School of Public and International Affairs at Virginia Tech. He has extensive research and consulting experience working on environmental policy and planning and on collaborative governance issues in North America, Europe, Asia, Africa, and the Middle East. Dr. Schenk received both a PhD in public policy and planning and a master's degree in city planning from the Massachusetts Institute of Technology, and a bachelor's degree in geography from the University of Guelph.

ANJU SETH, PHD, is the R. B. Pamplin Professor of Management in the Fisher College of Business at Virginia Tech. Seth takes a multidisciplinary approach to her teaching and research in the areas of corporate strategy, acquisitions, restructuring, strategic alliances and joint ventures, corporate governance and ethics, and globalization and strategy in emerging economies.

KATE SHERREN, PHD, is a professor in Dalhousie University's School for Resource and Environmental Studies, in Nova Scotia, Canada. Dr. Sherren's research combines social and spatial methods and focuses on change in multifunctional landscapes, cultural ecosystem services, climate adaptation, and sustainability transitions. She has worked in Canada, Australia, the United States, and the Falkland Islands across landscapes including farms, energy, wetlands, and coasts.

JISOO SIM is a landscape architect, urban researcher, and PhD candidate at Virginia Tech University. She holds a BLA and an MLA from Seoul National University. Her current research includes investigating the benefits of urban revitalization projects that transform urban infrastructure into urban parks, Big Data analysis for landscape architects, and using data visualization techniques, especially mapping, to interpret urban issues.

MARC J. STERN, PHD, is a professor in the Department of Forest Resources and Environmental Conservation at Virginia Tech University. He studies human dimensions of natural resource policy and management with a special interest in conservation. Stern received his PhD in Social Ecology from Yale University.

GENE L. THEODORI, PHD, is a professor of sociology at Sam Houston State University. He teaches, conducts basic and applied research, and writes professional and popular articles on rural and community development issues, energy and natural resource concerns, and related topics. A central feature of his work involves the design, implementation, and analysis of survey research. The findings from his research have been published in numerous articles, book chapters, research bulletins, and other professional reports.

JESSICA D. ULRICH-SCHAD, PHD, is an assistant professor of community and natural resource sociology in the Department of Sociology, Social Work, and Anthropology at Utah State University. Dr. Schad received her PhD in community and environmental sociology from the University of New Hampshire and completed a postdoc at Purdue University in the Natural Resource Social Science lab. Dr. Schad uses mixed methods to examine community and environmental issues in different types of rural places throughout the United States.

CAMERON T. WHITLEY, PHD, is an assistant professor of sociology at Western Washington University. He received his PhD in sociology from Michigan State University. His work is guided by a central question: How do our relationships with others inform our attitudes and behaviors? His most recent work looks at how our relationships with animals inform our perceptions of environmental risks such as hydraulic fracturing. His work has been published in numerous books and journals, including *Annual Review of Sociology*, *Sociology of Development*, *Political Behavior*, *Environmental Education Research*, *Clinical Chemistry*, *Teaching Sociology*, *Environmental Policy and Planning*, *Human Ecology*, *Risk Research*, and *Proceedings of the National Academy of Science*.

LAURA ZACHARY is an independent energy and climate consultant. Laura has worked on energy, climate, environmental, and education policy issues in the United States and abroad. She received a BA from Tufts University and an MPA from Princeton's Woodrow Wilson School of Public and International Affairs.

Index